International Series in Operations Research & Management Science

Volume 231

More information about this series at http://www.springer.com/series/6161

Adiel Teixeira de Almeida
Cristiano Alexandre Virgínio Cavalcante
Marcelo Hazin Alencar
Rodrigo José Pires Ferreira
Adiel Teixeira de Almeida-Filho
Thalles Vitelli Garcez

Multicriteria and Multiobjective Models for Risk, Reliability and Maintenance Decision Analysis

Springer

Adiel Teixeira de Almeida
Universidade Federal de Pernambuco
Recife, PE
Brazil

Cristiano Alexandre Virgínio Cavalcante
Universidade Federal de Pernambuco
Recife, PE
Brazil

Marcelo Hazin Alencar
Universidade Federal de Pernambuco
Recife, PE
Brazil

Rodrigo José Pires Ferreira
Universidade Federal de Pernambuco
Recife, PE
Brazil

Adiel Teixeira de Almeida-Filho
Universidade Federal de Pernambuco
Recife, PE
Brazil

Thalles Vitelli Garcez
Universidade Federal de Pernambuco
Recife, PE
Brazil

ISSN 0884-8289 ISSN 2214-7934 (electronic)
International Series in Operations Research & Management Science
ISBN 978-3-319-37157-3 ISBN 978-3-319-17969-8 (eBook)
DOI 10.1007/978-3-319-17969-8

Springer Cham Heidelberg New York Dordrecht London

Softcover reprint of the hardcover 1st edition 2015

Printed on acid-free paper

Springer International Publishing AG Switzerland is part of Springer Science+Business Media (www.springer.com)

To our families for their continuous support

Foreword

Any organization is interested in having a structured decision process for its strategic success. This is particularly relevant when the decision context involves technological risk, reliability or maintenance issues. In general these issues may often be associated with potential threats to human life (e.g. safety) and the environment. They may also affect the strategic results of any organization. All these matters may be integrated into a single decision problem in many systems, for example an electrical supply system. Service interruptions or accidents in this kind of system may affect health and other emergency services, traffic in big cities, air traffic control, and many other issues that society has become increasingly aware of as a result of media reports of major accidents that had or could well have had a very serious impact on human safety. These interruptions are often related to the decisions in a system involving Risk, Reliability and Maintenance (RRM). Usually, these decisions include more than one objective that need to be dealt with simultaneously, with appropriate support from multicriteria and multi-objective models. These multicriteria models become even more relevant for the example of electrical supply system with smart grids conception.

"Multicriteria and Multi-objective Models for Risk, Reliability and Maintenance Decision Analysis" is a book that enables the reader to have a better understanding of and guidelines on integrating important application areas of operations research and management science. This is done by discussing means of structured process for model building that incorporate RRM issues. This integration is based on the combination of concepts and foundations related to RRM areas within multicriteria methods.

The authors represent a group of active members of scientific societies in operations research and RRM areas. They set out to build a bridge between these areas with this book; They have had more than 20 years' experience of engaging on such research and have had many articles published both in journals of distinction in the areas of operations research and also in specialized journals related to risk, reliability and maintenance, since the 1990s. Many of these articles also consider real problems found in business organizations.

As the current IFORS (International Federation of Operational Research Societies) President, it is with great pleasure that I present a book that reports outstanding academic research results in operational research and management science, thus bridging these relevant areas in order to support the decision process related to issues of the utmost importance to society.

Nelson Maculan Filho
Prof Emeritus of Universidade Federal do Rio de Janeiro
President of IFORS (International Federation of Operational Research Societies)

Preface

Many decision problems have more than one objective that need to be dealt with simultaneously. Risk, Reliability and Maintenance (RRM) are contexts in which decision problems with multiple objectives have been on the increase in recent years. The importance of having a better structured decision process is essential for the success of any organization. Additionally, decisions on RRM matters may affect the strategic results of any organization, as well as, human life (e.g. safety) and the environment.

RRM influences society and organizations in many ways, since companies and governments must satisfy several expectations related to the everyday lifestyle inherent in modern society, such as safeguarding the safety of their employees, their customers and the community they are part of. Such a lifestyle includes new paradigms for judging what level of risk is acceptable and this requires multi-dimensional risks to be evaluated in order to meet society's and regulatory bodies' expectations. Reliability and maintenance have become more important also, since such expectations are extended to the demands that services are constantly available and that products are of a consistently high quality. Therefore, companies strive to reduce costs and simultaneously improve their performance with regard to meeting their strategic objectives. These are affected by reliability and maintenance and include implications for risk, namely that the analysis of risk and reliability demands a more conservative approach as do maintenance policies since failures may have serious implications regarding safety and environmental losses. As a result, MCDM/A approaches are becoming inevitable when modeling strategic problems that involve the RRM context.

This book integrates multiple criteria concepts and methods for problems within the RRM context. The concepts and foundations related to RRM are considered for this integration with multicriteria approaches. In the book, a general framework for building decision models is presented and this is illustrated in various chapters by discussing many different decision models related to the RRM context.

In general, a decision process or problem in the multicriteria context is related to the acronyms MCDM (Multi-Criteria Decision Making) and MCDA (Multi-Criteria Decision Aiding; also known as Multi-Criteria Decision Analysis). The distinctions between these acronyms are not emphasized in this text. Without loss of generality, the acronym MCDM/A is applied throughout the text to represent a variety of approaches associated with MCDM and MCDA (decision making, decision analysis and decision aiding).

The scope of the book is related to ways of how to integrate Applied Probability and Decision Making. In Applied Probability, this mainly includes: decision analysis and reliability theory, amongst other topics closely related to risk analysis and maintenance. In Decision Making, it includes a broad range of topics in MCDM/A. In addition to decision analysis, some of the topics related to Mathematical Programming area are briefly considered, such as multiobjective optimization, since methods related to these topics have been applied to the context of RRM.

The book addresses the needs of two specific audiences and these include practitioners and researchers of both areas:

- Those dealing with Risk analysis, Reliability and Maintenance areas, who are interested in using multicriteria decision methods;
- Those related to multiobjective and MCDM/A, who are interested in making applications in the contexts of RRM.

Those, who are dealing with decision problems related to the RRM context, in general need to improve their knowledge of multiobjective and multicriteria methods so they can build more appropriate decision models. Also, those dealing with multiobjective and multicriteria decision making area, require to improve their knowledge of the concepts and methods related to the contexts of RRM, so that they can approach decision problems on RRM in a more appropriate way.

The book addresses an innovative treatment for the decision making in RRM, thereby improving the integration of fundamental concepts from the areas of both RRM and decision making. This is accomplished by presenting an overview of the literature on decision making in RRM. Some pitfalls of decision models when applying them to RRM in practice are discussed and guidance on overcoming these drawbacks is offered. The procedure enables multicriteria models to be built for the RRM context, including guidance on choosing an appropriate multicriteria method for a particular problem faced in the RRM context. The book also includes many research advances in these topics. Most of the multicriteria decision models that are described are specific applications that have been influenced by this research and the advances in this field.

The book is not strictly for research and reference by researchers and practitioners. It has potential for use as an advanced textbook for one of the three topics: reliability, maintenance and risk management. That is, it could usefully complement a basic textbook on one of those topics.

The book is implicitly structured in three parts, with 12 chapters. The first part deals with MCDM/A concepts methods and decision processes (Chaps. 1 and 2). The second part corresponds to Chap. 3, in which the main concepts and foundations of RRM are presented. Then, comes the third part, which forms the greatest section of the book (Chap 4 to Chap. 12) and deals with specific decision problems in the RRM context approached with MCDM/A models.

Chap. 1 gives a first view on decision problems with multiple objectives, with a description of the basic elements needed to build decision models. This Chapter is directly integrated with Chap. 2, which focuses on the decision process and MCDM/A methods. Although the description and concepts are given in a general sense, they are focused on the main problems and situations found in the context that this book explores: risk, reliability and maintenance, although they can be applied to any other context. Therefore, an explanation is given as to why and how MCDM/A arises in the RRM context.

Chap. 2 deals with MCDM/A methods and the decision process. A procedure for building an MCDM/A decision model is presented. Some concerns on the choice of MCDM/A methods are presented, discussing the compensatory and non-compensatory approaches. Although this procedure may be applied to any context, some particular considerations are given to the RRM one. A few MCDM/A methods are presented, the focus being on deterministic additive methods (MAVT) and methods for aggregation in probabilistic context, with a focus on MAUT. Outranking methods are also presented, with some emphasis to ELECTRE and PROMETHEE methods.

Chap. 3 presents concepts of RRM. These concepts should be considered when building RRM decision models in order to indicate procedures and techniques that can be used to calculate and estimate consequences. This allows aspects related to the state of nature and particularities of RRM to be incorporated when modeling a decision problem. Chap. 3 includes techniques for dealing with risk analysis such as the HAZOP, FMEA, FTA, ETA, QRA and ALARP principle; cost effectiveness; and risk visualization. Reliability and maintenance aspects presented in Chap. 3 include random failure modeling, reliability and failure functions, maintenance and reliability interactions, FMEA/FMECA, redundant systems, repairable and non-repairable systems, maintenance goals and maintenance management techniques (TPM, RCM). Additionally Chap. 3 presents techniques for eliciting expert's prior knowledge.

Chaps. 4 to 12 present an integration of the first and second part when considering RRM decision problems structured within an MCDM/A approach, for which formulation and insights for decision problems are given. Chap. 4 presents a multidimensional risk analysis perspective by introducing a general structure for building a multidimensional risk analysis decision model. Based on the structure provided, Chap. 4 presents examples of multidimensional risk evaluation models for natural gas pipelines and an underground electricity distribution system. Other contexts are discussed, the purpose of which is to offer insights on how to evaluate multidimensional risks, such as in power electricity systems, for natural hazards, counter-terrorism and nuclear power.

Preventive maintenance decisions are presented in Chap. 5 with regard to how to go about selecting which is the most suitable time interval for scheduling preventive maintenance actions. This chapter explores the classical optimization approach for preventive maintenance modeling and gives insights on the implications of considering an MCDM/A approach by discussing illustrative applications of two kinds of MCDM/A approaches based on the general procedure for building MCDM/A models presented in Chap. 2.

Condition-based maintenance (CBM) is tackled in Chap. 6, including a discussion of MCDM/A models in CBM. An MCDM/A model is presented including delay time concepts followed by a case study conducted in a power distribution company, thereby illustrating the advantages of considering an MCDM/A perspective.

Chap. 7 presents maintenance outsourcing decisions regarding supplier and contract selection. Throughout this chapter, several criteria for such problems are discussed and five MCDM/A decision models are presented.

Spare part planning models are discussed in Chap. 8. General aspects of approaches to sizing spare parts are presented which gives insights into how an MCDM/A model considers the state of nature over reliability and maintainability, based on the probability of stockout and cost. Another MCDM/A decision model grounded on the same objectives is presented for sizing the need for multiple spare parts for which the case study uses a multiobjective genetic algorithm. Additionally, a spare parts model integrated with CBM is shown.

The allocation of redundancy is discussed in Chap. 9, and takes the combinatorial complexity of these problems into account. Therefore, multiobjective formulations for these problems, found in the literature are presented and the tradeoffs in redundancy allocation are emphasized. An MCDM/A model is presented for a standby system in the context of a telecommunications system of an electric power company with a 2-unit standby redundant system. The model takes interruption time and cost into account.

Design selection decisions are explored in Chap. 10 with a discussion on the roles of reliability, maintainability and risk in system design. Based on these aspects, this chapter includes an MCDM/A model for selecting the design of a car and an MCDM/A model for risk evaluation in design selection and gives illustrative applications.

Chap. 11 consists of MCDM/A models for priority assignment in maintenance planning. An MCDM/A model is presented within the RCM structure to establish critical failure modes considering a multidimensional perspective and this is followed by an illustrative example. The second MCDM/A model presented in this chapter considers the problem of identifying critical devices in an industrial plant. TPM aspects are also mentioned in this chapter and briefly discussed in order to emphasize potential MCDM/A problems that may be addressed.

Chap. 12 presents other RRM decision problems including the location of backup transformers, sequencing of maintenance activities, evaluating the risk of natural disasters, reliability in power systems, integrated production and maintenance scheduling, maintenance team sizing and reliability acceptance testing.

Depending on the reader's background and experience regarding MCDM/A and RRM concepts, a thorough understanding of the first and second parts of the book, respectively, may be required in order to understand the decision models presented in the third part (Chaps. 4 to 12). Otherwise, the reader may dip into Part 3 directly and choose to read any Chapter (Chaps. 4 to 12) without having read the first three Chapters. However, Chap. 2 is required, if the reader wants to use the procedure for building an MCDM/A decision model even though the reader has good knowledge of MCDM/A concepts.

We would like to thank our colleagues, students and professionals from industry, who jointly worked with us on modeling MCDM/A problems in the RRM context, integrated to the Center for Decision Systems and Information Development (CDSID). In addition, we are grateful to our sponsors (especially CNPq - the Brazilian Research Council) and the business organizations that have supported our research and activities since the 1990s. We would also like to thank the editors of Springer for their professional help and cooperation, and finally, but most of all, our families, who constantly supported and encouraged us in our research work.

Recife,
February, 2015

Adiel Teixeira de Almeida
Cristiano Alexandre Virgínio Cavalcante
Marcelo Hazin Alencar
Rodrigo José Pires Ferreira
Adiel Teixeira de Almeida-Filho
Thalles Vitelli Garcez

Contents

Acronyms

AHP	Analytic Hierarchy Process
ALARP	As Low As Reasonably Practicable
Aneel	Brazilian government agency responsible for regulating the generation of electrical power
ANP	Analytic Network Process
BLEVE	Boiling Liquid Expanding Vapor Explosion
CB	Cost benefit ratio
CBM	Condition-based maintenance
CBR	Case Based Reasoning
CDR	Composite dispatching rule
CDRNRGA	Non ranking genetic algorithm with composite dispatching rule
CDRNSGA-II	Non dominated sort genetic algorithm with composite dispatching rule
CRA	Comparative Risk Assessment
CSE	Concept Safety Evaluation
CVCE	Confined Vapor Cloud Explosion
DEA	Data Envelopment Analysis
DEC	Equivalent to System Average Interruption Duration Index
DM	Decision Maker
DSS	Decision Support System
DT	Delay Time
EC-JRC	European Commission - Joint Research Centre
EHS	Environmental Health Safety
ELECTRE	Elimination Et Choix Traduisant la Réalité
EPDC	Electric Power Distribution Company
ET	Event Tree
ETA	Event Tree Analysis
FAR	Fatality Accident Rate
FEC	Equivalent to System Average Interruption Frequency Index
FFA	Functional failure analysis
FMEA	Failure Modes and Effects Analysis
FMECA	Failure modes, effects, and criticality analysis
FT	Fault Tree
FTA	Fault Tree Analysis
GA	Genetic Algorithm
GD	Group Decision Making
GDN	Group Decision and Negotiation
GIS	Geographic Information System
GIT	Geo Information Technology
GPSIA	Genetic Pareto set identification algorithm
HAZID	Hazard Identification

HAZOP	Hazard and Operability Study
I	Indifference relation of preference
IEC	International Electrotechnical Commission
IEEE	Institute of Electrical and Electronics Engineers
ISO	International Organization for Standardization
J	Incomparability relation of preference
JIPM	Japan Institute of Plant Maintenance
LPP	Linear Programming Problems
M/M/s	A system where arrivals form a single queue, there are s servers and job service times are exponentially distributed
MACBETH	Measuring Attractiveness by a Categorical Based Evaluation Technique
MAU	Multi-Attribute Utility
MAUT	Multi Attribute Utility Theory
MAVT	Multi-Attribute Value Theory
MCDA	Multi-Criteria Decision Aiding; may also be applied to Multi-Criteria Decision Analysis
MCDM	Multi-Criteria Decision Making
MCDM/A	Indiscriminately applied to MCDM or MCDA
MOCBA	Multiobjective Computing Budget Allocation
MOEA	Multiobjective Evolutionary Algorithm
MOGA	Multiobjective Genetic Algorithm
MOLP	Multi-Objective Linear Problems
MOPSO	Multiobjective Particle Swarm Optimization
MTBF	Mean Time Between Failures
MTTF	Mean Time to Failure
MTTR	Mean Time to Repair
Natech	REPRESENTS a simultaneous occurrence of a natural disaster event and a technological accident, both requiring simultaneous response efforts
NCAP	New Car Assessment Program
NORSOK	NORSOK standards developed by the Norwegian petroleum industry
NPD	Norwegian Petroleum Directorate
NPRD	Non-electronic Parts Reliability Data
NRGA	Non ranking Genetic Algorithm
NSGA-II	Non dominated Sort Genetic Algorithm
OEE	Overall Equipment Effectiveness
OREDA	Offshore Reliability Data Handbook
P	Strict Preference Relation
PHA	Preliminary Hazard Analysis
PHM	Proportional Hazards Modelling
PM	Preventive Maintenance
PRA	Probabilistic Risk Assessment
PROMETHEE	Preference Ranking Organization Method for Enrichment Evaluation

PSA	Probabilistic Safety Assessment
PSM	Problem Structuring Methods
PSO	Particle Swarm Optimization
Q	Week Preference Relation
QRA	Quantitative Risk Analysis
RCM	Reliability Centered Maintenance
RDU	Rank-Dependent Utility
ROC	Rank Order Centroid
RPN	Risk Priority Number
RRM	Risk, Reliability and Maintenance
RUL	Residual Useful Life
S	Outranking Relation
SAFOP	Safety and Operability Study
SAIDI	System Average Interruption Duration Index
SAIFI	System Average Interruption Frequency Index
SEMOPS	Sequential Multiple-Objective Problem-Solving Technique
SJA	Safe Job Analysis
SMART	Simple Multi-Attribute Rating Technique
SMARTER	Simple Multi-Attribute Rating Technique Exploiting Ranks
SMARTS	Simple Multi-Attribute Rating Technique with Swing
SPEA2	Strength Pareto Evolutionary Algorithm
TPM	Total Productive Maintenance
TTR	Time To Repair
VCE	Vapor Cloud Explosion
VIP	Variable Interdependent Parameters
VTTF	Variance of Time to Failure

Chapter 1
Multiobjective and Multicriteria Problems and Decision Models

Abstract: The decision-making process for any organization may be a key factor for its success. Many decision problems have more than one objective that need to be dealt with simultaneously. This chapter introduces decision problems with multiple objectives, with a description of the basic elements needed to build decision models and focuses on multicriteria methods (MCDM; MCDA; MCDM/A), in which the DM's preference structure is considered. An overview for classification of MCDM/A methods is given, including a discussion on the DM's compensatory and non-compensatory rationality and on multi-objective and multicriteria approaches. The concepts and basic elements of MCDM/A methods are presented, including preference structures in a multi-attribute context, and intra-criterion and inter-criteria evaluation. The basic elements of a decision process for building decision models and the actors in this process are also presented. Differences between the descriptive, normative, prescriptive and constructivism decision approaches are discussed, considering the decision process. Although these concepts are presented in a general sense, this description deals mainly with the main context that this book explores: Risk, Reliability and Maintenance (RRM). Decision problems in a RRM context may affect the strategic results of any organization, as well as, human life (e.g. safety) and the environment. Therefore, an explanation is given as to why and how a MCDM/A arises in the RRM context. In particular, some peculiarities of service producing systems for MCDM/A models are presented, as well as for goods producing systems.

1.1 Introduction

In order to choose an alternative, from a set of possible alternatives, in a classical optimization problem, there is an objective function to be maximized or minimized, whether this function represents gains or losses, respectively. In a multiobjective or multicriteria problem, there is more than one objective to be dealt with. In many situations these objectives may be conflicting. These objectives are associated with the possible consequences (or outcomes) that will result from choosing an alternative. Therefore, these problems have more than one

A.T. de Almeida et al., *Multicriteria and Multiobjective Models for Risk, Reliability and Maintenance Decision Analysis*, International Series in Operations Research & Management Science 231, DOI 10.1007/978-3-319-17969-8_1

objective function to be dealt with simultaneously. In some particular situations, this means that these objectives are comprehensively optimized. Each objective is represented by a variable, in which its performance for a given alternative can be evaluated. This variable may be called a criterion or an attribute, depending on the multicriteria method used.

The acronyms MCDM (Multi-Criteria Decision Making) and MCDA (Multi-Criteria Decision Aiding) are applied to indicate a decision process or problem in the multicriteria context. MCDA may also be found as standing for Multi-Criteria Decision Analysis. Without loss of generality, the acronym MCDM/A is applied throughout the text to represent a number of approaches associated with MCDM and MCDA (decision making, decision analysis and decision aiding).

The perception of a decision process involving a tradeoff amongst several criteria was put forward since centuries ago.

A text of 1722, by Benjamin Franklin, is regularly quoted to indicate the nature of a multicriteria evaluation for a specific kind of decision problem, which consists of only one alternative, with either of two options: implement it or do not. He expressed this in a letter proposing a decision procedure (Hammond et al. 1998; Hammond et al. 1999; Figueira et al. 2005), as follows:

> "In the affair of so much importance to you, wherein you ask my advice, ... [...], my way is to divide half a sheet of paper by a line into two columns; writing over the one Pro, and over the other Con. [...] When I have thus got them all together in one view, I endeavor to estimate their respective weights; and where I find two, one on each side, that seem equal, I strike them both out. If I find a reason pro equal to some two reasons con, I strike out the three. If I judge some two reasons con, equal to three reasons pro, I strike out the five; and thus proceeding I find at length where the balance lies; and if, after a day or two of further consideration, nothing new that is of importance occurs on either side, I come to a determination accordingly."

Benjamin Franklin called this procedure prudential algebra. Much later on an MCDM/A method, called even swaps, was proposed based on this procedure (Hammond et al. 1998a; Hammond et al. 1999).

An identical perception of evaluation by tradeoff between two set of criteria for choosing a course of action, was made around 300 B.C., by Plato, a Greek philosopher, in the Protagoras dialogue. He proposed putting into the balance the two previous types of criteria Pro (pleasures) and the Con (pains), as follows:

> "I should reply: And do they differ in anything but in pleasure and pain? There can be no other measure of them. And do you, like a skilful weigher, put into the balance the pleasures and the pains, and their nearness and distance, and weigh them, and then say which outweighs the other. If you weigh pleasures against pleasures, you of course take the more and greater; or if you weigh pains against pains, you take the fewer and the less; or if pleasures against pains, then you choose that course of action in which the painful is exceeded by the pleasant, whether the distant by the near or the near by the distant; and you avoid that course of action in which the pleasant is exceeded by the painful. Would you not admit, my friends, that this is true? I am confident that they cannot deny this."

These quotations are just two of many others on situations reported long ago which offered this insight of making tradeoffs amongst criteria in order to evaluate alternatives in the decision process. The optimization conception is in these views, when considering the attempt to find the best action, is obtained by means of combining several objectives.

Historical views and perspectives for the MCDM/A area may be found in several texts (Koksalan et al. 2011; Edwards et al. 2007).

In this Chapter a first view is given on decision problems with multiple objectives, with a description of the basic elements needed for building decision models. This Chapter is directly integrated with Chap. 2, which focuses on the decision process, MCDM/A methods and multiobjective approaches. Then, an emphasis is given to the main problems and situations found in the context that this book explores: risk, reliability and maintenance (RRM), although they can be applied to any other context.

1.2 Multiobjective and Multicriteria Approaches

Most of the literature makes a distinction between the terms Multiobjective and Multicriteria. Therefore, one can say that a problem with multiple objectives can be approached by using either: MCDM/A method or a multi-objective optimization approach.

An MCDM/A method considers the preference structure of a decision maker (DM) and involves value judgment. The DM's preferences will be incorporated in the decision model in order to support the choice of the alternative, and by doing so, the multiple criteria will be analyzed simultaneously.

Multiobjective optimization approaches identify the Pareto frontier, the set of non-dominated alternatives, from the set of alternatives. An alternative A_1 is said to dominate another alternative A_2, if the following conditions hold: i) alternative A_1 is not worse than A_2 in all criteria, and ii) alternative A_1 is better than A_2 in at least one criterion.

The set of non-dominated alternatives consists of those which are not dominated by any other of the set of alternatives.

In this approach, the DM's preferences are not taken into consideration. This means that a specific final solution is not indicated, since a DM's preferences are not incorporated into the model for combining objectives.

On the other hand, using an MCDM/A method, the objectives are combined based on the DM's preferences. These preferences consist of the DM's subjective evaluation of the criteria. This subjectivity is an inherent part of the problem and cannot be avoided. Otherwise, it means that the model is related to any other problem, instead of the real problem faced by the DM. Thus, the methodological issues for dealing with this subjectivity have been one of the main purposes of research on MCDM/A.

1.3 Decision Models and Methods

The meaning of models and methods may vary amongst texts. In this text, an important distinction is made between MCDM/A models and MCDM/A methods, although slight variations may occur in our discussion because of particular contexts.

As is well known, a model is a simplification of a real situation and it is expected to deviate (err) to some extent from the real situation. Therefore, when building a model there is a conflict between its precision and its simplicity. This precision is related to how close the model is to the real situation (approximation of the model).

An MCDM/A model is a formal representation of a real MCDM/A problem faced by a DM. The MCDM/A model incorporates the DM's preference structure and particular issues for a specific decision problem. In general, an MCDM/A model is built based on an MCDM/A method.

An MCDM/A method consists of a methodological formulation, which can be applied so as to build specific MCDM/A models. A method may consist of a theoretical formulation based on a well-defined axiomatic structure.

The MCDM/A method has a more general characteristic and may be applied in order to build a class of MCDM/A models and may be applicable for a variety of situations related to preference structures. On the other hand, a decision model incorporates a preference structure of a specific DM. Some decision models may be built for a specific and unique problem, while others may be built for a more general and repetitive decision situation.

The use of the term model may appear to be an exception to the above concepts, when referring to the 'additive aggregation model', which indicates a group of MCDM/A methods. Here this term is associated with the kind of mathematical model applied for aggregating the criteria in a particular class of methods. The additive model for aggregation of criteria will be detailed in Chap. 2, but it is presented below in (1.1) so as to give a first view of an MCDM/A model.

$$v(a_i) = \sum_{j=1}^{n} k_j V_j(x_{ij}) \tag{1.1}$$

where:
$v(a_i)$ is the global value of the alternative a_i;
k_j is the parameter related to inter-criteria evaluation of criterion j; this parameter is named either as "weight" or "scale constant" of criterion *j*;
$V_j(x_{ij})$ is the value of consequence for criterion *j*;
x_{ij} is the consequence or outcome of alternative *i* for criterion *j*.

1.4 Decision Process

A model for decision process is given by Simon (1960), and consists of three stages. This model has been adapted, including the addition of new stages, by a number of posterior contributions, most of them from the area of information management and decision systems (Bidgoli 1989; Sprague and Watson 1989; Davis and Olson 1985; Thierauf 1982; Polmerol and Barba-Romero 2000).

Fig. 1.1 shows this updated model. Stages 1 to 3 are in the initial model proposed by Simon (1960) and consist of Intelligence, Design and Choice. Stages 4 and 5 were added later and are related to revising and implementing the decision process.

The intelligence stage sets out to search for decision situations, by monitoring the organization and its environment. This is not a conventional stage for most of the operational research procedures. In some ways, this stage is related to the view on structuring a problem given by Keeney (1992) with the Value Focusing Thinking (VFT) approach, with particular regard to identifying a decision situation. This stage is also correlated to the vision of strategic management, in which continuous monitoring and diagnosis of the organization and its environment has to be done in order to anticipate decision situations in a proactive way (de Almeida 2013).

Conventionally, most operational research procedures consider that there is already a decision problem to be faced and defining the problem is already part of working towards finding a solution (Ackoff and Sasieni 1968). Therefore, in most cases, the decision process starts with the second stage, that of design. This happens in general in most contexts, especially in the RRM context. However, even in the RRM context, the organization may derive great benefit by introducing a more strategic view for dealing with its decision process regarding risk management and maintenance. For instance, an inadequate maintenance model may affect the competitive position of any organization, when its clients are adversely affected by the effects of the unreliability of its products (goods or services).

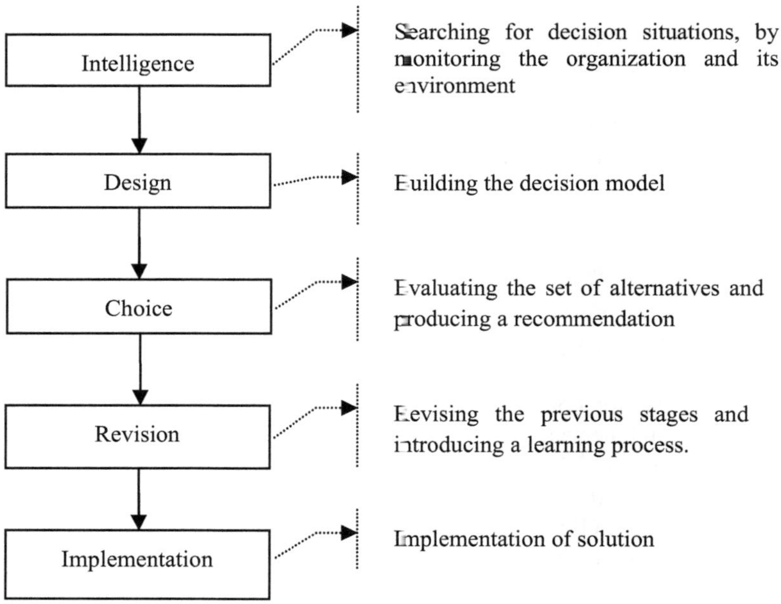

Fig. 1.1 Decision Process

The main focus of the design stage is on building the decision model. This stage includes generating alternatives and other ingredients of the decision model. In this stage the feasibility of alternatives are evaluated. Problem Structuring Methods (PSM) are very useful in this stage in order to ensure that the problem is clearly defined (Rosenhead and Mingers 2004; Eden 1988; Eden and Ackermann 2004). The mathematical model is worked out in this stage and the parameters of the model are estimated. The DM has an important role in this stage, with particular regard to information given through his/her preferences. Also, it is in the design stage that the MCDM/A method is chosen.

Therefore, this stage has a basic role in the decision process and the model designed has to be seen to guarantee that it is related as closely as possible to the real problem faced. As mentioned a model is an approximation of a real situation. There is a provocative aphorism about models related to this issue: "All models are wrong but some are useful" (Box 1979). In other words, the aphorism is saying that all models are approximations of the real situation. In the practical context of building models the following recommendation is relevant: "Remember that all models are wrong; the practical question is: how wrong do they have to be to not be useful?" (Box and Draper 1987).

In the choice stage the alternatives are evaluated according to the model built in order to produce a recommendation to the DM. The form of this recommendation depends on the problematic (Roy 1996), which may be, for instance, a selection of one of the alternative, ranking all alternatives, etc.

Before the recommendation is presented to the DM a revision stage is conducted, in order to evaluate the assumptions chosen and results obtained in earlier stages, and also to check for any possible inconsistencies. In this stage the model building process is evaluated, and takes a comprehensive view, before final confirmation that is given that the model is in an adequate state. Also, this stage incorporates an organizational learning process. Actually, this revision may be done at any time during this whole process, which may be based on a new perception about aspects dealt with in earlier steps (Davis and Olson 1985).

The implementation stage consists of applying the recommendation in the organization or in its environment. Communicating the recommendation is an important action in this stage.

In the decision process there are several actors who play different kinds of role in the decision process. The literature presents a few possible views on who these actors should be (Roy 1996; Vincke 1992; Belton and Stewart 2002; Figueira et al. 2005; Polmerol and Barba-Romero 2000), some of whom are considered in what follows. The decision maker (DM) plays the central role, but may be influenced by other actors. The other actors may include: an analyst, a client, experts, and stakeholders.

The decision analyst (most of the time simply referred to as 'analyst') gives methodological support to the DM in all stages of the decision process, and works on the problem structuring process and building the decision model.

The client is an actor who acts on behalf of the DM and interacts most of the time with the analyst, as a surrogate of the DM. In general this actor is a senior assistant of the DM, who is not available in many situations; or at least for many steps in the decision process. Perhaps this use of the term 'client' came into being as this person was seen as someone who sought the guidance of the analyst, who, in most cases, is an external consultant.

There are other actors, called stakeholders, who try to influence the DM's behavior in order to obtain a satisfactory result, for themselves or those whom they represent. In general, these stakeholders are affected by the decision that will be made by the DM.

The expert is an actor who has specialized knowledge of some part of the system, which is object of the decision process and who gives factual information to be incorporated within the model (de Almeida 2013). This information may be based on prior probabilities related to the state of nature, which represents variables not under the DM's control. This actor may be relevant for decision problems in the context of RRM, since this requires many probabilistic issues to be modeled, such as that done in the Bayesian Decision Theory framework (Raiffa 1968). This kind of actor is rarely mentioned in the MCDM/A literature, but is often present in the literature on Decision Analysis (or Decision Theory).

1.5 Basic Elements and Concepts of Multiobjective and Multicriteria Problems

This section briefly introduces basic ingredients and elements related to multicriteria problems and also relevant concepts that need to be reflected on the decision process related to MCDM/A.

1.5.1 Basic Ingredients and Related Concepts

The basic ingredients include the consequences and the set of alternatives. Concepts related to the family of criteria, the consequence matrix and the problematic are presented below.

A situation is a decision problem if the DM has at least two alternatives, one of which he/she must choose. The set of alternatives may be continuous or discrete. In organizations many managerial decision problems have a set of alternatives consisting of a discrete set of elements a_i, available to the DM. This set may be represented by $A = \{a_1, a_2, a_3, ..., a_n\}$. A continuous set of alternatives may be found, in several situations, such as in maintenance planning, in which the alternatives consist of the time interval t_p, within which a preventive maintenance action should be performed.

In some situations, a continuous set of alternatives may be adapted and presented as a discrete set of alternatives, when this is an adequate approximation for the problem. For instance, the time interval for preventive maintenance t_p, may be seen as calendar days, such that the set of alternatives becomes $A = \{d_1, d_2, d_3, ..., d_n\}$. For any organization, this model is more realistic, since there is no meaning in considering precisely a continuous time t_p, including any time of day or night. Making a choice of any day d_i is a reasonable approximation for the context of preventive maintenance, since a variation in 24 hours does not make a relevant difference in the consequences related to the decision problem.

The concept of problematic is related to the format of recommendation to be made for the set of alternatives, which is reflected in the algorithm to be applied and which will produce the desired result. There are a few types of problematic found in the literature (Roy 1996; Belton and Stewart 2002) and some of those, considered the most relevant for this text, are presented below:

- Problematic of choice - In this problematic the result consists of a chosen subset of alternatives, which should be as small as the procedure can make it. Normally it is desired to have only one alternative chosen, the optimal one. This is a particular situation of this problematic, called: optimization. If the sub-set chosen has more than one alternative, such alternatives are considered incomparable, since the procedure may not be able to find only one alternative. Whatever is the size of this subset, only one alternative is implemented in the end.

- Ranking problematic - In this problematic the alternatives of the set *A* are compared and ranked from the best to the worst.
- Sorting problematic - The alternatives of the set *A* are classified in categories or classes. These classes are specified in the model building process and have a certain order of preference.
- Portfolio problematic - In this problematic there is an interest in choosing a subset of the set *A*, in accordance with the objectives of the problem and subject to some constraints. Unlike the choice problematic, in the portfolio problematic, all alternatives of this subset are implemented in the end. This kind of problematic may be implemented based on the knapsack procedure. A typical example of this kind of problematic is the selection of projects for a portfolio, in which there is a combination of projects from which there is a global value of outcomes to be obtained and keeps within some constraint, such as a limit for the budget.

A fundamental ingredient for the model is the set of consequences, which consists of the outcomes to be obtained by the DM, when making the decision. These consequences are associated with the objectives. For each objective there is a set of possible consequences, which may be the result from the decision process.

The alternatives are evaluated by their consequences. In fact, given that this is the essential aspect of the decision process, the DM does not choose from amongst the alternatives. The choice is made from amongst the consequences, which are informed by the DM's preference structure. Based on this preference information, the model will choose the alternative that can supply the most desirable consequence, according to the DM's preferences.

At this point it is worth recalling an ancient vision regarding the preceding role of consequences for evaluating alternatives. It was presented by Pericles, around 430 B.C., in a Funeral Oration (Thucydides, History of the Peloponnesian War, II, 40):

> "We Athenians, in our own persons, take our decisions on policy and submit them to proper discussions: for we do not think that there is an incompatibility between words and deeds; the worst thing is to rush into action before the consequences have been properly debated. And this is another point where we differ from other people. We are capable at the same time of taking risks and of estimating them beforehand. Others are brave out of ignorance; and, when they stop to think, they begin to fear. But the man who can most truly be accounted brave is he who best knows the meaning of what is sweet in life and what is terrible, and then goes out undeterred to meet what is to come."

This has been quoted in many texts related to risk management. There are many decision problems, in which the consequences are presented in a probabilistic way or there is no information on the frequency of occurrence regarding the elements of the set of consequences. These situations involve decision problems under risk or under uncertainty.

Given the nature of the multicriteria problem, a vector of consequences is considered, since each dimension of this vector is related to each criterion.

For each alternative *i* there is a possible consequence X_{ij}, given the criterion *j*. Let us assume that the set of alternatives is discrete, then, a consequence matrix may be considered as illustrated in Table 1.1. This consequence may be represented by a deterministic or probabilistic variable. The Table 1.1 assumes the deterministic case, in which there is a specific outcome X_{ij}, for each combination of alternative and criterion. There are situations in which the consequence may be presented in a probabilistic way. For instance, for repair time *t*, the consequence may be represented by a probability density function *f(t)* over *t*.

Table 1.1 Consequence matrix

A	Criterion 1	Criterion 2	Criterion 3	....	Criterion j	...	Criterion n
A_1	x_{11}	x_{12}	x_{13}	...		...	x_{1n}
A_2	x_{21}	x_{22}	x_{23}	...		...	x_{2n}
...	...	...	...	...	...	...	...
a_i				...	x_{ij}	...	...
...	...	...	...	...	...	...	...
a_m	x_{m1}	x_{m2}	x_{m3}	...	...	...	x_{mn}

1.5.2 Preference Structures

The DM's preferences are evaluated by means of a preference modeling, considering basic concepts related to preference relations. These preference relations are binary relations applied to compare the elements of the set of consequences $X = \{x_1, x_2, x_3, ..., x_o\}$.

A binary relation *R* over a set $X = \{x_1, x_2, x_3, ..., x_o\}$ is a subset of the Cartesian product *RxR*. Let *x* and *y* be elements of *X,* then a binary relation is a set of ordered pairs *(x,y)*. This relation is represented by *xRy*. If the relation R between two elements (*x*, *y*) does not hold this can be represented as *not(xRy)*. Several properties may be considered for a binary relation *R* such as:
Reflexive, if *xRx*.
Symmetry, if $xRy \Rightarrow yRx$.
Asymmetry, if $xRy \Rightarrow not(yRx)$.
Transitivity, if *xRy* and $yRz \Rightarrow xRz$.

In preference modeling, a relation *R* is commonly called a preference relation. The main preference relations to be applied in this text are the following:

- Indifference (*I*) - *xIy* indicates that the DM is indifferent between the two elements *x* and *y*. Properties applied: reflexive and symmetry.
- Strict Preference (*P*) - *xPy* indicates that the DM clearly prefers the *x* to *y*. Property applied: asymmetry.

- Weak Preference (*Q*) - *xQy* indicates that there is some doubt if either the DM clearly prefers the *x* to *y* (*xPy*) or is indifferent between them (*xIy*), although it is clear that *not(yPx)*. Property applied: asymmetry.
- Incomparability (*J*) - *xJy* indicates that the DM is not able to compare the two elements. Any of the following situations may apply, but the DM can not differentiate amongst them: *xIy, xPy, yPx*. Properties applied: symmetry and not reflexive *(not(xJx))*.

A system of preferences or a preference structure is a collection of preference relations, applied to a set of consequences, such that, the two following conditions hold:

1. For each pair of elements *(x, y)* of *X*, at least one of the preference relations of the system of preferences is applied to *(x, y)*;
2. For each pair of elements (*x*, *y*) of *X*, if one of those preference relations is applied, no other may be applied.

Several preference structures are considered for preference modeling studies. The following preference structures are the ones most applied in practice:
Structure *(P,I)*;
Structure *(P,Q,I)*;
Structure *(P,Q,I,J)*.

Structure *(P,I)* has a symmetric preference relation (*I*) and the other relation is asymmetric. In this structure it is possible to obtain a complete pre-order or a complete order for the elements of *X*. In an order there are no ties (no relation *I*). A pre-order may have ties (existence of relation *I*). For a complete order there is no incomparability. The Structure *(P,I)* corresponds to the traditional preference model, with which many MCDM/A methods are associated. For instance, the additive model for aggregation of criteria, shown in (1.1) is related to this structure. Let *a* and *b* be elements of *X*, then, the following conditions hold for this structure:
$aPb \Rightarrow v(a) > v(b)$.
$aIb \Rightarrow v(a) = v(b)$.

Structure *(P,Q,I)* has a symmetric preference relation (*I*) and two asymmetric relations (*P,Q*). In this structure it is possible to obtain a complete pre-order for the elements of *X*. For this structure, the previous two conditions hold and the following may be added:
$aQb \Rightarrow v(a) \geq v(b)$.

Structure *(P,Q,I,J)* has the incomparability relation, which leads to partial pre-orders for the elements of *X*. This structure is relevant for situations in which the DM is not able to give full preference information; for instance the DM may not be able to compare two elements of *X*. This is not in agreement with one of the axioms for the model in (1.1), which is the first axiom of Utility Theory, and states that the DM is able to make a pre-order of all elements of *X*. This kind of situation has been pointed out by Roy (1996) and Simon (1955), who emphasizes that this may be relevant for MCDM/A situations, in which the DM has to face several dimensions in a multicriteria evaluation.

An evaluation of the DM's preference structure is essential for choosing an MCDM/A method and for building an MCDM/A model.

An arbitrary adoption of any preference structure with a convenient relation for elements of X, such as a complete pre-order or order, with no considerations for the DM's preference may be considered anti-ethical. A situation in which the DM has any doubt about applying the preference relation P is not a justification to assume the indifference relation I. For instance, if the DM declares that he/she is not able to distinguish whether xPy or yPx, and the analyst assumes that this means an indifference relation I between x and y, this may be a distortion in the process. Actually, a few elicitation procedures, for obtaining the preference information from the DM, may induce this kind of distortion. In this situation it should be considered whether indifference or incomparability relation should be applied.

1.5.3 Intra-Criterion Evaluation

Before considering the evaluation of consequences amongst criteria, an intra-criterion evaluation should be conducted. That is, the relative value (performance) according to the DM's preference over the outcomes for each criterion should be considered.

Each criterion represents an objective and can be more formally defined as a function g_j over the set of consequences for criterion j. Let us assume a discrete set of consequences X. This function $g_j(x)$ evaluates the performance obtained by any consequence x, according to the DM's preference. This function $g_j(x)$ may also be referred to as a value function $v_j(x)$, related to the consequence in the criterion j.

As in the previous discussion related to the decision process, in which a choice amongst consequences is involved rather than amongst alternatives, normally this value function is defined over the set of consequences. However, in some situations, this function may be related to the alternatives, such as in (1.1), since for each alternative there is a consequence as result of which this alternative receives its value in (1.1). Therefore, for the sake of simplification the value function may refer to alternatives or consequences, which does mean the concepts previously presented are violated.

Therefore, assuming a discrete and deterministic set of consequences, based on the consequences given for each alternative shown in Table 1.1, the value functions $v_j(x_j)$ for each criterion j may be obtained and applied over the consequences of each alternative i, so that a decision matrix may be obtained, replacing the elements shown in Table 1.1 by $v_j(x_{ij})$. This decision matrix is input for many MCDM/A methods, which include the intra-criterion evaluation.

In an intra-criterion evaluation a linear or a non-linear value function $v_j(x)$ may be obtained. Linear functions are quite common in MCDM/A problems, although the possibility of non-linear $v_j(x)$, should always be considered. For this reason, the consideration of normalization procedures is usual for intra-criterion evaluation, since for linear functions, all criteria and outcomes in Table 1.1 should have the same scale, in order to apply a model such as the additive shown in (1.1). In general this normalization uses a scale of between 0 and 1.

It is essential to understand the scales of each criterion and the restrictions of each MCDM/A method, with regard to this issue, since the kind of normalization may change the properties of the original scale, in which the outcomes are. These issues are discussed in Chap. 2.

1.5.4 Inter-Criteria Evaluation

Since the intra-criterion information is available, the following step can be that of evaluating the inter-criteria, in which all criteria are combined in order to have the global evaluation of all alternatives. For this evaluation an MCDM/A should be chosen and applied.

A classification of MCDM/A methods is presented in the next section and a description of a few methods is given in Chap. 2, but first the concept of a family of criteria has to be accounted for.

A family F of criteria $g_j(x_j)$ is the set $F = \{g_1(x_1), g_2(x_2), \ldots, g_m(x_m)\}$. The model building process should work for a consistent family of criteria (Roy 1996), in which a few properties has to be followed, such as: being capable of representing all objectives related to the decision problem and avoiding redundancies.

Since, for each criterion j, the value of the consequences $g_j(x_j)$ can be produced for all consequences x_j, then, the value of alternatives $g_j(a_i)$ can be obtained for each alternative a_i.

Given the family of criteria, a dominance relation D between two alternatives a and b is defined, considering all criteria g_j. Then, aDb if $g_j(a) \geq g_j(b)$, given all $j = 1, 2, 3, \ldots, m$, and since the inequality is strict (>) for at least one of the criterion j.

The use of the dominance relation could make the use of an MCDM/A method unnecessary. However, it is very rare for a solution to be found by applying the dominance relation. Since, in most situations, many alternatives will not be dominated by others, then, an MCDM/A method is required in order to evaluate the inter-criteria.

In Chap. 2 a description of a few MCDM/A methods are given and the following section gives an overview of their possible classifications.

1.6 Decision Approaches and Classification of MCDM/A Methods

There are four basic decision approaches, which represent perspectives for the decision process, and which are supported by many methods found in the literature.

These methods may be classified and grouped according to their characteristics. This grouping process enables common features of such methods to be understood and facilitates the process of choosing them so as to build particular decision models. Decision approaches on the other hand will give a perspective on the concepts and the organization of systematic knowledge that supports the decision process.

1.6.1 Decision Approaches

The literature differentiates amongst a few decision approaches, which are pointed out as perspectives for the study on the decision process. The literature on decision analysis considers three approaches: descriptive, normative and prescriptive (Bell 1988; Edwards et al. 2007). The literature on MCDM/A also considers a fourth perspective to the decision process: constructivism (Roy and Vanderpooten 1996).

The descriptive approach focuses on describing how people decide in a real situation, the concern being to describe how the DM makes judgments and choices in decision making. This approach is developed by the area of behavioral decision making (Edwards et al. 2007).

The normative approach focuses on rational choice, based on normative models, sustained by an axiomatic framework that aims to ensure a logical structure for decision making. The model in (1.1) is an example of such a normative model, which imposes a specific rational procedure which a DM may follow. The utility theory also provides a rational decision model for decisions under uncertainty.

The prescriptive approach consists of procedures that use a model from the normative perspective, and are structured to support a DM in the decision process. The prescriptive approach may use the results obtained in the descriptive approach, in order to deal with the limitation of human judgment. The errors and inconsistencies examined in the area of behavioral decision are studied in order to build procedures that can address a consistent way of interacting with DMs so as to build the preference modeling process and prescribe appropriate models.

The constructivism approach (Roy and Vanderpooten 1996) consists of an iterative process that uses a learning paradigm (Bouyssou et al. 2006), in which an analyst interacts with the DM with the support of some method, in order to construct the recommendation for the problem that the DM faces.

Whereas the prescriptive approach assumes that the DM has a well-defined preference structure (for instance a utility function to be elicited), in the constructive approach there is an interactive process that aims to help the DM reach a more thorough understanding of his/her preference structure.

1.6.2 Classification of MCDM/A Methods

There are many ways of classifying MCDM/A methods. As first mentioned, MCDM/A methods may be classified according to the action space, which can be either discrete or continuous. Both are of interest for the kind of decision problem analyzed in RRM, especially when a discrete set of alternatives is considered.

A common classification given in the literature (Roy 1996; Vincke 1992; Belton and Stewart 2002; Pardalos et al. 1995) for methods is that in which three types are considered:

- Unique criterion of synthesis methods
- Outranking methods
- Interactive methods

The unique criterion of synthesis methods are based on a process of an analytical combination of all criteria in order to produce a global evaluation or score for all alternatives and for this reason they are said to have a single criterion (global score) that synthesizes of all the criteria. The additive model in (1.1) is a common example of this kind of method and is the basis for many deterministic additive methods, such as AHP, SMARTS, MACBETH. These are methods for a deterministic set of consequences and may be referred to as Multi-Attribute Value Theory (Keeney and Raiffa 1976; Vincke 1992; Belton and Stewart 2002), for which the acronym MAVT is applied. Also, the Multi-Attribute Utility Theory (Keeney and Raiffa 1976), very well known by its acronym MAUT, is included in this group. Most of these methods use the preference structure *(P,I)*, and produce a complete pre-order.

Outranking methods do not use a unique criterion of synthesis, so many of these methods produce the final recommendation with no scores for alternatives. These methods uses the preference structure *(P,Q,I,J)*, considering the incomparability relation, and produce a partial pre-order. The main methods in this group are the ELECTRE and PROMETHEE methods (Roy 1996; Vincke 1992; Belton and Stewart 2002).

The unique criterion of synthesis methods and the outranking methods are representative of several discrete MCDM/A methods.

The interactive methods can be associated with discrete or continuous problems, although in the majority of cases this class of methods includes the Multi-Objective Linear Problems (MOLP). Pardalos et al. (1995) include mathematical programming methods as the third group of methods. A fourth group of methods is included in their classification for disaggregation methods, which consist of collecting information from the DM on global evaluation of a few alternatives for posterior inference on the parameters of an aggregation model. In the end, some of these methods are related to the unique criterion of synthesis methods.

1.6.3 Compensatory and Non-Compensatory Rationality

The methods may be also classified according to their form of compensation for aggregating the criteria, which may be considered a kind of rationality. In this case, two rationalities may be considered leading to: compensatory and non-compensatory methods (Roy 1996; Vincke 1992; Figueira et al. 2005). Bouyssou (1986) made remarks on the concepts related to compensation and non-compensation.

A number of methods may be included in the first type, for instance: MAUT for uncertainty situations and MAVT, such as the deterministic additive methods, including AHP, SMARTS, MACBETH, among many others, embracing basic elicitation procedures; for instance: tradeoff and swing methods (Figueira et al. 2005; Keeney and Raiffa 1976). For non-compensatory methods, lexicographical and outranking methods, such as PROMETHEE and ELECTRE are included in this group.

A preference relation P is non-compensatory if the preference between two elements x and y only depends on the subset of criteria in favor of x and y (Fishburn 1976). Let $P(x,y) = \{j: x_j P_j y_j\}$. That is, $P(x,y)$ is the collection of criteria for which $x_j P_j y_j$. Then:

$$\left.\begin{aligned} P(x,y) &= P(z,w) \\ P(y,x) &= \mathrm{P}(w,z) \end{aligned}\right\} \Rightarrow [xPy \Leftrightarrow zPw] \tag{1.2}$$

In this case, it does not matter what the level of the performance of x or y in each criterion is. The only information necessary is if one is higher or lower than the other.

That is, what the value is of the performance $(v_i(x_{ij}))$, in decision matrix, of an alternative for a particular criterion is not taken into account. It is enough to know if the level of performance $(v_i(a_j))$ of an alternative is higher of lower than another. That is, the only information needed is if $v_j(a_z) > v_i(a_y)$.This would mean that the performance of a_z is higher than the performance of a_y and a_z is preferred to a_y. This is the only information required in (1.2).

Conversely, for a compensatory relation P, it is not enough to know if the level of performance $(v_i(a_j))$ of an alternative is greater or less than another for criterion j. For the compensatory inter-criteria evaluation process, it matters what the value is of the performance $(v_i(a_j))$ for that criterion j, since that amount will be considered, in the aggregation model, as the opposite of a non-compensatory model. That is, for a compensatory method the disadvantage of one criterion may be compensated for by the advantage in another criterion, as can be done in the additive model in (1.1).

As remarked by Bouyssou (1986), a preference relation is compensatory if there are tradeoffs amongst criteria and it is non-compensatory otherwise.

There are many real situations in which the use of a non-compensatory rationality is found. Many examples may be found in sports and some of them are in voting systems.

For instance, in a game of volley-ball, the final result depends on the number of sets a team has won, rather than the total points it gets. The sets represent the criteria, with the same weight in the inter-criteria evaluation (de Almeida 2013). Table 1.2 shows an example of volley game between teams *A* and *B*. Team *A* wins three sets and is considered the winner, since the team *B* wins only two sets. It does not matter how many the teams get in each set. The winner of the set gets all the set value in the process. On the other hand, if a compensatory rationality is applied, then team B would be the winner, since it wins a total of 104 points, against the 93 points team A wins.

Table 1.2 A non-compensatory rationality in a volley-ball game

Team	*A*	*B*	Wins set
Set 1	25	23	*A*
Set 2	25	20	*A*
Set 3	11	25	*B*
Set 4	17	25	*B*
Set 5	15	11	*A*
Total points	93	104	

An interesting example is related to students on a course (Munda 2008), evaluated with grades in a scale from 0 to 10. A student receives grade 4 for mathematics and could compensate this grade, by obtaining a grade 10 in language, for instance, and therefore, passes the final evaluation. This is a compensatory procedure. Otherwise, if the system considers that each student should have a minimal performance in each subject, thereby not allowing compensation amongst different subjects, this evaluation system would be a non-compensatory one.

There is an interesting example in a voting system (de Almeida 2013), which concerns the presidential election in the United States of America (USA). In that system, each state has a symbolic weight, which is related to the number of senators and congress representatives it may have. This is proportional to the population of the state (there are a few exceptions that do not change the final result and for the sake of simplification, are not considered here). Then, the candidate running in the presidential election, who wins the majority of votes in a given state, keeps all the weight of that state. In other words, such a candidate wins all the electoral college votes of that state, no matter the number of electoral college votes that state has. For instance, California is a state with a high weight, and has 55 electoral college votes. The winner candidate in California gets all the 55 votes for the final process. Therefore, as in the non-compensatory process, and volley-ball game illustrated in the Table 1.2, it matters only if the candidate has the majority of votes cast in that state. At the end of the process, the winner candidate is the one who gets the states, whose votes sum up to the majority of weights.

In the presidential election of the USA, the states are equivalent to criteria and the number of votes obtained in each state corresponds to the score for that criterion. The combination of criteria, with their weights, plays the role described for the meaning of the weights in an outranking method (Vincke 1992), which are combined as a coalition of criteria in order to evaluate the best alternative. The winner is the one who gets the best coalition of criteria, with the greatest summation of criteria weights.

It is interesting to note that this non-compensatory rationality means that the presidential election in the USA is a system with a number of v elections, in which v = number of states.

1.7 MCDM/A Models in the Context of Risk, Reliability and Maintenance

The context of risk, reliability and maintenance (RRM) are the focus of this book, although all the concepts and methodological procedures of MCDM/A are applicable to any context in general. For this reason a few issues regarding RRM contexts are discussed below.

In a literature review on MCDM/A models in maintenance and reliability (de Almeida et al. 2015), more than 180 papers published between 1978 and 2013 were found, which had received more than 4,000 citations. In those studies many different criteria were found for modeling MCDM/A problems. Amongst the most common are cost, reliability, availability, time, weight, safety and risk.

Two issues are emphasized in this section regarding MCDM/A models in the RRM contexts:

- What happens when a decision model does not incorporate the DM's preferences;
- The need for MCDM/A models for different kinds of producing systems: services and goods.

The issue related to whether or not incorporate the DM's preferences within the decision model is discussed in the last Section.

There are important issues for MCDM/A models in RRM contexts, which are related to the peculiarities of two different kinds of producing systems: one for services and the other for goods, which have different frequencies of demand for MCDM/A models.

Whatever kind of product it may be, this distinction makes a great difference in the way that maintenance in general (and preventive maintenance in particular) is linked to the results of a business. For instance, a system that produces services has a feature related to simultaneousness (Slack et al. 2010). This means that at the time the system is producing the product itself, the customer is being served. Evidently, in such a context, when a failure in the system occurs, maintenance definitely has a direct and immediate impact on the competitiveness of the business (Almeida and Souza 2001). Therefore, preventive maintenance planning becomes a more strategic decision, linked to highest level of the hierarchical organizational structure. For the mentioned decision context mentioned above, the consequences are characterized by multiple and less tangible objectives, which may require support from an MCDM/A model.

1.7.1 Peculiarities of Service Producing Systems for MCDM/A Models

In service systems, the output is produced while the customer is being served. That is, the main feature of this system is its simultaneousness (Slack et al. 2010). Therefore, the perception of the quality of the service is being created as the client/user is being served, unlike in goods systems, in which the quality is linked to the characteristics of the product itself.

The objectives in service producing systems endeavor to reduce costs, when considered as part of a mix with other objectives, such as: availability, reliability of the system, time during which the system is interrupted and the quality of the service.

In service systems, the interruption of the system can be immediately perceived, since this affects its users. There are many examples of this kind of system: energy, telecommunications, health, transport, and other public services (security, defense, water supply).

For this kind of system, interruptions can lead to serious consequences. Actually, the domain of such consequences is not well defined when compared to the goods producing system. Another issue that has to be considered is related to the actors involved in the process. In the case of the service system, the number of people who are affected by the interruption may be huge. Also, the degree of impact may vary widely per person. Moreover, it is extremely difficult for a business organization to trace the totality of damage caused by the disruption of this kind of product, which is a service.

All things considered, it is easier to understand that failures in these systems are not only restricted to the financial dimension, so it is of paramount importance to have MCDM/A support, in order to provide the DM with a broader view about the problem, and to give to him/her the tools that best take into account the preferential aspects related to this multidimensional consequence space.

Furthermore, there is an increasing share of service products in the goods systems, so that the output of this kind of system turns out to be a combination of goods and services.

1.7.2 Peculiarities of Goods Producing Systems for MCDM/A Models

In systems that produce goods, losses due to machine downtime can be mitigated by increasing production beyond normal capacity or by taking some action to avoid downtime being noticed by clients. In general, failure entails production delays, re-works, inefficiencies, wastages, overtimes, and/or supply storage problems, which are easily converted into costs. This would make the problem change from being one that has multiple objectives to one that has the single objective of minimizing the total costs. That is why most decision models related to this context are not based on MCDM/A methods.

However, there are situations, in which, even for systems that produces goods, the decision context requires an MCDM/A model so that subjective issues can be, for evaluated. There are two main reasons for this:

- These are more strategic decision contexts which are linked to the highest level of a hierarchical organizational structure.
- Failures in the production system affect human or social issues, such as safety, and those to do with the environment.

Moreover, one should be concerned when no DM's preference is incorporated into the model, in the modeling process, as subsequently explained.

1.7.3 Models for RRM Contexts with no Preference Structure

Although most studies related to decisions in RRM contexts do not incorporate DM's preferences, this has been changing over the recent years. The review mentioned (de Almeida et al. 2015) shows that the increase in the number of studies and citations regarding MCDM/A models regarding this area is considerable. However, most studies on the decision process in RRM contexts still do not consider the DM's preferences.

Actually, a 'decision process' which does not include the DM's preferences is one in which no decision is being made. In such a situation, the model has whichever preference structure the analyst has introduced explicitly or implicitly, but this is not the DM's preference. This may be introduced within the model in many different ways, such as: arbitrarily or by chance.

In the former an arbitrary preference structure is explicitly (or almost that) incorporated within the model, in general following a decision previously made by someone else. Otherwise it may incorporate the analyst's perception of which would be the most appropriate preference structure for that context.

In the latter, some preference structure is implicitly incorporated within the model, at random, during the model building process. The analyst makes assumptions for simplifications or just applies what seems to be usual, following standard procedures, without properly considering the specific decision context.

For instance, in many situations the intra-criterion evaluation is skipped and a linear value (or utility) function is applied. This usually happens implicitly. That is, this is not made as an assumption for simplifying the model, in which process the approximation consequences are evaluated by the analyst and put forward to the DM. Actually, most models are built in such a way. In these cases, the characteristics of non-linearity, such as prone or averse behavior regarding risk, are not incorporated and may lead to a different solution, which is inappropriate. The model misinforms the actual decision that should be made. That is why it can be said that there is no decision being made.

References

Ackoff RL, Sasinieni MW (1968) Fundamentals of operations research. John Wiley & Sons, New York, p 455

Bell DE, Raiffa H, Tversky A (1988) Decision making: Descriptive, normative, and prescriptive interactions. Cambridge, UK: Cambridge University Press.

Belton V, Stewart TJ (2002) Multiple Criteria Decision Analysis. Kluwer Academic Publishers

Bidgoli H (1989) Decision support systems: principles and practice. West Pub. Co.

Bouyssou D (1986) Some remarks on the notion of compensation in MCDM. Eur J Oper Res 26(1):150–160

Bouyssou D, Marchant T, Pirlot M, Tsoukis A, Vincke P (2006) Evaluation and decision models with multiple criteria: Stepping stones for the analyst. Springer Science & Business Media

Box GEP (1979) Robustness in the strategy of scientific model building. Robustness Stat. pp 201–236

Box GEP, Draper NR (1987) Empirical model-building and response surfaces. John Wiley & Sons

Brans JP, Vincke Ph (1985) A preference ranking organization method: the Promethee method for multiple criteria decision making, Manage Sci 31:647–656

Davis CB, Olson MH (1985) Management Information Systems: Conceptual Foundations, Structure and Development. McGraw-Hill

de Almeida AT (2013) Processo de Decisão nas Organizações: Construindo Modelos de Decisão Multicritério (Decision Process in Organizaions: Building Multicriteria Decision Models), São Paulo: Editora Atlas

de Almeida AT, Ferreira RJP, Cavalcante CAV (2015) A review of multicriteria and multi-objective models in maintenance and reliability problems. IMA Journal of Management Mathematics 26(3):249–271

de Almeida AT, Souza FMC (2001) Gestão da Manutenção: na Direção da Competitividade (Maintenance Management: Toward Competitiveness) Editora Universitária da UFPE. Recife

de Almeida AT, Vetschera R (2012) A note on scale transformations in the PROMETHEE V method. Eur J Oper Res 219(1):198–200

de Almeida AT, Vetschera R, de Almeida J (2014) Scaling Issues in Additive Multicriteria Portfolio Analysis. In: Dargam F, Hernández JE, Zaraté P, et al. (eds) Decis. Support Syst. III - Impact Decis. Support Syst. Glob. Environ. SE - 12. Springer International Publishing, pp 131–140

Eden C (1988) Cognitive mapping. Eur J Oper Res 36:1–13
Eden C, Ackermann F (2004) SODA. The Principles. In: Rosenhead J, Mingers J (eds) Rational Analysis for a Problematic World Revisited. Second Edition, Chichester: John Wiley & Sons Ltd.
Figueira J, Greco S, Ehrgott M (eds) (2005) Multiple Criteria Decision Analysis: State of the Art Surveys. Springer Verlag, Boston, Dordrecht, London
Fishburn PC (1976) Noncompensatory preferences. Synthese 33:393–403
Hammond JS, Keeney RL, Raiffa H (1998) Even swaps: A rational method for making trade-offs. Harv Bus Rev 76(2):137–150.
Hammond JS, Keeney RL, Raiffa H (1999) Smart choices: A practical guide to making better decisions. Harvard Business Press
Keeney RL (1992) Value-focused thinking: a path to creative decisionmaking. Harvard University Press, London
Keeney RL (2002) Common Mistakes in Making Value Trade-Offs. Oper Res 50:935–945
Keeney RL, Raiffa H (1976) Decisions with multiple objectives: Preferences and Value Trade-Offs. Wiley Series in Probability and Mathematical Statistics. Wiley and Sons, New York
Koksalan M, Wallenius J, Zionts S (2011) Multiple Criteria Decision Making: From Early History to the 21st Century, World Scientific, New Jersey
Likert R (1932) A technique for the measurement of attitudes. Arch Psychol 22 140:55.
Munda G (2008) Social multi-criteria evaluation for a sustainable economy. Springer, Berlin
Pardalos PM, Siskos Y, Zopounidis C (eds) (1995) Advances in Multicriteria Analysis. Kluwer Academic Publishers
Polmerol J-C, Barba-Romero S (2000) Multicriterion Decision in Management: Principles and Practice. Kluwer
Raiffa H (1968) Decision analysis: introductory lectures on choices under uncertainty. Addison-Wesley, London
Rosenhead J, Mingers J (eds) (2004) Rational Analysis for a Problematic World Revisited. Second Edition, John Wiley & Sons Ltd.
Roy B (1996) Multicriteria Methodology for Decision Aiding. Springer US
Roy B, Słowiński R (2013) Questions guiding the choice of a multicriteria decision aiding method. EURO J Decis Process 1:69–97
Roy B, Vanderpooten D (1996) The European school of MCDA: Emergence, basic features and current works. J Multi-Criteria Decis Anal 5:22–38
Simon HA (1955) A Behavioral Model of Rational Choice. Q J Econ 69:99–118.
Simon HA (1960) The New Science of Management Decision. Harper & Row Publishers, Inc, New York
Simon, HA (1982) Models of Bounded Rationality. MIT Press
Slack N, Chambers S, Johnston R (2010) Operations management. Pearson Education
Sprague Jr RH, Watson HJ (eds) (1989) Decision Support Systems - Putting Theory into Practice, Prentice-Hall
Thierauf, RJ (1982) Decision support systems for effective planning and control - A case study approach. Prentice-Hall, Inc., Englewood Cliffs, New Jersey
Vincke P (1992) Multicriteria Decision-Aid. John Wiley & Sons, New York

Chapter 2
Multiobjective and Multicriteria Decision Processes and Methods

Abstract: An appropriate decision-making process is relevant for the strategic success of any organization. Most of the decision problems in these organizations have multiple objectives that have to be dealt with simultaneously. This chapter gives a brief description of a few multicriteria (MCDM; MCDA; MCDM/A) methods, including deterministic additive methods (MAVT), Multi-Attribute Utility Theory (MAUT), connected with Decision Theory, and outranking methods (ELECTRE and PROMETHEE). Additionally, Group Decision and Negotiation process is considered. A procedure for building an MCDM/A decision model is presented, which enables several factors to be incorporated such as: the DM's preference structure and experts' prior knowledge regarding the state of nature. The choice of the method is considered. Some concerns related to choosing an appropriate MCDM/A method are presented, including preference modeling with the evaluation of the DM's compensatory and non-compensatory rationality. This procedure enables an MCDM/A problem to be solved. Several issues concerning the implementation of this procedure are presented, such as: setting scales and normalizing criteria, time management in the scheduling of the decision process (including the procrastination process), and incorporating the intelligence stage of Simon's model into the procedure. Although this procedure may be applied to any context, some particular considerations are given to those of Risk, Reliability and Maintenance. For instance, a multidimensional risk analysis allows a broader view and may include the DM's behavior regarding risk (prone, neutral or averse). In the reliability and maintenance contexts, the models may include availability, maintainability, dependability, quality of repair and other aspects besides cost.

2.1 Introduction

In this Chapter two main issues are dealt with. First, considerations are given to building an MCDM/A model. Then, an overview of MCDM/A methods and multiobjective optimization approaches are set out.

There are many views for building decision models, since the first propositions of operational research area. First, some specific issues are emphasized in this subject in order to establish a basis for the process for building multicriteria

A.T. de Almeida et al., *Multicriteria and Multiobjective Models for Risk, Reliability and Maintenance Decision Analysis*, International Series in Operations Research & Management Science 231, DOI 10.1007/978-3-319-17969-8_2

models, which is subsequently presented. Then, regarding the MCDM/A models, a few concepts and basic issues are presented in order to give a general idea regarding the main concerns in this topic. Thereafter, a procedure is presented for dealing with how to tackle resolving MCDM/A decision problems, including the process for building the associated decision model. Also, a few basic issues related to the building of MCDM/A models in the RRM context are discussed, with some practical insights for this process.

The second topic consists of the describing a few MCDM/A methods, found to be amongst the most relevant for building MCDM/A models for the RRM (Risk, Reliability and Maintenance) context. There follows an overview of the main multiobjective optimization approaches, many of which are used in RRM decision models. Also, an overview of Group Decision and Negotiation (GDN) approaches is considered, since in some situations there is more than one DM.

2.2 Building MCDM/A Models

In the process for building models the main focus is on simplicity with a view to finding a degree of approximation that is good enough to make the model useful. Therefore, when aiming at making a model useful and simple to use several basic factors have to be borne in mind.

Bouyssou et al. (2006) point out that the use of formal models evokes the power of hermeneutics, associated with the facility with which a DM's preferences can be elicited. They state that the latter depends on the intellectual and cultural background of the DM. The analyst should be very cautions with regard to this issue.

On the other hand, the analyst should spend an additional effort in order to work on the DM's interpretation difficulties, which are commonly found in the interactions for preference modeling. Then again, the analyst should avoid the temptations of choosing easy approaches, that, although keeping away from these difficulties, deviates the model from the real problem, which it should be representing, first and foremost (de Almeida 2013a).

Wallenius (1975) states that DMs in general do not trust models, when they find them to be complex. Considering the observation from Bouyssou et al. (2006), it may be plausible that this resistance of a DM rather than that associated with the complexity, it is caused by the DM's intellectual background for dealing with the model. A certain complexity of the model may be acceptable, the better the DM's intellectual and cultural background are.

Building models is a creative process in nature, which involves intuition and other spontaneous actions by the analyst, some of them being inspirations driven in conjunction with the progression of the model (de Almeida 2013a).

In spite of the scientific basis of the models, the building process of which may follow several well-structured steps, as shown in sequence *a*, in Fig. 2.1, its creative side does not recommend a rigorously sequential procedure.

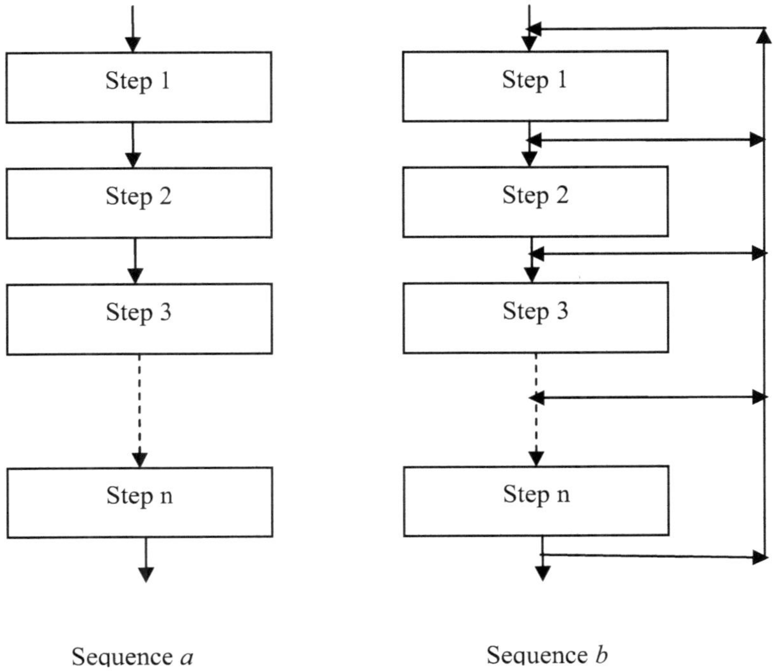

Fig. 2.1 Sequence of steps in the decision process

The rigid procedure, in sequence *a* with consecutive steps, leads to the same result, if a repetition is made in the process.

A different vision is shown in sequence *b* of Fig. 2.1, in which the building process follows a successive refinement procedure (Ackoff and Sasieni 1968). In this procedure, the analyst can return, at any time, from one step back to any other previous step, as often as necessary. This return may or not imply the revision of subsequent steps. This sequence consists of a recursive procedure.

The successive refinement procedure allows any step to be taken in a non conclusive way, so as to conclude it on returning back, after having a view and information from subsequent steps. This return makes it possible to enrich the process with better results for the whole process. Another benefit of this approach is that the creative modeling process is improved, since this flexibility produces an environment that is more susceptible to innovative results. In this process the analyst may get new insights at any time and return to a previous step. In contrast, the rigid approach of sequence *a* do not lead the creativity flows for innovation.

Moreover, it should be emphasized that this flexible and creative process does not hinder the support of the scientific foundations that any model should have. Also, the process for building models follows basic scientific patterns in order to avoid misconceptions.

To build decision models there is a strong support from PSM (Rosenhead and Mingers 2004), whose methods have become vital to understand decision problems, thus leading to a much closer connection between these problems and the models. By using PSM, the analyst has adequate support for organizing information from the actors of the decision process (Franco et al. 2004).

This link between the "real world" and a "model world" is discussed by Keisler and Noonan (2012). Fig. 2.2 illustrates these ideas, including an adaptation for considering Simon's model for a decision process.

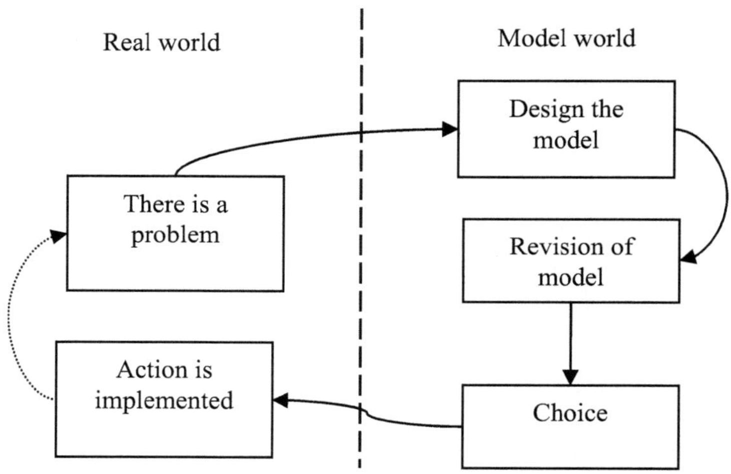

Fig. 2.2 Link between real world and a model world

Figure 2.1 shows that in the "real world", after recognizing the problem, a decision process is started, by building the decision model in the "model world", which will finally produce the implementation of an action. Comparing this view with Simons's model, the stages of design, revision and choice are in the "model world" and the two other stages are in the "real world". In this view, there is a possibility of returning back to reformulate the model, after implementing the action, since this can still provoke the step of problem recognition.

At this point, it should be observed that the model building process may lead to many possibilities of models, as illustrated in Fig. 2.3. In Fig. 2.3, at the beginning of the process, many models are possible. The models are represented by the black spheres. However, during the modeling process, many modeling decisions are taken, in which assumptions, choices of approaches and simplifications are introduced, leading to the elimination (filtering) of some possible models.

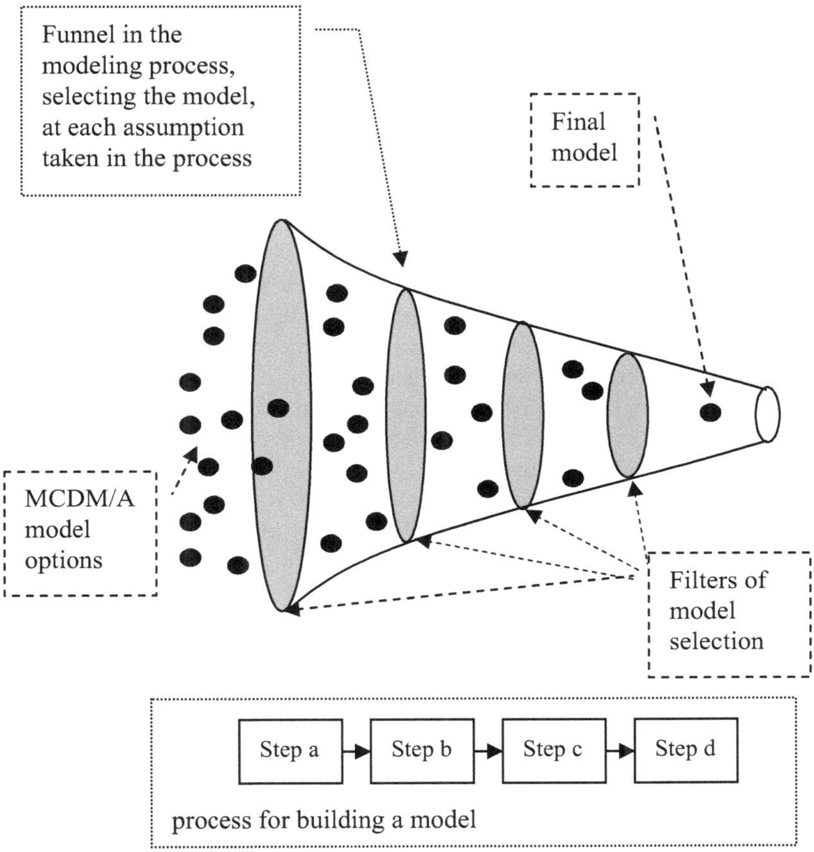

Fig. 2.3 Funnel in the model building process

The filters in Fig. 2.3 indicate that new assumptions or model definitions are taken, thus implying the elimination of some possible models, that would prevail with different assumptions. These modeling decisions also include the preference modeling information given by the DM. Therefore, during the process parameters are assigned to the MCDM/A model, thereby reducing the number of alternative models and leading the process to the final model, as indicated in Fig. 2.3. A similar illustration with a funnel is given by Slack et al. (1995), for a project management planning process.

It is interesting to observe that many models may not even be perceived by the analyst, who eliminates them, by taking directions in the process for building models. If the analyst has some kind of bias, then, this will be reflected in this elimination process and perhaps more useful models may not be taken into account.

There are many propositions or general views presented in the literature for building models in operational research, particularly when using PSM. A few of these propositions and views have been made for MCDM/A model building processes.

Roy (1996) presents a view with several stages for building an MCDM/A model, which includes: establishing the objectives and format of the recommendations; the analysis of consequences and development of criteria; comprehensive preference modeling and operational aggregation of performances; investigating and developing the recommendations.

Polmerol and Barba-Romero (2000) propose a few steps for MCDM/A model building, including: understanding and acceptance of the decision context; modeling alternatives and criteria; Discussion and model acceptance, refinements and evaluation of alternatives with a decision matrix; discussion on the choice of the method, gathering DM information; application of the method; recommendation and sensitivity analysis. They state that this procedure has a linear sequence, but can be done in a recursive way.

Belton and Stewart (2002) also present their view with the following steps: identification of the decision problem; problem structuring; model building; use of model to inform and challenge thinking; developing an action plan.

2.3 A Procedure for Resolving Problems and Building Multicriteria Models

In this section a procedure for building MCDM/A models is presented, based on Simons's model decision process, using the successive refinement procedure for resolution of MCDM/A problems and the basic ideas presented above.

The procedure for resolution of an MCDM/A problem includes the model building process, as shown in Fig. 2.4. The full arrows in Fig. 2.4 indicate the standard sequence to be followed in the process for building models and the dashed arrows indicate the possibilities of returning to a previous step, as allowed in the successive refinement process (for the sake of simplification, dashed arrows are only between two close steps, although the return can be done to any of the previous steps.

The procedure has three main phases, each one with several steps. The first two phases are related to the design stage of Simon's model. First, a preliminary phase is conducted, in which the main elements of the MCDM/A problem are approached and PSM may be applied for the problem structuring. The definitions in this first phase may influence definitively the whole process ahead. In this phase, many possible models may be eliminated, as illustrated in the filters shown in Fig. 2.3.

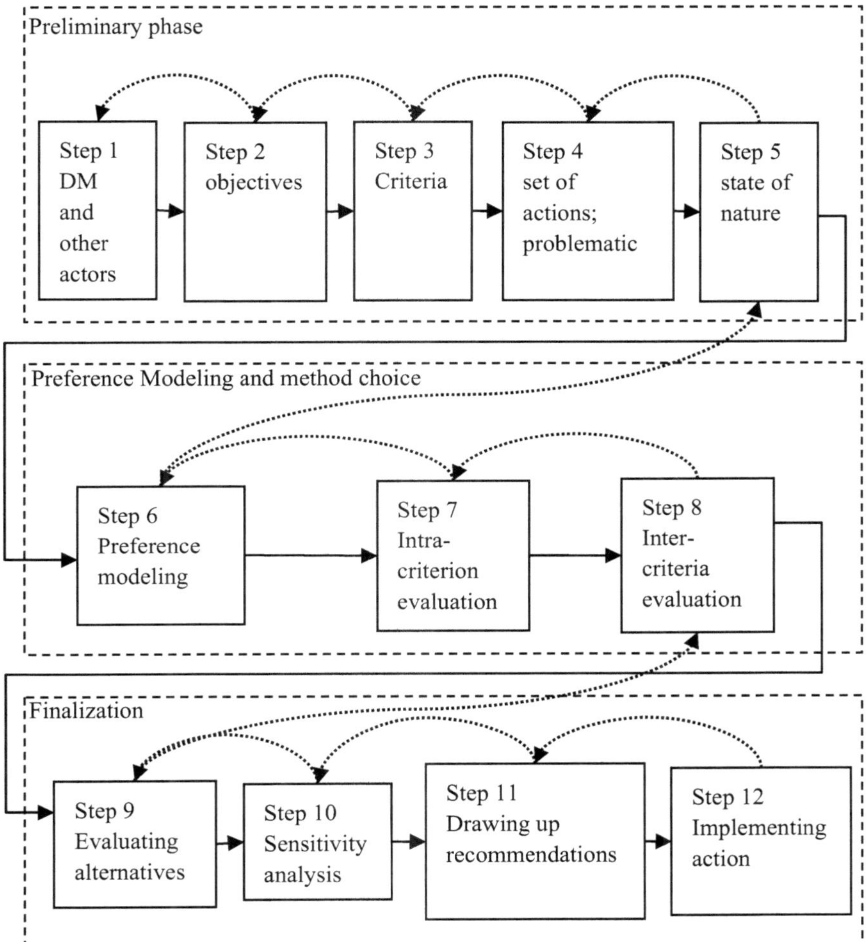

Fig. 2.4 Procedure for resolving an MCDM/A problem

In the second phase the preference modeling is conducted and the MCDM/A method is chosen. At the end of this second phase the decision model is ready to be applied in the third phase, meaning the end of the funnel, illustrated in Fig. 2.3. The second phase is the most flexible of all of them. In fact the three steps of this second phase may be done almost at the same time, exploring a richer insightful process. An already built MCDM/A model is an input to the third phase, although it still may be changed, due to the possibility of returning to review previous steps in the successive refinement process.

In the third phase, the choice and implementation stages of Simon's model are conducted, for the final resolution of the problem. However, it should be remembered that it is still possible to return and make revisions and changes in the built model. In this phase there is a key step of sensitivity analysis, in which this revision decision is evaluated.

The following sections present details regarding the conception and implementation of each step of this procedure.

2.3.1 Step 1 - Characterizing the DM and Other Actors

In this step is important to describe and typify the DM and other actors in the decision process. This procedure has an emphasis on decision problems with an individual DM, although adaptations may easily be conducted in order to contemplate the situation with a group of DMs.

In this step it has to be clear what the role of the analyst is going to be and the DM's participation should be. For instance, the DM may have a more direct or indirect involvement in the decision process. For the latter, another actor, often called the "client" may play some important roles in the process and may be very active in some of the steps of this procedure.

It is relevant to identify how other actors will take part in the process. It is important to characterize the role of each actor for each of the steps of this procedure.

Even for a situation of an individual DM, it is the DM who will decide if decision process may involve many other actors in some steps of the process in order to collect insights and a broad view regarding some particular issues to be included in the model. In this case, the analyst may play the role of a facilitator, who holds meetings with group of actors for a structured discussion of some issues. In general these meetings are supported by PSM approaches (Rosenhead and Mingers 2004; Eden 1988; Eden and Ackermann 2004; Ackermann and Eden 2004; Franco et al. 2004).

2.3.2 Step 2 - Identifying Objectives

This step may be considered the most important one, although this can be only stated in general terms. The most important step for this kind of decision process depends on the nature of the problem, which demands special attention to one these steps of the procedure. It may be the intrinsic nature of the problem may indicate that a particular step has the greatest influence in the quality of the final decision model. Therefore, only in general terms, it may be stated that this step is the most important, since the objectives are going to influence every step in this process.

Moreover, the identification of objectives may influence even the process of establishing the set of alternatives. This may be even more decisive, depending on the approach applied for creating alternatives. For instance, applying the Value Focused Thinking (VFT) approach the process for creating alternatives is very well associated with the structure given for proposing and organizing the objectives.

Actually, the PSM approaches (Rosenhead and Mingers 2004; Eden 1988; Eden and Ackermann 2004; Ackermann and Eden 2004; Franco et al. 2004) in general, amongst which VFT (Keeney 1992), are very useful for conducting this step.

A clear proposition of objectives may be obtained with the VFT approach, in which the objectives are characterized by three factors: the decision context, an object and a preference direction. The objectives are viewed in a hierarchical structure, including strategic objectives, fundamental objectives, and means objectives. The determination of a set of objectives in a decision frame is crucial, since they are the basis of any decision. The insight power of the process is reduced if the set of objectives are incomplete or vague (Keeney 1992).

2.3.3 Step 3 - Establishing Criteria

For each objective previously established a criterion or attribute has to be proposed, which will represent those objectives in the decision model. Therefore, the link between steps 2 and 3 are essential for the representation of the objectives in the whole decision model.

Keeney (1992) states that the attributes are related to the degree in which their associated objective is achieved. Therefore, each objective demands a variable in which this objective can have its degree of performance evaluated. This variable, usually called a criterion or an attribute, in MCDM/A, may also be called a measure of effectiveness or measure of performance.

A family of criteria F has to be established with some properties (Roy 1996). F cannot have redundancy; it must be exhaustive, since all objectives have to be present and represented by F; and it has to be consistent, in the sense that the DM's preferences over the criteria have to be coherent with the global evaluation of consequences.

A structured view for building attributes or criteria is given by Keeney (1992), considering three types: natural attributes, constructed attributes and proxy attributes.

The natural attributes have a common interpretation for all actors in the decision process, such as the cost, which is presented in monetary units. For the objective of minimizing human lives, a possible natural attribute is the number of fatalities per term (annual, for instance). The attributes should be associated with the decision context and must involve value judgments.

The constructed attributes are applied when it is not possible to use natural attributes. For instance, an objective that is concerned with improving the image of a business organization, requires such a type of attribute. Whereas the natural attributes may be used in any decision context, the constructed attributes are adequate only for a particular decision context, for which they have been built.

These attributes require the construction of a qualitative scale for evaluation of the associated objective. These attributes normally are on a discrete scale, which may be called subjective indices or subjective scales. A Table should be drawn up, indicating the meaning of each level of this scale, in a clear way (Keeney 1992). This description should indicate one or several impacts on consequences associated with each level, and specify the degree of achievement of that objective. It is quite common to reach a situation in which constructed attributes are necessary.

If the two previous attributes are not feasible, then a proxy attribute may be tried. This kind of attribute is an indirect measure of the associated objective. In general, the proxy attribute of a fundamental objective is the natural attribute of a mean objective that comprises that fundamental objective.

The criteria should have some properties: measurability, operationality, understandability (Keeney 1992). Measurability defines the objective with more details, thereby allowing the value judgment, necessary in the decision process. An attribute is operational if it describes the possible consequences and provides a common basis for value judgment, and is thus suitable for the intra-criterion evaluation. This property has a very close relationship with step 7, in which a return to this step, for refinement, may be necessary, if the criterion is not properly operational. Understandability means the attribute may not be ambiguous in the description of the consequences.

The criterion or attribute may be considered in two ways, regarding its variability and uncertainty: it may be deterministic or probabilistic. A deterministic criterion is assumed to have a constant level of performance or fixed outcome. A probabilistic criterion has a consequence x, which is a random variable and is specified in terms of its probability density function (pdf): $f(x)$. If a criterion is a random variable, with a not relevant variability it may be assumed to be deterministic. In this case, it is assumed that the standard deviation is so small, that the mean of the variable may represent the consequence x.

For instance, let us consider the time for delivering a product. If the criterion is assumed to be deterministic, then the establishment of the value function, in step 7, will be that of comparing delivering time such as of 2 or 3 hours, for instance. Another similar decision context, associated with the maintenance of electricity supply system, may consider the interruption time (t) of the energy supply. It is not plausible to assume that this kind of criterion is deterministic, since its variability is very high, thus it is clearly characterized as a random variable t. Therefore, the DM has to evaluate this criterion, considering its pdf $f(t)$, since, that is what the DM gets, as a consequence in the decision process.

Therefore, the evaluation to be conducted on this kind of criterion is related to comparing alternatives or consequences with different pdfs, as illustrated in Fig. 2.5.

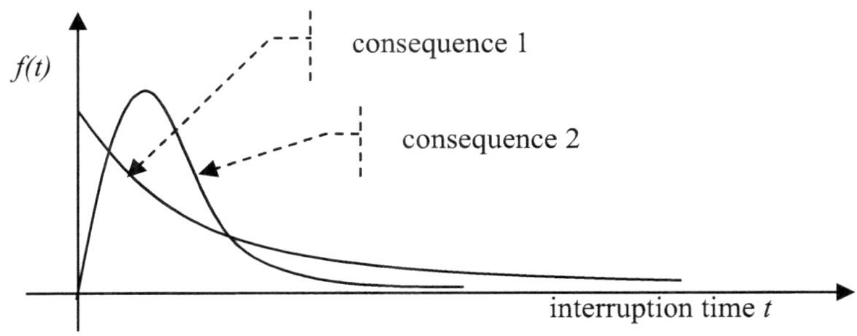

Fig. 2.5 Probabilistic consequences

In this case, the DM does not evaluate the difference in preference between 2 or 3 hours of interruption time (*t*) in the energy supply, since these two consequences do not really exist. Actually, the comparison would be between the consequences or alternatives shown in Fig. 2.5. Which of the two pdfs does the DM prefer? $f(t_1)$ or $f(t_2)$? That is, the DM evaluates the difference in preference between $f(t_1)$ and $f(t_2)$, shown in Fig. 2.5, related to the interruption time (*t*) in hours. This may appear more complicated, at first, although this is actually what the DM gets in the end in this kind of decision context. Regarding the complexity of the question to the DM, it should be pointed out that questions put to the DM in the elicitation procedures are much simpler.

Many problems in the RRM context have this probabilistic characteristic to be considered. A literature review on maintenance and reliability points out the nature of MCDM/A models in this context (de Almeida et al. 2015) and the plausibility of using deterministic representation for criteria, which is discussed in Sect. 2.3.15.

Thus, the model building process in this step may include a probabilistic modeling task for this kind of consequence, which goes together with the preference modeling.

Regarding uncertainty, a criterion or attribute may be found ambiguous in the representation of its value function, by the DM, and therefore fuzzy numbers (Pedrycz et al. 2011) could be used to represent them. In this case, a fuzzy approach may be considered for the decision model, which may influence the choice of the MCDM/A method. This should be properly evaluated in step 7.

2.3.4 Step 4 - Establishing the Set of Actions and Problematic

This step is related to the set of alternatives for solving the decision problem. There are four topics to be approached in this step: a) establishing the structure of the set of alternatives, b) establishing the problematic to be applied to this set, c) generating the alternatives; and d) establishing the matrix of consequences.

The structure of the set of alternatives has a direct connection with the choice of the MCDM/A method, since a discrete or continuous set implies completely different types of methods. For a discrete set of elements a_i, $A = \{a_1, a_2, a_3, \ldots, a_n\}$.

This issue also includes the determination of other features for the set *A*, which can be stable or evolutive (Vincke 1992). In the first case, it is known for the modeling process that the set *A* is fixed and does not change during the building process. For the latter, the analyst should be aware of the possibilities of changes during the decision process, which may represent some kind of constraint.

The set *A* can be globalized or fragmented. In the former, each element of *A* excludes other elements in the resolution process. In contrast, for a fragmented *A*, the elements may be combined for the resolution. A portfolio problematic may be associated to this kind of set. The use of this kind of set is illustrated in Chap. 10.

After establishing the structure of *A*, then the problematic to be applied to this set *A* has to be identified. The problematic may influence the kind of method, depending on the class of methods to be applied. Some methods may be applied in more than one problematic; the case of ranking problematic may include the solution for choice.

After establishment the previous conditions, the generation of the alternatives can proceed. This is one of the most creative tasks of the whole process. Analytical insight may be applied in this task, particularly those delineated by the VFT approach. In this approach the creation of alternatives is based on the value structure of the objectives. In general PSM can contribute in a considerable way to this task, and involve a group of experts supported with the guidance of a facilitator. Depending on the MCDM/A method chosen new alternatives may be included afterwards, even in an advanced stage of phase three, of finalization. Some MCDM/A methods assume a fixed set of alternatives and make pairwise comparisons, for instance. Other MCDM/A methods, build the model and the preference modeling in a consequence space and may introduce new alternatives later on.

At this stage, with the criteria and the set of alternatives established the matrix of consequences can be presented, which consists of the information shown in Table 1.1. For some problems this matrix can be built very easily, since the association of alternatives with the corresponding outcome for each criterion can be made straightforward.

However, for other decision problems this association may not be so straightforward for some of the criteria. In some cases, the outcome to be achieved by the alternatives has to be worked out in more complicated procedures.

This possible complexity for establishing the matrix of consequences could justify that this task would be done as a separate step. However, this association of each alternative with the outcomes corresponds to the very definition of the alternatives, including how they are detailed and specified.

2.3.5 Step 5 - Identifying the State of Nature

The state of nature corresponds to one of the ingredients of decision theory (Raiffa 1968; Berger 1985; Edwards et al. 2007; Goodwin and Wright 2004).

The state of nature consists of factors in the system that are not under the DM's control and may change randomly, influencing the outcomes of the decision process. A variable θ may represent the state of nature and may be a discrete or continuous set of elements.

For instance, in a decision problem related to capital investment, regarding new technologies or machines in an industrial unit, the alternatives are a discrete set of elements a_i, $A = \{a_1, a_2, a_3, \ldots, a_n\}$, which is a factor under DM's control. On the other hand, the demand for the product is the state of nature θ_s, in this problem, which is not under the DM's control. Depending on the nature of the product, it may be represented by a discrete set of states of nature, $\Theta = \{\theta_1, \theta_2, \theta_3, \ldots, \theta_t\}$, such as for units of computers. Otherwise, the set of θ is continuous, for instance: liters of juice.

One should be careful with this ingredient of a decision problem, which is in some situations may be understood as a consequence and represented as a criterion within the model. This could be a critical modeling error, and affect the decision process substantially, including a preference modeling on θ. Natural consequences of this kind of problem may lead to two criteria: *C*: the total cost of the technology (considering the purchasing and operational costs); and *I*: the image of the enterprise as a confident supplier for its costumers.

This ingredient θ is integrated in the model, by its association with the consequences. A consequence function (Berger 1985) makes this association and may be represented by $P(x|\theta, a)$, which for a probabilistic association, such as in the example of machine purchase. $P(x|\theta, a)$ means the probability of obtaining the consequence x, given that the state of nature is θ and the DM chooses the alternative a.

For a discrete representation of θ_s, considering the consequences *C* and *I*, Table 2.1 shows the decision matrix with the state of nature. In this case, the θ_s may represent different scenarios for demand.

Table 2.1 Consequence matrix with the state of nature θs

A	θ_1	θ_2	θ_3		θ_s		θ_t
A1	$(C,I)_{11}$	$(C,I)_{12}$	$(C,I)_{13}$	...		...	$(C,I)_{1t}$
A2	$(C,I)_{21}$	$(C,I)_{22}$	$(C,I)_{23}$	...		...	$(C,I)_{2t}$
...	...	...	...	..	...	...	...
Ai				..	$(C,I)_{is}$	...	...
...	...	...	...	..	...	...	...
Am	$(C,I)_{m1}$	$(C,I)_{m2}$	$(C,I)_{m3}$	..	...	...	$(C,I)_{mt}$

The modeling process with this ingredient is approached by decision theory (Raiffa 1968; Berger 1985), which includes MAUT. The decision model may incorporate prior probabilities $\pi(\theta)$ on θ. Otherwise, the decision is conducted under an uncertainty approach, using an appropriate procedure such as MinMax (Raiffa 1968; Berger 1985).

Thus, if prior probabilities $\pi(\theta)$ are incorporated, a probabilistic modeling task complements the preference modeling. In probabilistic modeling, the analyst applies an elicitation procedure so as to obtain the $\pi(\theta)$. This procedure is usually applied to an expert on the behavior of θ.

2.3.6 Step 6 - Preference Modeling

This is the first step of the second phase of this procedure In this phase the model is built and the MCDM/A is chosen, although both may be changed by returning to previous steps.

This step is very connected to the next two steps and all of them are considerable relevant for choosing the final model, according to the funnel view, given before.

The preference structure should be evaluated in this step. For instance, the preference structure *(P,I)* should be checked with the DM, evaluating if this structure is appropriate for representing the DM's preference. If it is, a traditional aggregation model may be applied, such as the additive model.

However, if *(P,I)* is not adequate, then other structures should be checked, such as the preference structure *(P,Q,I,J)*, in which the incomparability relation is considered.

The analyst may start this process by checking some basic properties of the preference structure *(P,I)*, such as the transitivity and if the DM is able to make a complete pre-order or order in the consequence space. These properties are essential to the structure *(P,I)* and can easily be evaluated with the DM, by checking the relations *P* and *I* on the consequences. This format is more conceptual than operational and could be checked as a preliminary procedure, since these questions in many cases are included in the elicitation procedures of step 8.

That is, these steps 6, 7 and 8, of the second phase, may be conducted in a very flexible sequence, even simultaneously and integrated. This process should be conducted under a non-structured approach, in the sense of this management information systems concept (Bidgoli 1989; Sprague and Watson 1989; Davis and Olson 1985; Thierauf 1982). That is, the non-structured approach is due to the extremely interactive nature of the process, which depends on the DM's characteristics and availability. The process is recursive, with many moves forwards and backwards. This is beyond the view of successive refinement shown in Fig. 2.1. For instance, the evaluation of relations *P* and *I* on the consequences, at step 6, could be done as an anticipation of the elicitation process of steps 7 and 8.

Also, for some decision contexts, the three steps of this phase may be conducted in a sequential way, with no repetitions or returns. Considering the nature of the preference modeling process, everything depends on the DM and decision context.

An important issue to be evaluated in this step is the assessment of rationality regarding compensation amongst criteria, which can be shown in Fig. 3.6.

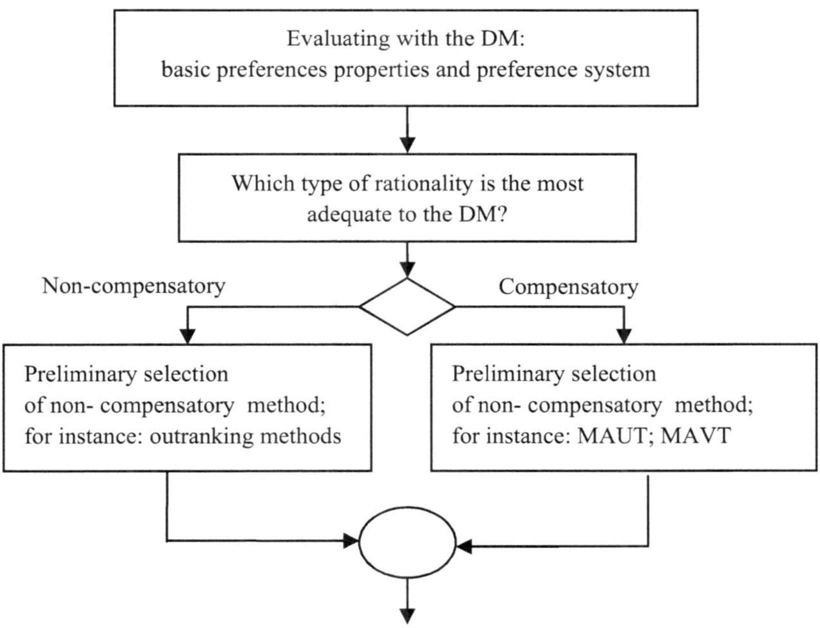

Fig. 2.6 Evaluation of compensatory rationality

This evaluation of compensation is a question for which the number of studies is still very limited and are of a preliminary nature. Therefore, this evaluation may be subjected to some improvisation, since, everything depends on the context, afterwards. This is an important question when choosing the MCDM/A method, since the main classifications of these methods divide into two representative groups: compensatory and non-compensatory methods.

Unfortunately, in many situation when modeling MCDM/A problems, this issue is not even considered. The preference modeling process in most situations is limited to steps 7 and 8 (only this step in many cases), only to parameterizing a model, with a method that has been already chosen, since the very beginning. This is similar to that proverb in which a hammer (method) is always applied, when it is considered that any problem is a nail.

The notion of compensatory and non-compensatory rationality has already been presented and it is related to the Fishburn (1976) concept.

Therefore, after the evaluation proposed by the model illustrated in Fig. 2.6, then, the choice of the MCDM/A method is partially made. Partially, because the final evaluation of methods, in steps 7 and 8 (mainly in step 8), are based on an initial method already chosen in a first round. For instance, if the compensatory rationality is indicated, a method related to the additive model is a natural starting point. Then, the properties of this first method are evaluated, before making a final choice.

2.3.7 Step 7 - Conducting an Intra-Criterion Evaluation

This intra-criterion evaluation consists of eliciting the value function $v_j(x)$ (may be referred to as $g_i(x)$), related to the value of different performances of outcomes in the criterion j. The information given in the decision matrix should be produced in this step.

This intra-criterion evaluation depends on the preliminary selection of an MCDM/A method, in the previous section. On the other hand, the results of this step may influence a revision on the pre-selection of the kind of MCDM/A method made in the step 6.

Regarding the influence of the previous step, if a non-compensatory method is found to be the most appropriate, then, an ordinal evaluation for the consequences may be enough. Therefore, the intra-criteria evaluation may not necessary, if the preferences of consequences in each criterion j are already ordered. In such a case, only a normalization for a common scale may be necessary, which is not often the case.

For a non-compensatory method, such as an outranking method, the indifference and preference threshold consists of an intra-criterion evaluation and is conducted in this step. Also the veto and discordance threshold, commonly part of the ELECTRE method, are evaluated in this step. It should be observed that an interval scale may be required, depending on the formulation required for veto and discordance.

For a compensatory method, such as the unique criterion of synthesis type of method, a cardinal evaluation of outcomes should be considered and so, an elicitation procedure should be applied for obtaining the value function $v_j(x)$. This procedure may produce either: linear or non-linear value functions $v_j(x)$.

For probabilistic consequences, usually the terminology applied is utility function $u_j(x)$, since the value function is usually a term applied for deterministic consequences. Therefore, one of the available utility function elicitation procedures (Keeney and Raiffa 1976; Raiffa 1968; Berger 1985) is applied to obtain $u_j(x)$. These procedures consider lotteries, in which a probabilistic consequence is considered, in order to place choice questions between consequences to the DM. These procedures identify the DM's behavior regarding risk, which may be classified into: neutral, averse, or prone to risk. For a neutral risk behavior, the $u_j(x)$ is a linear function. For both averse and prone risk $u_j(x)$ is a non-linear function. In the elicitation procedure, $u_j(x)$ is obtained in a scale of 0 to 1. Therefore, no normalization procedures are necessary for a linear function $u_j(x)$. It should be observed that the utility function $u_j(x)$ is given in an interval scale.

For deterministic consequences, there are a few procedures available (Belton and Stewart 2002), in which approximations may be made very easily and partial information may be applied to approach the value functions $v_j(x)$.

First of all, it should be evaluated if the value functions $v_j(x)$ are either: linear or non-linear. For linear $v_j(x)$, one of the normalization procedures should be applied, verifying the compatibility of scales for the MCDM/A method and the inter-criteria evaluation procedure applied. For some of the inter-criteria elicitation procedures related to the additive model, the interval scale is considered.

In many practical situations a linear function for $v_j(x)$ may be found to be the most appropriate. Even when a non-linear $v_j(x)$ is indicated, there are many situations in which a linear function can be applied as a good approximation, as has been pointed out by Edwards and Barron (1994), highlighting that a deviation in a model may be better than an elicitation error. A deviation in a model means the use of a linear model instead of a non-linear one.

At this point, this can illustrate the advantages of the flexible process proposed, with the possibilities of returning revise of previous steps. The linear approximation for a non-linear that may be indicated for $v_j(x)$, can have its impact evaluated at the sensitivity analysis step, when the impact of variations in this function $v_j(x)$, may be considered. If variations in $v_j(x)$, change the final recommendation, then, a return to this step in order to replace $v_j(x)$ with a non-linear function may be made.

Step 7 can be affected by the way in which step 3 has been conducted, since the type of attribute (natural, constructed or proxy) may change the process in this step, and in some cases, it can already bring in the intra-criterion evaluation. This is very often the case for the constructed attribute. This may include even the non-linearity of the scale in some cases.

The intra-criterion evaluation may require specific issues depending on the kind of problematic applied; for instance: sorting or portfolio.

If a sorting problematic is applied, then, this step includes the evaluation of the profiles for the categories, in which the alternatives will be classified. These profiles involve an intra-criterion evaluation for the bounds of each category.

For a portfolio problematic the scales of the value function $v_j(x)$ should be considered very carefully. For instance, when using an outranking method, such as

PROMETHEE V, it has been shown that the necessary transformation in the scales requires additional evaluation (de Almeida and Vetschera 2012; Vetschera and de Almeida 2012). For the unique criterion of synthesis methods, based on the additive model, the value function $v_j(x)$ should use a ratio scale instead of an interval scale, which is used by many of the elicitation procedures (de Almeida et al. 2014).

2.3.8 Step 8 - Conducting an Inter-Criteria Evaluation

In this step, the choice of the MCDM/A is made, at the beginning or it may already have been made. The inter-criteria evaluation in this step leads to the parameters of the MCDM/A model, involving the elicitation procedure for the criteria weights. This evaluation depends strongly on the kind of method chosen. Since the meaning of weights changes for different methods, the elicitation procedure depends on the method.

Regarding the additive model, the meaning of the weights, normally called scale constants k_j, does not involve only the importance of the criteria and their elicitation is related to the scales of the value function $v_j(x)$ in each criterion. Actually, there are quite a few MCDM/A methods related to the additive model for aggregation of criteria, in which the main differences amongst them are related to the elicitation procedure applied for k_j.

For the additive model there are also indirect procedures, in which an inference is made, based on the DM's global evaluation of some alternatives. This kind of method is usually classified as a disaggregation method.

Regarding outranking methods, the elicitation of weights is completely different from that for compensatory methods. In this case, the meaning of weights is closely related to the importance of criteria and can be obtained considering this issue.

In the group of methods classified as interactive methods, in which MOLP methods are included, the intra-criteria evaluation is worked out by an interactive process involving dialog with the DM and a system, in general a DSS (Decision Support System). The DM gives preference information at each dialog action, which is alternated by computation action by the system. The DM views the problem by considering the consequence space related to the decision context in question.

There are also many adaptations of classical elicitation procedures for the additive model, in which partial information is required, using interactive procedures.

For probabilistic consequences, using MAUT, there are very well structured elicitation procedures for obtaining the scale constants for aggregation of the utility functions of the criteria (Keeney and Raiffa 1976).

This step concludes the second phase of the process, with two important results:

- the decision model has been built;
- the MCDM/A method has been chosen.

Now, the third phase is started in order to resolve the problem, recalling that a return to and revision of previous steps may be made and the model may change.

2.3.9 Step 9 - Evaluating Alternatives

This is the first step of the third phase of the procedure, the finalization. In this step the set of alternatives is evaluated, according to the problematic proposed. The decision model is finally applied.

This step is straightforward and consists basically of applying an algorithm in the decision model in order to evaluating the set of alternatives.

This step will rarely produce a situation that requires a return to a previous step and the successive refinement has no place in this step, although this may be represented in the model as a vague possibility.

The output of this step is still not enough for an evaluation, required for revision of previous steps. Actually the final result concerning the alternatives has its final consolidation in the next step.

2.3.10 Step 10 - Conducting a Sensitivity Analysis

The result of step 9 consists of a preliminary recommendation, which must be confronted with an analysis of the robustness of the process, regarding variations on the parameters of the model and its input data. This step may indicate that the recommendation is either: robust or sensitive to the input data or to the model features. Also, this step may show that the results in step 9, should be reevaluated, after a revision in previous steps, due to some of the assumptions or input data, or even any inadequate simplification in the model, for instance in the elicitation process.

That is, this step checks to what extent the result of step 9, the model output, is sensitive to variations on the input data and parameters of the model. Regarding data, any organization may have imprecise data with a varied degree of approximation, the impact of which can be tested in this step. Also the process for building the model may have some degree of approximation, and the impact of this can be evaluated in the sensitivity analysis.

Regarding the kind of solution given by the model for each problematic, the sensitivity analysis checks different questions and may require different procedures.

For each problematic, the following questions (changes in the model output) are checked:

- For the choice problematic, the output may present alternatives other than those of step 9, as a solution for the problem. If so, it is desirable to evaluate: how many alternatives are presented; in which alternatives this happens; and in which frequency this happens.
- For the ranking problematic, the output may change the position of some alternatives in the ranking. If so, it is desirable to evaluate: how often this happens; in which alternatives it happens; and the significance of these changes.
- For the sorting problematic, the output may present some alternatives in a class other than that found in step 9. If so, it is desirable to evaluate: how often this happens; in which alternatives it happens; and the significance of these changes.
- For the portfolio problematic, the output may present portfolios other than that of step 9, as a solution. If so, it is desirable to evaluate: how many portfolios are presented; and in which frequency this happens.

If no changes are observed this indicates that the model is robust for that particular set of input data. It may happen that a model appears to be robust for a set of input data and the opposite may happen for another set of input data. It is important to check the model and its parameters and also the input data.

If changes happen in the model output, then, it is necessary to investigate how unacceptable this is. Also, the particular input data or parameter that influences this change is an important piece of information. This may be useful in order to evaluate if the model should be revised, returning to some previous step. At this point, it is worthy remembering that there is not a right model; there are useful models.

The sensitivity analysis may be conducted based on either: analytical analysis of the mathematical structure of the model or numerical analysis on the model, by changing the input data. In spite of simplifications of the model, the complexity of a model may require a numerical analysis.

Many procedures for sensitivity analysis are available in the literature and are not detailed in this text, the main focus of which is to discuss the role of this procedure in the model building process. Therefore, for this focus, the following two kinds of sensitivity analysis are considered:

- for the evaluation of the overall model in a comprehensive process, including all parameters and input data, at once.
- for a particular analysis of a specific parameter or input data.

The former procedure consists of an evaluation of the overall model, by changing, simultaneously, a subset or all sets of input data and parameters of the model. The Monte Carlo simulation procedure may be applied in this case. In this procedure a random generation of the subset or all set of data is made and applied in the model to check the results. This procedure is repeated a number of times (may be hundreds of thousands of times) in order to compare the frequency at which the output changes, considering the problematic in question. Other information to be considered is how significant these changes are, by applying some statistical hypothesis tests, as demonstrated in Daher and de Almeida (2012).

The changes in each piece of data are established, according to a range around the nominal value considered for the model and applied in step 9. The range is specified according to the considerations for assumptions and approximations, given to that particular piece of data in the modeling process. In general, a percentage around the nominal value is applied; for instance, plus and minus 30%, 20%, or 10 %. A probability distribution should be applied for the random generation of data, according to the nature of imprecision observed in the modeling process; for instance, uniform, triangular, normal probability distributions may be applied.

This first procedure consists of an overall evaluation of the model and may indicate whether of not there is a need to continue to the second procedure. The result of this procedure is included in the recommendation to be given to the DM, which is worked out in step 11.

The second procedure is very simple to implement and consists of changing the particular variables of awareness. Each variable is evaluated one at a time, in order to check its specific impact in the model. This procedure may have an important managerial role in the process of building the model. During this process a decision may be made to simplify some step in the procedure. This may be motivated by the time available being limited or the high costs of collecting information (preferential or factual data).

These simplifications may be made on the following issues:

- general assumptions for the model;
- the elicitation process, with approximations in parameters; for instance in the criteria weights;
- assumptions regarding specific analytical structures inside the model;
- using partial information for approximate estimation of input data or model parameters.

For instance, let us suppose that the elicitation process, in step 8, has considered approximations in the criteria weights, due to limitations on the DM's time. Then, in this procedure, the particular impact of changes in the weights may be evaluated, in order to check whether or not approximations in the criteria weights were adequate. If there is no relevant variation, then the simplification in the model may be considered harmless and the results may be accepted. Otherwise, an evaluation should be made of the possibility of returning to step 8 and repeating the elicitation process.

The DM may consider that other solutions produced in this step are equivalent to the nominal solution presented at step 9 and therefore, the results of model may be accepted. The performance proximity of alternatives may lead to such a situation.

A second example may be given for input data, such as estimates for the cost of implementing each alternative. Estimates of costs for implementing projects are obtained with a high level of approximation in many situations. In this case this second procedure for the sensitivity analysis may indicate if the impact of such approximation is relevant and should be reevaluated, by returning to step 3.

The results obtained in this step are as relevant as the solution given in step 9. The DM should know not only the alternative indicated by the model, but also the impact of model simplifications on this result.

There is moreover, an insightful consideration of this step for the whole process for building models. Since the sensitivity analysis can indicate how the model simplifications can affect the results, then, this possibility may influence decisions that the analyst will make as to on simplifying the modeling process. That is, the possibility for successive refinement may indicate that any step, which is cost or time consuming may be conducted with approximations in a preliminary way, and is expected to be repeated, after evaluating the impact of these approximations on the result of step 9.

This may reveal that a rigorous procedure for some steps may be useless, in the context of a building process for producing a useful model, as a simplification of the reality. Therefore, the analyst has to be careful when evaluating the DM and the organizational contexts, when building models.

2.3.11 Step 11 - Drawing up Recommendations

After the conclusion of the last step, if no return to revise previous steps is necessary, then, the finalization is approached in this step by analyzing the final results and producing the report for the DM, with the final recommendations.

The two previous steps produce the main topics to be included in the recommendations to be given to the DM. Also, the main considerations on assumptions and simplifications on the model should be included in the report to the DM.

That is, the DM is not given only the solution indicated in step 9. This is only part of the recommendation. The DM has to be aware of the simplifications in the modeling process and its impact in the solution proposed. This kind of report may be useful for future evaluations regarding results to be achieved by implementing alternatives.

A good report indicates to the DM the extent to which the solution can be trusted. The DM should be advised on the nature of the models. The DM should understand that there is no right model and the usefulness of the model is the main issue to be evaluated.

2.3.12 Step 12 - Implementing actions

Finally, after the DM has received the recommendation and accepted the proposed solution, then, its implementation process can start. This may be either: simple and immediate or complex and time consuming. The latter situation, may require special attention. Also, the way in which the decision is taken may influence the implementation process (Brunsson 2007).

A complex implementation process may be as complex as the decision process and may take much more time to accomplish than the decision process itself. In such situations, occasionally the implementation process may be conducted by an actor other than the DM, who may be afraid of changes in the expected outcomes.

For instance, the implementation process for decisions related to public policy may be so complex and require so much time to be spent on them, that the complex solution may change in format as time goes by, leading to outcomes that are different from those expected at the time of the decision process.

Possible changes in the expected outcomes may happen, when the actor conducting the implementation introduces modifications in the process that may alter the format of the solution and its expected outcomes. In these cases, the DM may be concerned with controlling the content of the solution, although in some cases this cannot be done. The analyst should be aware of this, since this may influence the DM's perception on the relations between the consequences and the alternatives, if the latter may be changed, during the implementation.

There is another issue of time, which is related to the time at which the implementation process should be started. That is, the deadline for starting the process may be considerable, compared with the time for the decision process. This may appear to be controversial, since the time given for producing the recommendation may be short, thus leading to a stressed model building process, and at the end, a longer time is available before starting the implementation. This may be required, when the organization needs to announce the decision made and there is still some time available before initiating the action.

In this situation, a procrastination process may be introduced in this step. The procrastination process consists of introducing and managing a delay before implementing the solution, so that a re-evaluation of the decision may take place. The procrastination (Partnoy 2012) takes place under the allegation that it is more important to take the correct action, than to take it sooner. In this case, it would be wise to procrastinate, taking time to think over the chosen solution. This thinking time may allow the decision made to be revised and thereby to gain new insights from the whole process already conducted.

For some situations, managing this delay is more important than other steps of the decision process. This may suggest the introduction of a sub-step in step 12, in which the implementation delay is managed.

Prudence is of the utmost importance in a procrastination process, since a delay beyond the deadline may bring terrible consequences, even making the chosen solution unfeasible in some situations.

Regarding the scheduling of the decision process (de Almeida 2013a), there are two main deadlines to be taken into account in the 12 steps of this procedure:

- the deadline for choosing a final solution and having a recommendation, in step 11;
- the deadline for starting the implementation process.

The whole scheduling and the time managing process may be illustrated in Fig. 2.7. The first above mentioned deadline has its main effect in phases 1 and 2. The times for working out phases 1 and 2 are related to building the decision model, as illustrated in the final part of the funnel of Fig. 2.7, in which the model is built (chosen). In these phases the deadline is a constraint that obliges the analyst to simplify the model in phases 1 and 2. A greater deadline allows a more cautious process for building models, resulting in a more elaborated model. On the other hand, the first two steps of phase 3 are more technical and take their own time. In this phase, the analyst is concerned with the application of the model.

The dosage of time management is an important issue, since it suggests a balance between two opposite and damage tendencies: streamlining the process too much and detaining the process for unnecessary improvements.

The second deadline is related to the final step, since the decision has already been made. The concerns with the first deadline are over and concentration is on the deadline for starting the action. At this time, the procrastination process may be introduced and this deadline has to be managed carefully. Here the deadline allows a delay in order to give the DM the opportunity to review the decision made, before its implementation. This process is much related to the organizational context and the DM should be very prudent with this delay management.

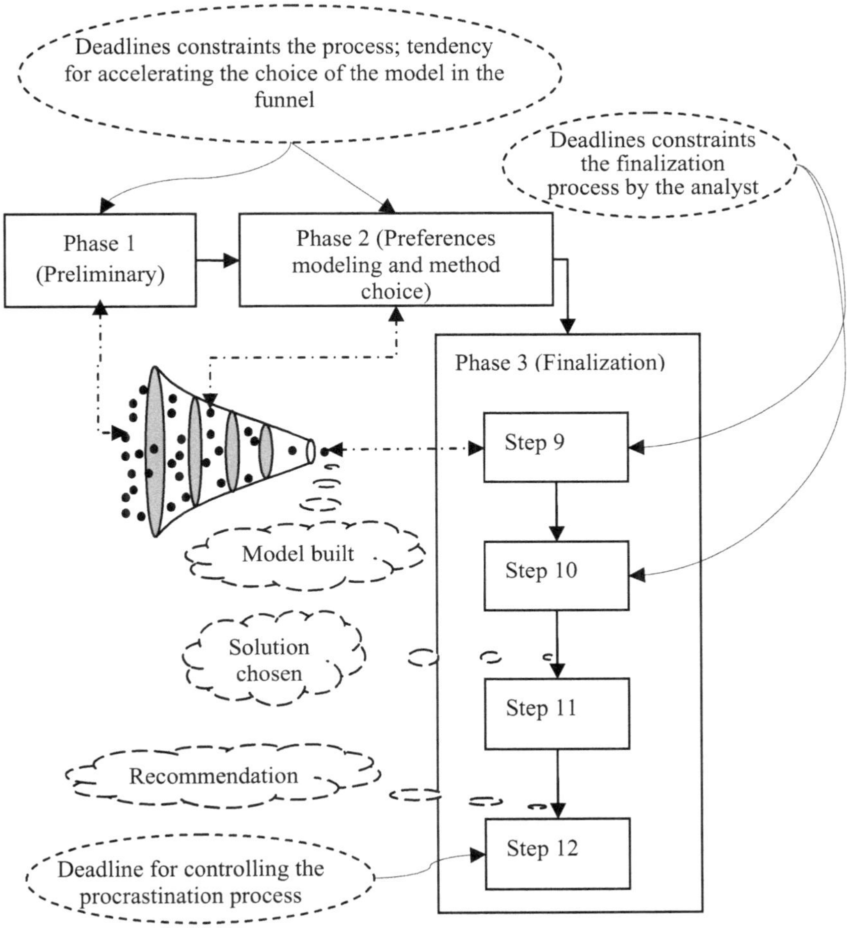

Fig. 2.7 Time managing in the scheduling of the decision process

2.3.13 The Issue of Scales and Normalization of Criteria

Just as in the preference modeling so too in the inter-criterion evaluation, the performance of the consequences may be expressed in terms of numbers. These numbers are presented in a given scale. The scale on which a criterion is presented may define the possibilities for choosing an MCDM/A method. For instance, if the scale of information given in the consequence matrix or in the decision matrix gives only ordinal information one can identify that a given consequence may be greater or lesser than other, but by how much cannot be measured. In such a case, the additive model in (1.1) may not be applied. Therefore, the scales impose constraints for the kind of method to be applied.

Familiarity with these scales and their associated normalization procedures (Polmerol and Barba-Romero 2000; Munda 2008) are important issues for dealing with MCDM/A problems.

First of all, two kinds of scales may be considered: a) a numerical scale; and b) a verbal scale. Amongst the numerical scales the following are the main interest of this text: ratio scale, interval scale, and ordinal scale.

The ordinal scale is the one that has a minimal degree of information. In this scale the numbers only represent the order to be assigned to the elements in a set. They do not have cardinality in the sense that one can say 4 is twice as much as 2. Basic arithmetic operations, such as summation, are not allowed when using this scale. If a decision problem is presented in such a way that some of the criteria are presented in the ordinal scale, then an ordinal method should be applied. A careful application of another method is possible, considering an approximation, in which case one should be careful, when drawing conclusions from the results.

Many verbal and numerical scales are applied for outcomes of criteria, represented by subjective scales, which in the end present information that is only consistent with an ordinal scale. Actually, most pieces of information collected from a DM, by subjective evaluation, using a verbal or numerical scale, are not consistent with a cardinal scale, unless, an adequate procedure is applied to ensure that they are.

The ratio scale is the scale with the greatest degree of information. As suggested by the name, in this scale the cardinality is in the ratio between two numbers. For instance, the weight of an object is presented on this scale. This means that 4 kg is twice as much as 2 kg. The ratio scale has unity and e origin, represented by the zero of the scale, which means absence of property. That is, 0 kg means absence of weight. In this scale a transformation of the following type may be done and the scale properties are maintained: $y = ax$, with $a > 0$. In this transformation the origin is kept and the unity is changed. That is what happens when the weight scale is changed between kg and g. Length and the time are other examples of ratio scales.

In the interval scale, the cardinality is in the interval between two outcomes. In this scale, the following linear transformation may be applied, keeping the properties of the interval scale: $y = ax + b$, with $a > 0$. In this transformation the unity and the origin are changed, respectively by a and b. In this scale the zero does not have the same meaning as in the ratio scale. The zero means just the minimum value of the scale (as is usual in MCDM/A problems). Temperature is an example of an interval scale. In this scale, considering the Celsius scale, one cannot say that 40°C is twice as much than 20°C. On the other hand, one can say that passing from 30°C to 10°C is twice as much as passing from 40°C to 30°C. The above linear transformation may be applied for temperature, so that on changing from Celsius (x) to Fahrenheit (y), one can apply $y = (9/5)x + 32)$.

Verbal scales are applied in many MCDM/A problems and can be transformed into a numerical scale in order to be incorporated into a decision model. This scale may be ordinal or cardinal (ratio or interval), depending on the elicitation procedure applied. However, a simple process of asking a DM to declare a verbal scale for a set of consequences in most cases will produce an ordinal evaluation. A verbal scale that is very often applied is the Likert scale (Likert 1932), in which the number of levels for evaluation is limited to five (there are variations, such as a four-level scale), due to the limited human cognitive capacity for making evaluation in a scale of many levels, such as a ten-level scale, from 1 to 10, which is often applied inadequately.

The type of scale for the consequences of a criterion, as represented in the consequence matrix, causes constraints for choosing an MCDM/A method. Also, the type of scale for a value function $v_j(x)$, shown in the decision matrix is chosen according to the necessary degree of information required and the kind of transformation to be done.

An interval scale is applied in many MCDM/A methods, such as in Utility Theory, and it is in its axiomatic structure. This scale presents a piece of information which has a particular relevance for comparing two alternatives. It shows how much performance is added from one alternative to another. In many situations the DM wants to know, how much is added to go from one position to another. Of course the ratio scale also has interval cardinality and, therefore, gives the same information as the interval scale.

The interval or ratio scale are both applied for methods, such as those based on the additive model in (1.1). The interval scale includes an additional feature that may lead it to be the scale preferred by many of those methods, based on the additive model. In this scale, the minimum value (x_{min}) of an outcome for a criterion j is set to be zero, so that the value of $v_j(x_{min}) = 0$. Since the maximum (x_{max}) outcome is set to be 1, so that the value of $v_j(x_{max}) = 1$, in this scale the range ($x_{max} - x_{min}$) is reduced to a minimum, for the scale 0 to 1. In contrast, for the ratio scale the range ($x_{max} - 0$) is higher, for this scale of 0 to 1. This makes the interval scale more precise for estimating subjective values in the preference modeling process.

There is a specific situation for the model in (1.1), in which the interval scale is not adequate. When using MCDM/A in the portfolio problematic, the interval scale may not be applied, since it induces a wrong solution due to a size effect caused by this scale. In this case a ratio scale should be applied (de Almeida et al. 2014). For other MCDM/A methods similar situations occur (de Almeida and Vetschera 2012) and additional procedures should be implemented.

If the value functions $v_j(x)$ obtained in the intra-criterion evaluation are linear, then, the information produced in the decision matrix can be obtained by a normalization procedure. It should be observed that the term normalization in MCDM/A does not have the same meaning as it has in statistical procedures of normalization.

A normalization procedure consists of carrying out a scale transformation so as to change all criteria to the same scale, since some methods, such as the additive model in (1.1), require this in order to work out the aggregation process. These procedures may change the unity or the origin of the original scale.

There is a close relationship between setting managerial indices (or managerial indicators) and the scales and their normalization process for a criterion. If these indices have to show the level of performance in objectives they should be associated with the DM's preferences.

In MCDM/A methods, in general, this transformation for normalization is made to a scale of 0 to 1. In this case the least preferred (x_{min}) and the most preferred (x_{max}) consequence have the values 0 and 1, respectively.

A few normalization procedures are presented below, considering the discrete set of consequences such as that presented for Table 1.1 (consequence matrix), and an increasing preference with the value of x:

- Procedure 1: $v_j(x) = (x - x_{min})/(x_{max} - x_{min})$.
- Procedure 2: $v_j(x) = x/x_{max}$.
- Procedure 3: $v_j(x) = x/\sum_i x_i$.

For all procedures the values of $v_j(x)$ are obtained in the interval $0 \leq v_j(x) \leq 1$.

Procedure 1 uses an interval scale and the values of $v_j(x)$ may be interpreted as the percentage of the range ($x_{max} - x_{min}$). In this procedure the zero means the minimum value x_{min}. Of course, this procedure does not maintain the proportionality of x. That is, the relation $v_j(x_k)/v_j(x_l)$ may not be the same as that of x_k/x_l.

Procedure 2 maintains the proportionality of x, uses a ratio scale and the values of $v_j(x)$ may be interpreted as the percentage of the maximum value of X (x_{max}), indicating the distance to the leader alternative in the consequence matrix. In this procedure the zero means $x = 0$.

Procedure 3 maintains the proportionality of x and uses a ratio scale. The values of $v_j(x)$ may be interpreted as the percentage of the summation of all consequences of X (x_i), indicating the distance to the leader alternative in the consequence matrix. In this procedure the zero means $x = 0$. This procedure is widely applied when normalizing weights of criteria.

2.3.14 Other Issues for Building MCDM/A Models

This section deals with a few specific issues for building MCDM/A models, such as psychological traps, the choice of the method, compensation of criteria, and the intelligence stage of Simon's model.

Psychological Traps

There are some psychological traps, discussed in the behavioral decision making literature that can affect the quality of the information obtained from the DM, during the elicitation procedures for preference modeling. This is relevant, since the DM's preferences to be included in the model are items of subjective-based information. Simon (1982) discusses the limitation on rationality that people in general have.

A few of these psychological traps are briefly presented below (Hammond et al. 1998a):

- Anchoring - People tend to give a strong weight to information received (impressions, estimates, data) just before making any subjective evaluation. This should be considered in the way that preference questions are put to the DM or factual questions to an expert.
- Status Quo - There is a tendency of choosing actions that maintain the Status Quo. This may lead to confirm and repeat past decisions.
- Estimating and Forecasting - In general people are skilled at making estimates about time, distance, etc, in a deterministic way. However, making these estimates considering uncertainty is different. On the other hand, DMs usually have to make such kinds of estimates for their decisions.
- Overconfidence - DMs tend to be overconfident about their own accuracy, thus naturally guiding them to errors of judgment in preference elicitation procedures. This is one of the traps that affect the DM's ability to assess probabilities adequately.

With regard to the estimating and forecasting trap, Hammond et al (1998a) state that DMs rarely get clear feedback about the accuracy of those estimates they have to make. The feature of successive refinement in the decision procedure described above can minimize this situation, combined with the results of the sensitivity analysis, although this does not improve the accuracy for future estimates.

The way in which questions are put forward to the DM may induce errs, in any of these traps. For instance, the more choices the elicitation procedure gives to the DM, the more chance there is that the status quo will be chosen (Hammond et al. 1998a).

Suggestions to deal with these difficulties are given by Hammond et al (1998a).They also present other psychological traps, which include: confirming evidence, framing, and prudence.

The Choice of the MCDM/A Method

In the literature there are not many studies dealing with the choice of a proper MCDM/A method for a decision problem. However, this seems to be changing. The concern with the matching between the method and problem has increased and may be influencing adaptations in classical methods and even the development and use of hybrid methods. The latter require many cautions, since the integration of different axiomatic structures may lead to serious errors. A few studies deal with this matter. Roy and Słowinki (2013) put several questions for guiding the choice of a method.

The above procedure for building an MCDM/A model gives substantial emphasis to this issue of choosing the MCDM/A method, particularly concerned with the matching with the decision problem, which is the central issue in this matter. Phase two of that procedure is devoted to this topic.

Several factors should be observed for the choice of method, which are closely related to the context of the model building process, and may include:

- The nature of problem analyzed, which is the central feature in the whole process;
- The context in which the problem is faced, which includes organizational issues, and the time available for the decision to be made;
- The DM's preference structure;

Unfortunately, the analyst's preference on the method may play an important role in this process. This may bring ethical considerations to the process. Rauschmayer et al. (2009) discuss the ethical issues in the modeling process. They state that the choice of the method and its parameterization is not neutral and may bring an ethical problem if:

- Distortions in the results are made for interests other than the DM's and the organizational one, in which the problem is faced.
- The assumptions are not shared with the DM.
- The assumptions are selected in a malicious way

It should be noted that the second issue above is carefully considered in step 11 of the above procedure, since all this information should be included in the recommendation report.

One of the main issues in the choice of an MCDM/A method is the evaluation of the DM's preference structure with regard to compensatory and non-compensatory rationality, as highlighted in step 6 of the procedure for building MCDM/A models. Simon (1955) pointed out the importance of this issue, before many of the MCDM/A methods had been developed. Bouyssou (1986) made remarks on the concepts and notion of compensation and non-compensation and discussed a few axiomatic issues.

According to Vincke (1992) the choice of a method for aggregating criteria, such as the additive method, for instance, is equivalent to choosing the type of compensation amongst those criteria. Roy and Słowinki (2013) are concerned with this issue, in the context of choosing a method, when they put the following question "Is the compensation of bad performances on some criteria by good ones on other criteria acceptable?".

Although step 6 of the procedure for building an MCDM/A model includes this evaluation of the DM's willingness or otherwise to make compensation, no details are given on how to deal with this. Indeed, there is still much research work to be conducted on the evaluation of the DM's willingness to make compensations, even though this is an extremely relevant factor for the choice of methods.

The Intelligence Stage of Simon in the Procedure for Building Models

The foregoing procedure for building an MCDM/A model does not include the intelligence stage of Simon's model for the decision process (Simon 1960). This procedure assumes that there is already a problem that has been identified at the start of the design stage of Simon's model. Fig 2.8 shows how this intelligence stage can be integrated with the procedure described above for building a decision model.

This intelligence stage requires a continuous monitoring process on the status of the organization or the decision context, in which attention to the decision process is established, and also its external environment.

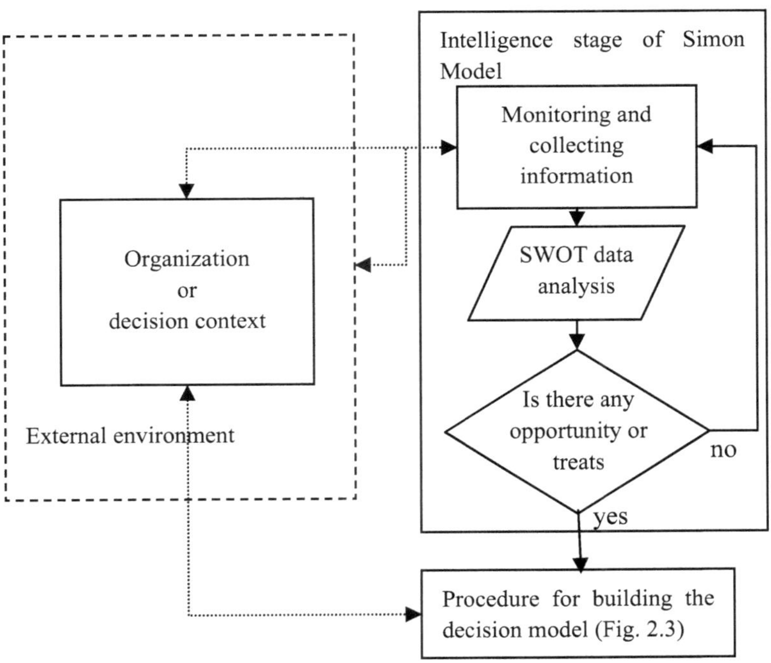

Fig. 2.8 Integrating Simon's intelligence stage

This monitoring process may, at any moment, indicate a situation requiring attention and then data collected are analyzed, in order to identify whether or not there is a problem to be solved, which may include an opportunity to be explored. If so, then, the above procedure is initialized.

This monitoring process is very well associated to the strategic management process, in which the diagnosis analysis of the internal and external environment of the organization is conducted. Also, the VFT approach proposed by Keeney (1992) can be considered in the model shown in Fig. 2.8. Using the VFT approach, the specification of values would guide the monitoring process.

2.3.15 Insights for Building MCDM/A Models in the RRM Context

In an MCDM/A model for the RRM Context, uncertainty is usually a certain thing. That is, a decision under a certainty situation may be possible only as a simplification of the model. Also, this may be justified either: when the variability of the random variable is not considerable or when the use of quantiles of the probability distribution for the variables, such as criteria, may be applied as a good approximation.

For the former, a deterministic approximation is quite useful and justifiable. The mean of the random variable can be applied, since the standard deviation is assumed to be too small.

For the latter formulation, a deterministic approach is usually applied, although there are many concerns to be taken into account with that approximation. An alternative to this procedure is the disaggregation of the criterion into two: the mean and the standard deviation of the random variable. The analyst should evaluate very carefully, which of these possibilities the DM can better understand. Even, the choice of the quantile should be considered the best option for the DM's understanding; for instance, the quantile could be either: 90% or 80% of the distribution.

It should be noticed that deterministic MCDM/A methods are largely applied in reliability and maintenance contexts. Table 2.2 derived from a literature review shows the percentage use of different MCDM/A approaches in maintenance and reliability problems (de Almeida et al. 2015).

Table 2.2 MCDM/A approaches applied in reliability and maintenance research

Method	Percentage
Pareto Front	48.39
MAUT	10.22
AHP	9.68
MACBETH or other MAVT	8.60
Goal Programming	3.23
ELECTRE	2.69
PROMETHEE	2.15
TOPSIS	1.08

As can be observed, in most cases, it seems that a deterministic model is applied, since it is not clear what amount of probabilistic adaptations is conducted in these methods. One can wonder how much this is related to either: a simplification of model itself or a bias in the analyst's choice.

This issue is relevant, since reliability and maintenance contexts are very closely related to risk considerations by their very concepts. An interesting reference on uncertainties in MCDM/A (Stewart 2005) shows different meanings for uncertainty and how to deal with them, including a few guidelines for practitioners. Also, many issues related to a risk analysis of uncertain systems are considered by Cox (2009). For instance, he discusses the limitations of some quantitative risk assessment, such as frequency, which is often applied to explain risk, yet which does not contain enough information for a clear decision to be made.

MCDM/A Models in the Risk Context

With regard to the risk context, there is a variety of concepts in the literature on risk and also on its perception (Chap. 3 deals with this topic). Some of them consider only the probability for a specific context. However, if a decision is being made then the consequences should be considered. Also, the model should incorporate the DM's preferences over these consequences. In fact, a 'decision process', in which the DM's preference is not considered is not a process in which a decision is actually being made, as discussed at the end of Chap. 1.

According to Cox (2012), the application of utility functions rather than simple risk formulas – consisting of terms such as exposure, probability and consequence - allows a DM's risk attitudes to take into account, thereby improving the effectiveness of the decision making process to reduce risks. Cox (2009) discusses many issues related to the decision process in the risk context, including the limitations of risk assessment using risk matrices and a normative decision framework.

Another classical problem within the risk context is the direct association between the quality of a decision and the actual consequence obtained at the end. In fact, at the time in which the decision is being made, the DM cannot assure the best consequence, since there are uncertainties in the process. Therefore, only expectations can be evaluated when making the decision. In general, this is something difficult for many DMs to understood and the analyst should be aware of how to deal with this by clarifying all these issues to the DM, instead of using inadequate models for simplifying what is going to be shown. These clarifications should be made in step 11, when drawing up the recommendations to the DM.

Interpretation of an MCDM/A Model or Utility Function Scores

There are many concerns in the literature with regard to interpreting the scores for the alternatives given by utility functions. This concern is extended in general to any MCDM/A method that gives final scores for alternatives, thus representing a global evaluation, based on the aggregation of multiple criteria. However, these numbers can be interpreted according to the properties of the scale, for each particular method, in order to compare alternatives.

If the method uses a ratio scale, it is relatively easy to produce a comparison of alternatives, considering the ratio of their scores. For instance, in a choice problematic, a first alternative may be twice as good as the second one, or it could be 20% better than the second one.

Even for a specific scale, such as the ratio scale, the meaning of this ratio may be explained, by taking the rationality behind the method into account. For instance, in the PROMETHE II method, the scores are based on the summation of criteria weights, within a non-compensatory rationality.

With regard to the interval scale, which is applied for the utility function of many of the MAVT methods, the alternatives may be compared based on the properties of this scale.

The interval scale allows an incremental comparison between alternatives. That is, the differences of the scores of the alternatives are considered. However, a ratio may also be considered between two differences, as shown in Chap. 4 (see Equations (4.13) and (4.14)). Therefore, a difference ratio *DR* may be applied to interpret the values in relation to the alternatives, so that: $DR = \frac{v(a_p) - v(a_{p+1})}{v(a_{p+1}) - v(a_{p+2})}$, in which p represents the position in the ranking obtained by alternative a_p and $v(a_p)$ represents the score of the alternative. By analyzing these *DR* results, the DM can perceive the distance between the pairs of alternatives. This is illustrated in Table 2.3.

Table 2.3 Analysis of scores of an MCDM/A method with an interval scale

Alternative *i*	Position *(p)* of the alternative	Value or Utility	Interval	Ratio of intervals *(DR)*
A2	*1*	0.70	0.10	0.77
A5	2	0.60	0.13	6.50
A1	3	0.47	0.02	0.40
A3	4	0.45	0.05	1.00
A7	5	0.40	0.05	5.00
A8	6	0.35	0.01	0.04
A4	7	0.34	0.24	-----
A6	8	0.10	-----	-----

Table 2.3 presents the position of the alternatives in the second column, their scores in the third column and their comparisons by the increments of the scores in the fourth column. The fifth column shows the *DR*, from which it can be observed

that the increment of the scores from A_1 to A_5 is 6.50 times greater than that from A_3 to A_1.

Another possible way to explain these results to the DM is to consider the ratio of differences between two alternatives and the whole range, given by the range between the best and the worst scores. This difference can be expressed as a percentage of the whole range. That is, in Table 2.3, the whole range is $v(A_2)-v(A_6)=0.70-0.10=0.60$. Therefore, the difference in scores between alternatives A_5 and A_1 is 22% of the whole range, while for alternatives A_1 and A_3 it is 3%.

Applications of these indices are given in Chap 4. The analyst may use any one of these indices, after evaluating which of them is the most appropriate for a given DM to understand.

Paradoxes and Behavioral Concerns Related to Risk Evaluation

With regard to the use the expected utility function for models in the risk context, there are a few paradoxes with which the analyst should be aware of. These paradoxes have been analyzed by behavioral decision making studies, in the descriptive perspective context.

There are other approaches that deal with some particular situations, such as Rank-Dependent Utility (RDU) and Prospective Theory (Edwards et al. 2007; Wakker 2010).

There are many situations regarding risk which cannot be easily integrated into decision models. The kind of event known as a 'black swan', related to the so called 'black swan theory' may be an example of such a situation. This event is related to a kind of occurrence that is very unexpected (very low probability), with very undesirable consequences. These are rare events, which result in a great damage. In general, their evaluation is not well accepted in the expected value principle, since the multiplication of the value of such great damage is excessively reduced by the value of an extremely low probability.

On the other hand, although many concerns with the use of the expected utility function are clamorously announced in part of the literature, the analyst should be aware that in many situations these behavioral issues do not matter for many practical problems. It is necessary to understand their meaning and to evaluate them when they are relevant. Unfortunately, in many situations, these matters are inappropriately announced in order to justify other less adequate approaches.

2.4 Multicriteria Decision Methods

A brief overview of MCDM/A methods is given in this section with emphasis on those most often found in practical application, balanced with the most appropriate ones for the RRM context.

First the methods related to unique criterion of synthesis are presented, then some outranking methods are introduced. Interactive methods, related to MOLP are very briefly mentioned, since most of the problems in the RRM contexts are non-linear problems. The next section deals with heuristics and evolutionary multiobjective algorithms for dealing with multiobjective models.

2.4.1 Deterministic Additive Aggregation Methods

This is one of the most applied models for aggregating criteria and it is usually classified as MAVT (Belton and Stewart 2002), being part of the group of methods of unique criterion of synthesis. MAVT is distinguished because it considers deterministic consequences, whereas MAUT (see next subsection) deals with probabilistic consequences (Keeney and Raiffa 1976).

The additive model, also called a weighted sum model, is recalled from (1.1) and reintroduced below for prompt reference in (2.1), in which the global value ($v(x_i)$) is considered for a consequence vector $x_i = (x_{i1}, x_{i2}, ..., x_{in})$, for the alternative i, which is the same as the global value $v(a_i)$ for alternative a_i, as indicated in (1.1).

$$v(x_i) = \sum_{j=1}^{n} k_j v_j(x_{ij}) \tag{2.1}$$

where:
k_j is the scale constant (weights) for attribute or criterion j.
$v_j(x_{ij})$ is the value of consequence for criterion j, for the alternative i.
x_{ij} is the consequence or outcome of alternative i for criterion j.

The scale constant is usually normalized as follows:

$$\sum_{j=1}^{n} k_j = 1. \tag{2.2}$$

Properties for the Additive Model

The additive model has a few properties that should be checked before making a decision on its application. For practical modeling purposes the main properties are briefly described.

This model follows the preference structure *(P,I),* in which it is possible to obtain a complete pre-order or a complete order. For two consequences x_z and x_y, the following conditions hold for this structure: a) $x_y P x_z \Rightarrow v(x_y) > v(x_z)$;

b) $x_y I x_z \Rightarrow v(x_y) = v(x_z)$. Therefore, one of the assumptions of this model is that the DM is able to compare all consequences and order them. Also the transitivity property holds for the preference relation R, whether it is P or I, so that for three consequences x_w, x_y and x_z, if $x_w R x_y$ and $x_y R x_z \Rightarrow x_w R x_z$.

Another property of this model is the mutual preference independence condition amongst the criteria (Keeney and Raiffa 1976). Let Y and Z be two criteria, the preference independence between Y and Z occurs if and only if the conditional preference in the Y space (intra-criteria evaluation given, different levels of y, such as y' and y''), given a certain level of $z = z'$, does not depend on the level of z. That is, $(y',z')P(y'',z') \Leftrightarrow (y',z)P(y'',z)$, for all z, y' and y''.

This property may be formally presented in the following formulation (Vincke 1992). Let a, b, c and d be four vector of consequences in a consequence space with two criteria Y and Z. Then, Y and Z are preferentially independent if the following condition holds: If for criterion Y, $v_y(a) = v_y(b)$, and $v_y(c) = v_y(d)$, and for criterion Z, $v_z(a) = v_z(c)$, and $v_z(b) = v_z(d)$, then, $aPb \Rightarrow cPd$. This is illustrated in Fig. 2.9.

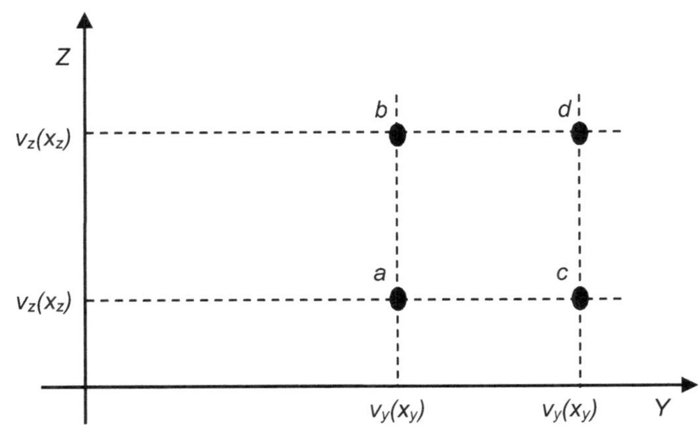

Fig. 2.9 Preference independence condition

Therefore, the validation of this model should be done by confirming that the DM's preference structure is according to these properties. In some practical situations a DM may refuse to follow the final recommendation based on this kind of model, when a violation of one of these properties occurs and there alternatives close to the solution, in which it is obvious for a global evaluation that a property is violated. The DM may not be able to perceive which property is being violated, in such cases, but can recognize the inconsistency of the final result. Although, the DM can distinguish this kind of inconsistency only in an obvious situation, this shows that this may not be an issue to be ignored.

Therefore, these properties should be evaluated very carefully, before making a decision of going through them. Of course, the additive model may be applied, as a typical simplification procedure for model building, where some property is not consistent with the DM's preference. However, the analyst should evaluate carefully to what extent this is inconsistent with the DM's preference.

Regarding the preference independence property, it has been observed that in most practical situations this property is not violated. This may explain, in part, the broad dissemination of the use of this model, although the other properties should also be considered. Yet, regarding the preference independence, Keeney (1992) points out that the preference dependence may indicate that a criterion may be missing. In this case, a revision of steps 2 and 3 of the above procedure may allow a better structuring of the problem.

Also, practical applications have shown that the violation of this property is more likely to happen for a large range of consequences. For a small range of consequences, the mutual preference independence is more likely to hold. This has an interesting relation with the kind of scale applied to a criterion. For instance, a ratio scale tends to be larger than an interval scale. Therefore, one should be careful, when changing from an interval scale to a ratio scale, for a problematic of portfolio that requires the latter (de Almeida et al. 2014).

Elicitation Procedures for Scale Constants

There are many elicitation procedures in the literature for the elicitation of the scale constants (Weber and Borcherding 1993). Amongst these are the tradeoff and the swing procedures which are described below.

The tradeoff procedure is presented in detail by Keeney and Raiffa (1976). Weber and Borcherding (1993) consider that this is the procedure with the strongest theoretical foundation.

This procedure is classified as an indirect procedure (Weber and Borcherding 1993), since the determination of the scale constants is based on inference from information given by the DM. It is also classified as an algebraic procedure, since it calculates the n scale constants from a set of n-1 judgments often using a simple system of equations, which also includes (2.1).

This procedure is based on a sequence of structured questions (Keeney and Raiffa 1976) put to the DM, in order to obtain preference information, based on choices between two consequences. A first group of questions obtains the ordering of the scale constants, then, other questions prepare the DM to understand better the consequence space and finally, the DM makes choices between pairs of consequences related to neighboring criteria, in order to make the tradeoffs for the equations for the algebraic process.

Thus, the procedure is based on the DM making a comparison on two consequences $x^b = (x_1, x_2, ..., x_j, ..., x_n)$, which is a vector with the consequences x_j for each criterion j. These consequences have the best outcome b_j, for one of the criteria and the worst outcome w_j for the other criteria. For instance, $x^2 = (w_1, b_2, ..., w_j, ..., w_n)$ has the best outcome for the criterion $j = 2$, and $x^3 = (w_1, w_2, b_3, ..., w_j, ..., w_n)$ has the best outcome for $j = 3$. If the DM's preference is such that x^3Px^2, then, $v(x^3) > v(x^2)$. Based on (2.1), the value of $v(x^b) = k_b$, since $v(b_j) = 1$ *and* $v(w_j) = 0$. Therefore, if x^3Px^2, then, $k_3 > k_2$. Using these kinds of questions, the order of the scale constants is obtained.

Next, another pair of consequences is compared in order to find indifference between them, by decreasing the value of the outcome b_j for criterion j which is the preferred one. For instance, for x^3Px^2, the consequence b_3, of x^3, has the outcome decreased to the level of x_3, such that $x^{3'}Ix^2$, in which $x^{3'} = (w_1, w_2, x_3, ..., w_j, ..., w_n)$. If the DM can specify the outcome $x^{3'}$, such that $x^{3'}Ix^2$, then, $v(x^{3'}) = v(x^2)$. Since, $v(x^b) = k_b$ and $v(x^{b'}) = k_bv_b(x_b)$, by applying (2.1), this leads to $k_3v_3(x_3) = k_2$. This equation is related to one of the $n\text{-}1$ judgments for the system of equations necessary in this procedure, in order to obtain all the scale constants k_j.

A critical judgment in this procedure is adjustment of the outcome in order to obtain the indifference between the two consequences above (Weber and Borcherding 1993).

The swing procedure is included in the SMARTS method (Edwards and Barron 1994). This procedure is classified as an algebraic procedure and also as a direct procedure (Weber and Borcherding 1993), since the determination of the scale constants are based on direct information given by the DM, taking the range of the consequences into consideration.

This procedure is also based on a sequence of structured question (Edwards and Barron 1994). The first question considers the following consequence $w = (w_1, w_2, ..., w_j, ..., w_n)$., in which all criteria have the worst outcome. Then, the DM is asked to choose one of the j criterion to improve the outcome of w_j to the best outcome b_j. That is, the DM may choose a criterion to 'swing' from the worst to the best outcome. This indicates criterion j for which the scale constant k_j has the greatest value. Then, the DM is asked to choose the next criterion, and so on. At the end the scale constants of the criteria are ordered. Then, in another step, the criterion with the largest value of scale constant is arbitrarily assigned 100 points. The other criteria are assigned points expressed as percentages of the criterion with the largest scale constant value, considering their range. Finally, these percentages are normalized to produce the final scale constants.

Avoiding Misinterpretations Regarding the Scale Constants

There is a quite commonly disseminated misconception (for additive models) of associating the meaning of the scale constants with the degree of importance of the criteria. This represents a source of one of the main modeling mistakes when the additive model is used.

In the additive model, this parameter cannot be determined as weights, considering only the degree of importance of the criterion, which may be appropriate in other methods, such as in outranking methods. Although the value of a scale constant of a criterion may be associated with its importance, there are other issues to be considered. The value of a scale constant is also related to the scale range of the consequences for the criterion (Edwards and Barron 1994). For instance, in a decision problem for purchasing a product, in which any five criteria are considered, including the price, one could state that the price is the most important criterion, thus with the largest weight. However, if the outcomes related to price are in a very narrow range of consequences, let us say between $ 99,990 for the best price and $ 100,005, for the worst price, it does not seem relevant to assign the highest weight to such a criterion. This is even clearer considering the additive model in (2.1) and the most usual normalization procedure for the value function such that the worse outcome is set to 0 and the best outcome is set to 1.

Actually, the scale constants are substitution rates between the criteria (Keeney and Raiffa 1976; Vincke 1992; Belton and Stewart 2002). Keeney and Raiffa (1976) point out that it might happen that a criterion may have a scale constant larger than any other and yet it has less importance. Several practical examples are discussed on this issue by Keeney and Raiffa (1976) and Keeney (1992).

Finally, one should be aware that changing the normalization procedure or using different scales (for instance: a ratio or an interval scale) for the value function completely affects the set of values established for the criteria weights (or scale constants). In such a case a new set of values for the criteria weights should be computed. Of course this is valid for the additive model, although it is not valid for other methods, such as the outranking methods.

Some MAVT Additive MCDM/A Methods

There are quite a few methods incorporating the additive model. The main difference amongst them is in the elicitation procedures of the parameters, including both the intra-criterion and inter-criteria evaluations, with emphasis on the scale constants.

In many situations the use of the additive model is straightforward with the use of one of the classical elicitation procedures, there being no explicit consideration of an MCDM/A method. In other cases, an MCDM/A method is considered.

One of the most applied methods that incorporates the additive model is SMARTS (Simple Multi-Attribute Rating Technique with Swing), in which the swing procedure is applied (Edwards and Barron 1994). SMARTER (Simple Multi-Attribute Rating Technique Exploiting Ranks) is a related method that applies the first step of ordering the scale constants of the criteria and then, uses a surrogate weight. In these methods the value function for each criterion is assumed to be linear (Edwards and Barron 1994).

The AHP (Analytic Hierarchy Process) presents a particular procedure for preference modeling, considering the possibility of a hierarchical structure of objectives (Saaty 1980). The method uses the additive aggregation model, and collects information based on pairwise comparison of alternatives. In the literature there are some complaints that this method does not follow some of the properties of the additive model and a few other concerns, such as: the possibility of order reversal, and the interpretation for the criteria weights (Belton and Stewart 2002; Howard 1992). Howard (1992) points out that it is widely applied, since it does not demand much effort from the DM.

Macbeth (Measuring Attractiveness by a Categorical Based Evaluation Technique) is a method based on a qualitative evaluation on the difference of attractivity (Bana and Costa et al. 2005). They say that this method seeks to be concerned with constructing the value of outcomes, but does not force the DM to produce a direct numerical representation of preferences. The DM gives some preference information that is applied to build a numerical scale, based on a set of Linear Programming Problems (LPP).

The even swaps are based on the procedure proposed by Benjamin Franklin for the tradeoff on choosing whether or not implement an action (Hammond et al. 1998a; Hammond et al. 1999).

Additive-Veto Model

The compensatory nature of the additive model may recommend an alternative with a very low outcome level in one of the criteria, which is compensated by high outcome levels in one of more of the other criteria. However, it may happen that the DM may prefer not to select such a kind of alternative, whatever the criterion with low performance is. Thus, additive-veto models (de Almeida 2013b) may solve this problem by vetoing alternatives in such situations.

Numerical simulation in such kinds of situations has shown that it may not be rare for alternatives from a set of alternatives have this kind of characteristic (de Almeida 2013b), namely, one in which a very low outcome level in one of the criteria is compensated by high outcome levels in other criteria, thus ranking this alternative in a high position. This means that, depending on the DM's preference structure, if the DM is not willing to accept such a kind of alternative, then, a veto of the best alternative should occur in the additive model.

Roy and Słowinki (2013) discuss the choice of MCDM/A methods, considering several questions, such as this kind of compensation of bad performances in some criteria by good ones in other criteria. They pointed out that the acceptability of this situation should be evaluated for a compensatory method.

Additive Models for the Portfolio Problematic

The use of additive models for the portfolio problematic demands some concerns with the scale to be applied, since there is a size effect that causes the wrong solution to be selected in the interval scale, which is the one most applied for elicitation procedures (de Almeida et al. 2014).

The portfolio problematic in the additive model is based on the selection of a portfolio p_r that maximizes the value $V(p_r)$ as given in (2.3).

$$V(p_r)=\sum_{i=1}^{m}\left(x_i\sum_{j=1}^{n}k_j v_j(a_i)\right) \tag{2.3}$$

subject to some constraints, such as a budget constraint of $\sum_{i=1}^{m} x_i c_i \leq B$.

where:

$p_r = [a_1, \ldots, a_m]$ is the portfolio, which is a vector with the items (projects) a_i.

$$x_i = \begin{cases} 1 & \text{if the item (project) } x_i \text{ is included in the portfolio} \\ 0 & \text{if the item (project) } x_i \text{ is not included in the portfolio} \end{cases}.$$

C represents the vector of item costs, $C = [c_1, c_2, \ldots, c_m]^T$.

B is the budget or the limit for total cost C.

For portfolio selection, based on additive models, as in (2.3), the interval scale may not be applied. It has an impact on the result due to the size effect of the portfolio in this kind of scale, thus causing the wrong portfolio to be selected. What has been proved to be most appropriate is the ratio scale for this kind of problem (de Almeida et al. 2014). Most weight elicitation procedures are based on the interval scale that sets the worst outcome to zero, whereas using a ratio scale for the portfolio selection, the weights to be applied with the scale should be changed. The transformation of these scales can be seen at de Almeida et al. (2014).

Methods Based on Partial Information for Elicitation of Weights

Many behavioral studies have been conducted in order to evaluate the consistencies of the elicitation procedures. Borcherding et al (1991) have reported on inconsistencies of 50% and 67% of the time, when using ratio swing tradeoff procedures.

There has been some justification for using procedures with partial information instead of those elicitation procedures with complete information, since the elicitation of weighs can be time-consuming and controversial (Kirkwood and Sarin 1985; Kirkwood and Corner 1993) and because the DM may not be able to respond specifically to tradeoff questions (Kirkwood and Sarin 1985).

A few approaches have been proposed to deal with the model in (1.1) using partial information. One of the ways of dealing with this is to use surrogate weights. SMARTER (Edwards and Barron 1994) uses this idea, based on the partial information of the order of criteria weights. Another procedure (Danielson et al, 2014) increases the precision for surrogate weights by adding numerically imprecise cardinal information into rank-order methods, such as the ROC (Rank Order Centroid), also applied in SMARTER.

Other approaches collect more information and use procedures based on decision rules, formulating linear programming problems (LPP) or simulation procedures in order to analyze the alternatives. Among these approaches are: PAIRS (Salo and Hämäläinen, 1992), which uses interval judgments; VIP Analysis (Dias and Climaco, 2000), based on the progressive reduction of the number of alternatives; PRIME (Salo; Hämäläinen, 2001) which uses preference information based on swing method or holistic information; and RICH (Salo and Punkka, 2005) which uses incomplete ordinal preference statements. Mustajoki and Hamalainen (2005) integrate preference elicitation in the partial information framework for the SMART/SWING method.

A flexible elicitation procedure adapts the tradeoff elicitation procedure by using partial information in an interactive way, and conducts analysis by means of a set of LPPs (de Almeida 2014a; de Almeida 2014b).

2.4.2 MAUT

MAUT has been developed for MCDM/A problems, from Utility Theory (von Neumann and Morgenstern 1944), keeping its axiomatic structure (Keeney and Raiffa 1976). According to Edwards and Barron (1994), Howard Raiffa presented the fundamental insight for MAUT in 1968, pointing out that there would be more than one reason to value an object. Raiffa (1968) presented a few considerations for a multicriteria view in the context of health problems.

This approach gives one of the most classical MCDM/A methods, in which the most widely applied aggregation approach has been the additive model, for which the axiomatic structure of the theory indicates a number of properties to be considered. As mentioned, the main difference from the MAUT additive model to the model in the previous section is that the probabilistic consequence is approached in the utility function $u_j(x_j)$ for each criterion j.

The decision models with MAUT may include the framework of Decision Theory (Raiffa 1968; Berger 1985; Edwards et al. 2007), also called as Decision Analysis, which may consider the Bayesian approach to dealing with uncertainties, incorporating prior probabilities. Therefore, the uncertainties on the state of nature (θ) may be obtained from experts, in the form of prior probabilities $\pi(\theta)$. Thus, θ is an additional ingredient to be considered with MAUT, although this may not be explicit in some models.

For each θ_s chosen by nature and each action a_i chosen by the DM, a consequence x may be obtained, according to a consequence function (Berger 1985)

$P(x \mid \theta, a)$, which shows the probabilistic association amongst these ingredients, meaning the probability of obtaining x, given θ and a.

Thus, the model building process with MAUT incorporates a probabilistic modeling task for these ingredients, which complements the preference modeling. This probabilistic modeling task, in general, may involve another actor in the decision process, namely an expert. Usually the expert brings knowledge on the probabilistic behavior of the state of nature, so that the analyst applies elicitation procedures for obtaining $\pi(\theta)$, as subjective probabilities.

Therefore, when applying MAUT, the final model consists of a multi-attribute utility (MAU) function $u(x_1, x_2, ..., x_n) = f[u_1(x_1), u_2(x_2), ..., u_n(x_n)]$, to be maximized by the choice of an alternative probabilistically associated with the consequences $(x_1, x_2, ..., x_n)$. This corresponds to the expected utility function for the consequences under consideration.

From now on, the main elements of MAUT are going to be presented considering the case of two criteria x and y leading to the MAU function: $u(x, y) = f[u_1(x), u_2(y)]$.

The choices in Utility Theory consider the concept of lottery, which represents a probabilistic consequence. For instance, a lottery with two consequences is represented by $[A, p; B, 1–p]$, which means the possibility of obtaining one of two consequences A or B, where p is the probability of obtaining A, and 1-p is the probability of obtaining B.

There has been a set of axioms for Utility Theory, ever since its first formulation (von Neumann and Morgenstern 1944, Raiffa 1968; Keeney and Raiffa 1976; Berger 1985), which are applied to MAUT.

Just as in the additive model for MAVT, in MAUT the models follow the preference structure (P,I). Therefore, the first axiom is related to the ability of the DM to compare all consequences and order them. The second axiom is the transitivity preference relations P and I. These two axioms are implicitly related to probabilistic consequences, so they may apply for lotteries. The other axioms are explicitly related to lotteries. Let the lotteries with the consequences A, B and C and the probabilities p and q, then, there are the two following axioms:

- If APB, then there is a probability p, $0<p\leq 1$, so that for any C, $[A,p; C,1–p]P [B,p;C,1–p]$. This is also applied to indifference relation I.
- If $APBPC$, then, there are p and q, $0<q<p<1$, so that $[A,p;C,1–p]PBP[A,q;C,1–q]$.

Consequence Space

The whole evaluation process for the utility function is made over the consequence space, with which the DM should be familiar. Fig. 2.10 shows the consequence space for two criteria x and y.

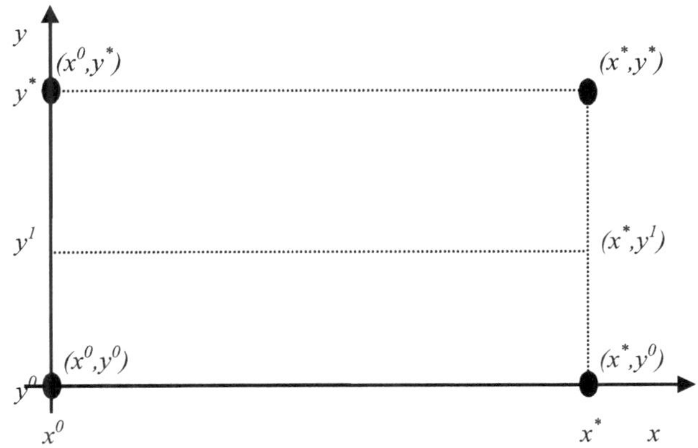

Fig. 2.10 Consequence space for two criteria

In the consequence space shown in Fig. 2.10, for each criterion, the most desirable outcomes are x^* and y^*, while the least desirable outcomes are x^0 and y^0. For the whole space the points (x^*,y^*) and (x^0,y^0) represented respectively the most and least desirable outcomes for the multi-attribute space. The scale for the utility is arbitrarily set in the interval 0 to 1, so that $u(x^*,y^*) = 1$, $u(x^0,y^0) = 0$, $u_j(x^*) = 1$, $u_j(y^*) = 1$, $u_j(x^0) = 0$ and $u_j(y^0) = 0$.

Elicitation of the Conditional Utility Function

The utility function $u_j(x_j)$ for each criterion j, related to the intra-criterion evaluation, is assessed considering a conditional utility function of criterion j, which is conditioned to a fixed level of the outcomes in other criteria. For instance, on the x axis of Fig. 2.10, there is a conditional utility function of criterion x, given a fixed level of the outcome for criterion $y = y^0$.

The intra-criterion evaluation consists of eliciting this single dimensional utility function $u_j(x_j)$. There are several procedures for this elicitation (Raiffa 1968; Keeney and Raiffa 1976; Berger 1985), many of them use the concept of certain equivalent of a lottery. This certain equivalent is the consequence B for which there is a probability p, such that the DM is indifferent between B and a lottery $[A, p; C, 1\text{–}p]$, with consequences A and C.

In general, the consequences of the lottery are the least and the most desirable, so that the probability $p = u(B)$. Since, $u(x^*) = 1$ and $u(x^0) = 0$, and considering the indifference between B and $[x^*,p;x^0,1\text{–}p]$, then $u(B)=pu(x^*)+(1\text{-}p)u(x^0)$. From this, it follows that $u(B) = p$.

Therefore, the elicitation procedure consists of obtaining the indifference between this kind of lottery and the consequences x, so that the utility function

$u(x)$ can be obtained. Detailed elicitation procedures are provided in Keeney and Raiffa (1976).

Elicitation of the MAU Function

For the elicitation of the MAU function $u(x_1,x_2,\ldots,x_n)=f[u_1(x_1),u_2(x_2),\ldots,u_n(x_n)]$, after obtaining the conditional utility function of each criterion, then the elicitation procedure is conducted for global utility. Let the two criteria be x and y and the consequence space in Fig. 2.10. Then, the elicitation seeks to obtain $u(x,y)=f[u_x(x),u_y(y)]$.

For the elicitation of the MAU function there are a few structured procedures (Keeney and Raiffa 1976). The main process, described below, is based on a prescriptive approach, in which preference conditions are evaluated with the DM, and based on these, analytical functions may be applied to $u(x,y)$.

The two main concepts of preference conditions considered for this purpose are: the additive independence condition and the utility independence condition.

If the mutual additive independence condition is found between x and y, in the DM's preference structure, then the additive model, $u(x,y)=k_xu(x)+k_yu(y)$, may be applied. (2.4) gives a more general model for n criteria.

$$u(x) = \sum_{j=1}^{n} k_j u_j(x_j) \tag{2.4}$$

where:
k_j is the scale constant for attribute or criterion j;
$u_j(x_j)$ is the utility function for criterion j;
x_j is the consequence or outcome for criterion j.

The scale constant k_j is usually normalized as in (2.1).

If the mutual utility independence condition is found between x and y, in the DM's preference structure, then the multilinear model, $u(x,y)=k_xu(x)+k_yu(y)+k_{xy}u(x)u(y)$, may be applied. Similar to (2.4), a generalization may be made for a model with n criteria.

The Utility Independence Condition

This independence preferential condition is associated with the context of utility functions. This concept may be understood considering the consequence space of Fig. 2.10. Criterion x is said to be utility independent of criterion y, if the conditional utility function $u(x,y^0)$ is strategically equivalent to any other utility of x, whatever the outcome for y is. The utility $u(x,y^0)$ is the utility for x, given that $y=y^0$. This means that the certain equivalent of the lottery $[(x^*,y^0),p;(x^0,y^0),1-p]$, whatever the value of p is, is the same for any other lottery $[(x^*,y),p;(x^0,y),1-p]$, whatever the outcome for y is.

It is interesting to note that for the strategically equivalent utility function $u(x,y^0)$, a utility $u(x,y)$ may be found by a linear transformation, such as $u(x,y)=a(y)u(x,y^0)+b(y)$, where: $a(y)>0$ and $b(y)>0$ are constants, established for any outcome for y.

Therefore, as shown with this utility independent condition, the utility function $u(x,y)$ depends only on the particular level of the outcome in criterion y, even so, by a linear transformation. More details on this concept are given by Keeney and Raiffa (1976).

The Additive Independence Condition

This independence condition imposes stronger constraints on the additive model. Let the following consequences of the space in (x,y) be: A, B, C and D, respectively corresponding to (x^1,y^1), (x^1,y^2), (x^2,y^2), (x^2,y^1), as illustrated in Fig. 2.11.

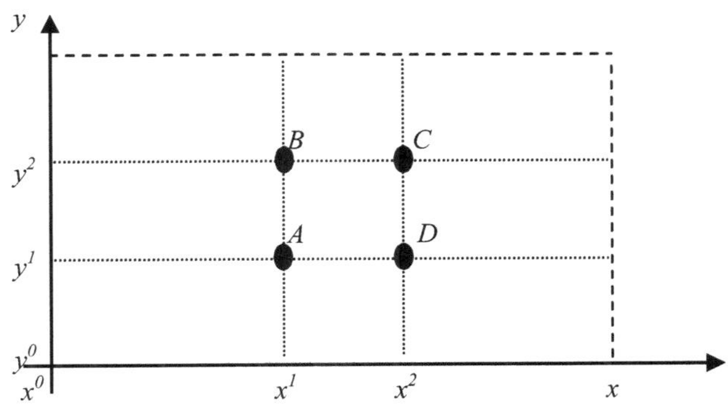

Fig. 2.11 Additive independence condition

The additive independence condition holds if the DM is indifferent between the following lotteries: $[A,0.5;C,0.5]$ and $[B,0.5;D,0.5]$, whatever x and y are, in the consequences A, B, C and D. Since, the same probability $p = 0.5$ is applied to both consequences in these lotteries, its representation may be simplified as follows: $[A,C]$ and $[B,D]$.

Considering the indifference between two lotteries similar to those in Fig. 2.11, such as those of the values of (x,y) for consequences A, B, C and D, being $[(x^0,y^0),(x,y)]$ and $[(x^0,y),(x,y^0)]$, then the utility of the lotteries has the same value. Thus, $0.5u(x^0,y^0)+0.5u(x,y)=0.5u(x^0,y)+0.5u(x,y^0)$. Given, the normalized scale for the extreme values x and y, then $u(x,y)=u(x,y^0)+u(x^0,y)$.

It can be seen that $u(x,y^0)$ and $u(x^0,y)$ can be obtained based on the scale constants k_j, such that: $u(x,y^0)=k_x u_x(x)$ and $u(x^0,y)=k_y u_y(y)$. This leads to the format of (2.4). This concept and its development are given in detail by Keeney and Raiffa (1976).

Elicitation of the Scale Constants

A complete and detailed procedure for the elicitation of the MAU function is given by Keeney and Raiffa (1976). The elicitation of the scale constants k_j is based on the analytical model obtained, associated with the independence conditions.

For instance, the scale constants k_j for the additive model on the two criteria x and y correspond to the utility of the two specific consequences (x^*,y^0) and (x^0,y^*), shown in Fig. 2.10. That is, $k_x=u(x^*,y^0)$ and $k_y=u(x^0,y^*)$.

Therefore, the elicitation of k_x consists of finding the probability p for which (x^*,y^0) is the certain equivalent to the lottery $[(x^*,y^*),p;(x^0,y^0),1-p]$. A similar evaluation may be made for k_y.

Again, as can be seen the scale constants k_j for an MAU function are not simply the relative degree of importance of the criterion. They are related to the scale, considering the limits for x and y, since the lottery $[(x^*,y^*),p;(x^0,y^0),1-p]$ is the basis for their elicitation.

Rank-Dependent Utility and Prospective Theory

There are quite a few paradoxes related to the use of the expected utility function, which are presented in the literature. Many of these paradoxes have been analyzed in a descriptive perspective within the context of behavioral decision making.

In many situations Rank-Dependent Utility (RDU) and Prospective Theory (Edwards et al. 2007) have been considered as ways of dealing with such situations (Wakker 2010).

MCDM/A models based on MAUT may be adapted with Rank-Dependent Utility and Prospective Theory views on modeling risk preferences, which may have particular relevance for the RRM context.

2.4.3 Outranking Methods

This kind of method has a completely different rationality from the methods in the two previous subsections. These methods are non-compensatory and may be applied to a preference structure (P,Q,I,J). The possibility of the incomparability relation is one of the issues distinguished in this kind of method, and therefore, only partial pre-orders may be obtained.

Therefore, unlike MAVT and MAUT, this kind of method may be applied in a situation for which the DM's preferences are not in agreement with the first two properties. That is, the DM is not able to compare all consequences and order them. Also, the transitivity property may not be followed.

This section presents some basic elements of these methods and then, introduces an overview on the two most widely applied outranking methods: ELECTRE and PROMETEE.

These methods are based on pairwise comparison of the alternatives, by exploring an outranking relation between the pairs of alternatives.

There is an important difference between outranking methods and those of MAVT and MAUT that impacts the preference modeling process, namely the different meaning for the inter-criteria parameters, which may be called weights. The meaning of criteria weights corresponds directly to the degree of importance of the criteria, for outranking methods.

This notion of importance amongst criteria may be compared with votes in a voting process (Roy 1996; Vincke 1992). Let there be two subsets of criteria *G* and *H* and two alternatives *a* and *b*. If the subset of criteria in *G* is more important (has more votes) than the criteria in the subset in *H*, and the following conditions hold (Vincke 1992):

- *a* is better than *b* for all criteria in the subset *G*;
- *b* is better than *a* for all criteria in the subset *H*; and
- *a* and *b* are indifferent for any other criteria;

Then: *a* is globally better than *b*.

If this importance (or votes) can be represented by the criteria weights, the comparison between the subsets of criteria *G* and *H* can be based on the summation of these weights.

That is, the summation of weights for criteria in favor of *a* is greater than those in favor of *b*. This means that *a* makes a better coalition of criteria than *b*.

These methods are worked out in two main steps (Roy 1996; Vincke 1992):

- Building the outranking relation, by comparing all pair of alternatives in the set of alternatives;
- Exploiting the outranking relation by applying an algorithm or procedure for solving the problem, according to each particular problematic.

These methods may work with different kinds of criteria, depending on their intra-criterion characteristics. In a true criterion there is no threshold. For a pseudo criterion there are thresholds that may be one of the following or both: an indifference threshold and a preference threshold.

The outranking relation *S*, is applied over all pairs of alternatives of the set of alternatives, such as *a* and *b*. Therefore, *aSb* means that alternative *a* outranks alternative *b*, which means that *a* is at least as good as *b*.

ELECTRE Methods

In the ELECTRE (*Elimination Et Choix Traduisant la Réalité*) methods the outranking relation *aSb*, between two alternatives *a* and *b*, is based on concordance and discordance concepts, on which the DM gives preference information in the form of thresholds.

The family of ELECTRE methods includes the following methods, which differs from the problematic and the kind of criteria (Roy 1996; Vincke 1992):

- The ELECTRE I method is applied for a choice problematic, considering true criteria;
- The ELECTRE IS method, which is applied for a choice problematic, considering pseudo criteria;
- The ELECTRE II method, which is applied for a ranking problematic, considering true criteria;
- The ELECTRE III method, which is applied for a ranking problematic, considering pseudo criteria;
- The ELECTRE IV method, which is applied for a ranking problematic, considering pseudo criteria;
- The ELECTRE TRI method, which is applied for a sorting problematic, considering pseudo criteria.

The ELECTRE I method is subsequently described in order to illustrate the basic approach followed by these methods. The other methods have some differences in the parameters for the step of building the outranking relation and are at their most different in the step of exploiting the outranking relation, according to their problematic.

For building the outranking relation, ELECTRE I uses the concepts of concordance and discordance. The former indicates if a considerable subset of criteria is in favor of an outranking relation *S* between two alternatives. The latter, may disagree with this relation *S*, even if the concordance is in agreement.

Therefore, when evaluating the outranking relation *aSb,* between two alternatives *a* and *b,* the following indices are applied: the concordance index $C(a,b)$ and the discordance index $D(a,b)$.

The concordance index $C(a,b)$ is given by (2.5).

$$C(a,b) = \sum_{j:g_j(a)\geq g_j(b)} w_j \tag{2.5}$$

where:
w_j is the weight for criterion j; the weights are normalized, such that $\sum_j w_j = 1$.

$g_j(a)$ and $g_j(b)$ is the value of the outcome for criterion j, respectively for alternatives a and b.

There are a few different formulations for the discordance index (Roy 1996; Vincke 1992; Belton and Stewart 2002). *D(a,b)* may be given by (2.6):

$$D(a,b) = \max\left(\frac{g_j(b) - g_j(a)}{\max[g_j(c) - g_j(d)]}\right) \quad , \forall j \mid g_j(b) > g_j(a); \forall j,c,d \,. \tag{2.6}$$

A concordance threshold c' and discordance threshold d' should be specified by the DM in order to build the outranking relation. The outranking relation *aSb* between *a* and *b*, is established by (2.7).

$$aSb \text{ if and only if } \begin{cases} C(a,b) \geq c' \\ D(a,b) \leq d' \end{cases} \tag{2.7}$$

Having obtained these formulations and parameters, this step for building of the outranking relation can be finalized, by applying (2.7) for all pair of alternatives. It may happen with a pair of alternatives that *aSb* and *bSa*. In this case there is a circuit and these alternatives are considered indifferent.

The second step of exploiting the outranking relation can now be worked out. For the ELECTRE I method, the purpose of this step is to obtain the kernel, which is the subset of alternatives, in which each of its elements is not outranked by any other in the kernel. If only one alternative is found in the kernel, the choice problematic reaches its particular case of optimization. Otherwise, the alternatives in the kernel have been found to be incomparable.

More details on ELECTRE methods may be found in many basic texts on MCDM/A methods (Roy 1996; Vincke 1992; Belton and Stewart 2002; Figueira et al. 2005).

PROMETHEE Methods

PROMETHEE (Preference Ranking Organization Method for Enrichment Evaluation) is a group of outranking methods, based on a valued outranking relation (Brans and Vincke 1985; Vincke 1992; Belton and Stewart 2002).

In PROMETHEE methods the DM does not have to specify information on concordance and discordance regarding the outranking relation. The DM provides the information on the criteria weights and on the intra-criterion evaluation, related to the indifference or preference thresholds, if any of them are considered.

This group of methods uses the following formulation for the first step of building the outranking relation, thereby establishing the outranking degree $\pi(a,b)$, for each pair of alternatives *a* and *b*, from (2.8).

$$\pi(a,b) = \sum_{j=1}^{n} w_j F_j(a,b) \qquad (2.8)$$

where:
w_j is the weight for criterion j; the weights are normalized, such that $\sum_j w_j = 1$.

$F_j(a,b)$ is a function of the difference $[g_j(a)\text{-}g_j(b)]$ of the outcomes of the alternatives for criterion j.

The method has six different patterns for this function $F_j(a,b)$. In the basic form for $F_j(a,b)$, it does not use either of indifference or preference thresholds for criterion j. In this case, $F_j(a,b) = 1$, if $g_j(a) > g_j(b)$ and $F_j(a,b) = 0$, otherwise. Thus, the outranking degree $\pi(a,b)$, is the summation of all criteria weights for those criteria, in which a has a better performance than b.

The other five forms for $F_j(a,b)$ considers indifference or preference thresholds, or both, for criterion j. In these five patterns for $F_j(a,b)$, it has a value between 0 and 1, for criterion j, when the difference $[g_j(a)\text{-}g_j(b)]$ is in the range of the indifference or preference thresholds. In this range, the outranking degree $\pi(a,b)$, adds a partial value of the weights of criterion j, in which a has a better performance than b, as can be seen in (2.8).

These forms for $F_j(a,b)$, are chosen by the DM, in the context of the intra-criterion evaluation, and includes the specification of values related to the indifference or preference thresholds, for that criterion j.

The matrix with the values of the outranking degree $\pi(a,b)$ for each pair of alternatives can be available now, thus concluding the first step.

For the second step of exploiting the outranking relation, each alternative a is evaluated based on the outgoing flow $\phi^+(a)$ and on the ingoing flow $\phi^-(a)$.

The outgoing flow $\phi^+(a)$ indicates the advantage of the alternative a over all other alternatives b in the set of alternatives A. $\phi^+(a)$ is obtained from (2.9).

$$\phi^+(a) = \sum_{b \in A} \pi(a,b) \qquad (2.9)$$

where, $n\text{-}1$ gives a normalized scale between 0 and 1, since n is the number of criteria.

The ingoing flow $\phi^-(a)$ indicates the disadvantage of the alternative a compared with all other alternatives b in the set of alternatives A. $\phi^-(a)$ is obtained from (2.10):

$$\phi^-(a) = \sum_{b \in A} \pi(b,a). \qquad (2.10)$$

Another index for evaluation of the alternatives is the liquid flow $\phi(a)$, given by (2.11), which is obtained in a scale of *-1* to *1*.

$$\phi(a)=\phi^{+}(a)-\phi^{-}(a) \tag{2.11}$$

Now the second step of exploiting the outranking relation, may be concluded by using these indices on specific procedures for each problematic.

In the PROMETHEE I method two pre-orders are built, based on (2.9) and (2.10), which indicate the relations of preference *(P)*, indifference *(I)* and incomparability *(J)* between the pairs of alternatives of set A (Brans and Vincke 1985; Belton and Stewart 2002). Therefore, PROMETHEE I outputs a partial pre-order of the elements of A.

The PROMETHEE II method is based on the liquid flow $\phi(a)$ from (2.11), in which each alternative has a score. Therefore, PROMETHEE II outputs a complete pre-order on the elements of A.

The family of PROMETHEE methods includes other methods: PROMETHEE III and IV, for a stochastic situation; PROMETHEE V for a portfolio problematic, as discussed in the following sub-section; and PROMETHEE VI, when the DM specifies a range for each criterion weight, instead of a precise value of weight.

PROMETHEE V for Portfolio Problematic

The PROMETHEE V method (Brans and Mareschal 1992) is applied for selecting portfolios using a non-compensatory method for evaluating of alternatives in a model similar to that in (2.3). The only difference is in computing the value of the portfolio $V(p_r)$, which is based on the application of PROMETHEE II for scoring the items a_i (projects) as values $v_i(a_i)$.

There is also a problem of scale with this method, although different from that with the additive model. In this case PROMETHEE II presents positive and negative scores for $v_i(a_i) = \phi(a_i)$ to be applied in (2.3). Therefore, to work in the maximization model, the negative scores have to be transformed into positive scores, thereby changing the properties of the ratio scale (Vetschera and de Almeida 2012).

This transformation has a similar effect, with the possibility of selecting the wrong portfolio. Contrary to the case of the additive model in (2.3), the ratio scale cannot be applied in the PROMETHEE V. In order to overcome this problem, an analysis should be conducted based on the concept of a c-optimal portfolio (Vetschera and de Almeida 2012; de Almeida and Vetschera 2012).

2.4.5 Other MCDM/A Methods

There are other approaches and concepts that may be seen either as specific methods or tools that can be applied in any method, such as those presented above. Belton and Stewart (2002) consider the latter option for fuzzy sets and rough sets. A comprehensive view of fuzzy approaches for modeling MCDM/A problems is given by Pedrycz et al. (2011), while the rough sets approach is briefly described in the next subsection.

There are a few approaches classified as disaggregation methods, which are based on holistic (or global) evaluation by the DM, followed by a subsequent step of inference of the parameters of an aggregation model. Pardalos et al. (1995) consider these approaches as a fourth group of methods in their classification.

Some of these approaches, such as the UTA method (Jacquet-Lagréze and Siskos 1982), are related to the single criterion of synthesis methods. However, inference procedures proposed for the ELECTE TRI method use the same process of collecting information from the DM on global evaluation for posterior inference of the parameters of inter-criteria evaluation. The preference learning approach (Slowinski et al. 2012) uses a similar process.

Rough Sets

This is a kind of MCDM/A method based on preference learning. These methods consider the DM's preferences by evaluating a set of decision rules discovered from preference data, which can be elicited previously from the DM and afterwards used as an input to establish comparisons among the set of alternatives (Slowinski et al. 2012).

Rough sets theory has been widely used as an MCDM/A approach based on preference learning (Pawlak and Slowinski 1994; Greco et al. 2001; Greco et al. 2002; Slowinski et al. 2012).The preference learning approach seeks to avoid the elicitation of model parameters, such as importance weights or scale constants and others related to thresholds. It uses information from previous preferences stated by a DM to establish preference relations among the alternatives based on this input by assuming that the sample of statements gathered from the DM is enough to establish decision rules for evaluating the set of alternatives.

This approach may be applied to evaluating risk conditions, for which decision rules may be built, grounded on preferential information given by the DM. That is, rough sets could be applied in a similar way to the problem of territorial risk evaluation (Cailloux et al. 2013), based on ELECTRE TRI method.

2.4.6 Mathematical Programming Methods

Several mathematical programming techniques have been proposed to solve multiobjective problems, such as involving linear (MOLP - Multi-Objective Linear Programming) and nonlinear programming principles. There is a broad range of relevant literature on this topic (Korhonen 2009; Korhonen 2005; Korhonen and Wallenius 2010; Steuer 1986; Ehrgott 2006; Miettinen 1999; Coello et al. 2007).

Basically, a mathematical programming for solving a multiobjective problem can be approached in the following ways:

- By considering a preference structure in advance so as to solve the problem by some approach, such as: transforming multiple objective functions into a single objective function, solving by an interactive process, and so forth.
- By identifying the non-dominated solutions which together form the set of Pareto optimal outcomes (more commonly referred to as the Pareto front), without taking the DM's preferences into account.

The latter is discussed in the next section. The former may consider the DM's preferences by either: collecting information or taking assumptions. In terms of articulating the DM's preferences, three classes can be defined: a priori, posteriori and progressive articulation of the preferences. Some of these approaches are listed in Table 2.4.

Table 2.4 Summary of MCDA representative methods

Articulation of preferences	MCDA methods
A priori	Global Criterion Method (Osyczka 1984); Goal Programming (Charnes and Cooper 1961); Goal-Attainment Method (Chen and Liu 1994); Lexicographic Method (Rao 1984); Min-Max Optimization (Osyczka 1984); Surrogate Worth Trade-Off (Haimes et al. 1975).
A posteriori	Weighted Sum; ε-constraint Method (Miettinen 1999).
Progressive	STEP Method (Benayoun et al. 1971); (SEMOPS) Sequential Multiobjective Problem Solving Method (Duckstein et al. 1975)

2.5 Multiobjective Optimization

Multiobjective optimization approaches are related to complex problems and have spread to the research fields of heuristics and evolutionary algorithms. Two possible reasons for this evolution are that problems have become more complex and the ability of these approaches to find Pareto solutions promptly. In terms of complexity, some problems are classified as NP-Hard and exact methods have not been successful in finding non-dominated solutions. Therefore, some heuristics and evolutionary multiobjective algorithms are described.

Multiobjective optimization is based on Pareto-front analysis. In multiobjective optimization the notion of optimum was generalized by Vilfredo Pareto (in 1896). It can be said that a vector of decision variables, x^*, is Pareto optimal if there is no other vector of decision variables, x, such that $f_i(x) \leq f_i(x^*)$ for all $i = 1, \ldots, k$ and $f_i(x) < f_i(x^*)$ for at least one j (Coello et al. 2007).

In multiobjective optimization, all objectives are considered important and all non-dominated solutions should be found. Thereafter, higher-level information, generally on non-technical, qualitative and experience-driven matters, can be used to compare non-dominated solutions before making a choice. This principle is defined as an ideal multiobjective optimization procedure (Deb 2001).

Several studies are focused only on determining the non-dominated solutions, assuming that all non-dominated solutions are equally optimum, or that the DM will provide information on his/her preferences after he/she learns what the Pareto front is. These assumptions make sense in complex problems where finding non-dominated solutions is an independent and hard task.

In terms of multiobjective evolutionary algorithms, there are some algorithms that do not incorporate the concept of Pareto dominance in their selection mechanism. These are considered first generation methods. They started to become obsolete in the literature because some algorithms started to rank the population based on Pareto dominance, which are second generation methods (Coello et al. 2007). It is important to point out that, in general, multiobjective optimization based on evolutionary algorithms concentrates its efforts on the first step of the MCDM/A problem: identifying the Pareto front. Main multiobjective evolutionary algorithms of these generations are represented in Table 2.5.

Table 2.5 First and second generations of multiobjective evolutionary algorithms (MOEAs)

MOEAs Generation	Methods
First Generation	GA with Aggregating Functions
	VEGA - (Schaffer 1985)
	MOGA - (Fonseca and Fleming 1993)
	NSGA - (Srinivas and Deb 1994)
	NPGA - (Horn et al. 1994)
	NPGA 2 - (Erickson et al. 2001)
Second Generation	SPEA and SPEA2 - (Zitzler and Thiele 1999)
	NSGA-II - (Deb et al. 2002)
	PAES - (Knowles and Corne 2000)
	PESA and PESA II - (Corne et al. 2000)
	micro-GA - (Coello Coello and Toscano Pulido 2001)

2.6 Group Decision and Negotiation

In many decision processes there is more than one DM. In such situations a group decision model or a negotiation process has to be applied in order to come to a final solution. Therefore, a brief overview is given of Group Decision and Negotiation (GDN) methods and processes, particularly of those aspects most closely related to MCDM/A models. The GDN area covers decision problems with multiple DMs, over a wide range of topics such as: Conflict Analysis (Fraser and Hipel 1984; Keith et al. 1993; Kilgour and Keith 2005), web-based negotiation support systems (Kersten and Noronha 1999), evolutionary systems design (Shakun M.F 1988), connectedness (Shakun 2010), formal consciousness (Shakun 2006) and fair division (Brams and Taylor 1996).

As stated by Kilgour and Eden (2010) negotiation and group decision contain both unity and diversity. Regarding the latter, some of the scholars in the field of GDN understand that it is appropriate to distinguish between Group Decision (GD) making and negotiation. Kilgour and Eden (2010) explain that in this view GD making is related to a decision problem shared by more than one DM, who must make a choice, for which all DMs will have some responsibility. On the other hand, a negotiation is seen as a process in which two or more DMs, acting in an independent way, may either: make a collective choice, or not do so. For the latter, one (or more) of the DMs may give up taking further part in the decision process and walk away.

Additionally, it can be considered that a GD process involves an analytical procedure in order to aggregate the preferences of the individual DMs, which results in a kind of collective representation of the preferences of the group. With regard to negotiation, this involves a process of interaction between DMs, in order to find a collective solution for the problem of their mutual interest.

As to using the analytical procedure in order to aggregate the DMs' preferences, the process for building models pays great attention to following rules of rationality, related to a normative perspective. Also, there are some concerns about dealing with some paradoxes, as shown by the descriptive perspective. As for the negotiation process, the interaction between people invokes other concerns, such as the accuracy of their communication process.

These issues show some diversity between GD making and a negotiation process. However, there are some elements of unity between them. For instance, most negotiation processes are grounded in analytical results and endeavor to ensure the rationality and fairness of the collective choice. Also, the building process for the GD model involves agreements with the group of DMs, regarding several issues and parameters of the model, especially when the problem also involves multiple objectives, leading to an integrated MCDM/A and GD model. Therefore, in order to build GD models, some interaction processes may be necessary between the DMs. The process for building GD models will depend on the available time of these DMs, and most of all, on how simultaneously their availability can be made. Also, it should be considered issues related to the distributive and integrative models (Kersten 2001).

Given the very close relationship between GD making and the MCDM/A modeling process, a brief description of some aspects of this topic is given below. Although some studies suggest that MCDM/A models may be straightforwardly applied for GD aggregation, one should be aware that aggregating people's preferences is completely different from aggregating criteria that represent the objectives of an individual. The area of GDN brings contributions to the concerns to be dealt with when integrating DMs' preferences.

2.6.1 Aggregation of DMs' Preferences or Experts' Knowledge

While most studies on GD making are related to the aggregation of DMs' preferences, others are associated with experts' knowledge. These two GD procedures are related to aggregating or integrating two substantially distinct situations. These two kinds of aggregating process have differences in their foundations. Unfortunately, in some studies this distinction is not clear and may lead to misconceptions and mislead the decision modeling process. That is, using an inappropriate foundation to build a decision model will produce a wrong model and thereby lead to an unsuitable solution.

The aggregation of DMs' preferences is related to consequences value (Leyva-Lopez and Fernandez-Gonzalez 2003; Morais and de Almeida 2012). On the other hand, the aggregation of experts' knowledge is associated with some specific subject.

In the former, the process does not seek the true solution. Instead, the process seeks the most appropriate solution, considering the DMs' preferences. The foundations for the aggregation process are concerned with aspects such as rationality and preference elicitation. This kind of aggregation process considers the differences in objectives between DMs, and takes into account elements associated with preferences, such as the DMs' tradeoffs and the possibilities of compromising; in other words, the extent to which a DM is wiling to make concessions in order to reach a final group decision. In this case DMs do not change their preferences. Instead, they make concessions, always according to their preferences.

In the latter, the process is focused on seeking the true about some particular situation, based on the experts' knowledge. The foundations for this kind of aggregation process are concerned with aspects such as experts' knowledge and their accuracy on evaluating variables in a system. This process considers the differences in perception among experts, taking into account elements associated with knowledge, such as the experts' different backgrounds and experiences. The experts are not supposed to keep their initial opinion on a subject, unless their knowledge gives grounds for doing so. An expert may change his/her opinion on a subject, since they can learn something new from other experts. That is why many studies are focused on searching for consensus regarding the experts' perceptions of that particular topic.

Regardless of these differences on these two kinds of aggregation, some models are built in order to tackle these two issues mutually, since both are present in many GDN problems.

Many fuzzy approaches are applied to this kind of problem (Ekel et al. 2008; Pedrycz et al. 2011), and deal with factors such as ambiguity and uncertainties the experts have as to describing their perception on the variables that are being evaluated.

There is a particular kind of situation related to experts' aggregation of probabilities, which is related to prior probabilities $\pi(\theta)$ on the state of nature θ. There are many studies in the literature on Decision Theory (or Decision Analysis) related to the elicitation of prior probabilities (Raiffa 1968; Berger 1985) and the aggregation of a group of experts' prior probabilities (Edwards et al. 2007). At the end of Chap. 3 there are more details about this topic.

The following subsection gives a brief description of types of group decision aggregations regarding DMs' preferences.

2.6.2 Types of Group Decision Aggregations

Regarding the aggregation of DMs, different actors may play specific roles. For instance, instead of an analyst, a facilitator or a mediator may act in some situations. For instance, a facilitator may act so as to intensify the interaction process between DMs or among other actors in the decision process. With regard to DMs, the way in which they act and are available for the interaction in the decision process, for a particular problem, plays an important role when classifying the types of GD aggregation.

The GD aggregation process consists of reducing the set of individual DMs' preferences to a collective DMs' preference. There are some situations in which one of the actors in the GD process is a supra-DM. This supra-DM makes decision on final issues, in general, related to global evaluations in the process, such as evaluating the other DMs' choices. The supra-DM may have a hierarchical position above the other DMs in the organization's structure. Keeney (1976) considers two types of GD process, with regard to DMs' interrelationships: the 'benevolent dictator problem' and the 'participatory group problem'. The former is related to the situation regarding a supra-DM and in the latter, the group acts jointly in the GD process, with the same power.

Whether or not a supra-DM is present in the process, two kinds of GD aggregation general procedures may be considered (Kim and Ahn 1999; Leyva-López and Fernández-González 2003; Dias and Clímaco 2005):

- Aggregation of DMs' initial preferences.
- Aggregation of DMs' individual choices, which means the ranking of alternatives by each DM's;

These two GD aggregation procedures are illustrated in Fig. 2.12, with the first kind on the left-hand side and the second on the right-hand side. With regard to the first steps of preparation for the GD process, there is an integration in the former procedure, whereas in the latter, the process is completely separate for each DM.

In the former the DMs provide their initial preferences in an integrated way, in which the aggregation process is considered from the very beginning. Then, the process produces the final choices for the set of alternatives. This may be given as a simple ordinal ranking of the alternatives or may include a cardinal score for each alternative, depending on the method applied, which is the same for all DMs. The same criteria are considered for all DMs, but the intra-criterion and inter - criteria evaluations may be different. In most models the former is the same and the main difference is in the analysis of the criteria weights.

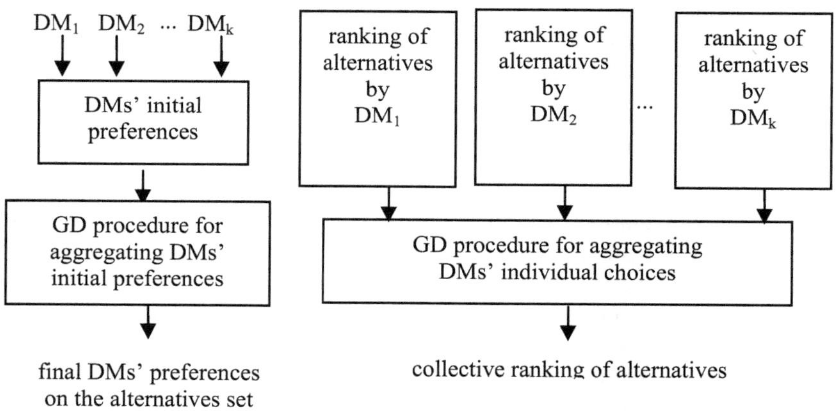

Fig. 2.12 Types of GD aggregation procedures

In the latter, each DM provides his/her individual ranking of alternatives. That is, the individual DMs' choices produce the final ranking of alternatives or other results if another problematic, such as choice or sorting, is applied, although in these cases information on scores of the alternatives is not expected to be produced, in general. These may be produced by completely different methods, with different criteria for each DM. It does not matter which objective each DM considers. The only information that matters is the final individual evaluation of each alternative by each DM. With regard to the GD process, if a ranking of alternatives is produced by each DM, then the GD procedure may be conducted by using a voting procedure, which is based on the foundations of Social Choice Theory (Nurmi 1987; Nurmi 2002).

References

Ackermann F, Eden C (2001) SODA - journey making and mapping in practice. In: Rosenhead J, Mingers J (eds) Rational Analysis in a Problematic World Revisited, 2nd ed. John Wiley & Sons Inc., United Kingdom, pp 43–61

Ackoff RL, Sasinieni MW (1968) Fundamentals of operations research. John Wiley & Sons, New York, p 455

Bana e Costa C, De Corte J-M, Vansnick J-C (2005) On the Mathematical Foundation of MACBETH. Mult. Criteria Decis. Anal. State Art Surv. SE - 10. Springer New York, pp 409–437

Belton V, Stewart TJ (2002) Multiple Criteria Decision Analysis. Kluwer Academic Publishers

Benayoun R, de Montgolfier J, Tergny J, Laritchev O (1971) Linear programming with multiple objective functions: Step method (stem). Math Program 1(1):366–375

Berger JO (1985) Statistical decision theory and Bayesian analysis. Springer Science & Business Media, New York

Bidgoli H (1989) Decision support systems: principles and practice. West Pub. Co.

Borcherding K, Eppel T, von Winterfeldt D (1991) Comparison of Weighting Judgments in Multiattribute Utility Measurement. Manage Sci 37:1603–1619

Bouyssou D (1986) Some remarks on the notion of compensation in MCDM. Eur J Oper Res 26(1):150–160

Bouyssou D, Marchant T, Pirlot M, Tsoukis A, Vincke P (2006) Evaluation and decision models with multiple criteria: Stepping stones for the analyst. Springer Science & Business Media

Brams SJ, Taylor AD (1996) Fair Division: from cake-cutting to dispute resolution. Cambridge University Press, New York

Brans JP, Mareschal B (1992) PROMETHEE V: MCDM Problems with Segmentation Constraints. INFOR 30(2):85-96

Brans JP, Vincke Ph (1985) A preference ranking organization method: the Promethee method for multiple criteria decision making, Manage Sci 31:647–656

Brunsson N (2007) The consequences of decision-making. Oxford University Press New York, NY

Cailloux O, Mayag B, Meyer P, Mousseau V (2013) Operational tools to build a multicriteria territorial risk scale with multiple stakeholders. Reliab Eng Syst Saf 120:88–97

Charnes A, Cooper WW (1961) Management Models and Industrial Applications of Linear Programming. John Wiley & Sons

Chen Y-L, Liu C-C (1994) Multiobjective VAr planning using the goal-attainment method. IEE Proc. - Gener. Transm. Distrib. IET, pp 227–232

Coello CAC, Toscano Pulido G (2001) A Micro-Genetic Algorithm for Multiobjective Optimization. In: Zitzler E, Deb K, Thiele L, Coello Coello CA, Corne D (eds) First International Conference on Evolutionary Multi-Criterion Optimization: 126-140. Springer-Verlag. Lecture Notes in Computer Science No. 1993

Coello CC, Lamont GB, Van Veldhuizen DA (2007) Evolutionary algorithms for solving multi-objective problems. Springer Science & Business Media

Corne DW, Jerram NR, Knowles JD, Oates MJ (2001) PESA-II: Region based Selection in Evolutionary Multiobjective Optimization. In: Spector L, Goodman ED, Wu A, Langdon WB, Voigt HM, Gen M, Sen S, Dorigo M, Pezeshk S, Garzon MH, Burke E (eds) Proceedings of the Genetic and Evolutionary Computation Conference (GECCO'2001): 283-290, San Francisco, California, Morgan Kaufmann Publishers

Cox LA Jr (2009) Risk analysis of complex and uncertain systems. Springer Science & Business Media, New York

Cox LA Jr (2012) Evaluating and Improving Risk Formulas for Allocating Limited Budgets to Expensive Risk-Reduction Opportunities. Risk Anal 32(7):1244–1252

Daher S, de Almeida A (2012) The Use of Ranking Veto Concept to Mitigate the Compensatory Effects of Additive Aggregation in Group Decisions on a Water Utility Automation Investment. Group Decis Negot 21(2):185–204

Danielson M, Ekenberg L, Larsson A, Riabacke M (2014) Weighting under ambiguous preferences and imprecise differences. Int J Comput Int Sys 7(1):105-112

Davis CB, Olson MH (1985) Management Information Systems: Conceptual Foundations, Structure and Development. McGraw-Hill

de Almeida A, Vetschera R, de Almeida J (2014) Scaling Issues in Additive Multicriteria Portfolio Analysis. In: Dargam F, Hernández JE, Zaraté P, et al. (eds) Decis. Support Syst. III - Impact Decis. Support Syst. Glob. Environ. SE - 12. Springer International Publishing, pp 131–140

de Almeida AT (2013a) Processo de Decisão nas Organizações: Construindo Modelos de Decisão Multicritério (Decision Process in Organizaions: Building Multicriteria Decision Models), São Paulo: Editora Atlas

de Almeida AT (2013b) Additive-veto models for choice and ranking multicriteria decision problems. Asia-Pacific J Oper Res 30(6):1-20

de Almeida AT, Almeida JA, Costa, APCS, ALmeida-Filho AT (2014b) A New Method for Evaluation of Criteria Weights in Additive Models by Interactive Flexible Elicitation. Working paper, CDSID

de Almeida AT, Costa, APCS, Almeida JA, Almeida-Filho AT (2014a) A DSS for Resolving Evaluation of Criteria by Interactive Flexible Elicitation Procedure, In: Dargam F, Hernández J, Zaraté P, Liu S, Ribeiro R, Delibasic B, Papathanasiou J. Decision Support Systems III - Impact of Decision Support Systems for Global Environments". LNBIP 184 (Lecture Notes in Business Information Processing), Springer. 157–166

de Almeida AT, Ferreira RJP, Cavalcante CAV (2015) A review of multicriteria and multi-objective models in maintenance and reliability problems. IMA Journal of Management Mathematics 26(3):249–271

de Almeida AT, Vetschera R (2012) A note on scale transformations in the PROMETHEE V method. Eur J Oper Res 219:198–200

Deb K, Pratap A, Agarwal S, Meyarivan T (2002) A fast and elitist multiobjective genetic algorithm: NSGA-II. IEEE Trans Evol Comput 6:182–197

Dias LC, Climaco JN (2000). Additive aggregation with variable interdependent parameters: The VIP analysis software. J Oper Res Soc 51:1070–1082

Dias LC, Clímaco JN (2005) Dealing with imprecise information in group multicriteria decisions: A methodology and a GDSS architecture. Eur J Oper Res 160(2) 291–307

Duckstein L, Monarchi D, Kisicl CC (1975) Interactive Multi-Objective Decision Making Under Uncertainty. Theor Decis Pract Hodder Stoughtor, London 128–147.

Eden C (1988) Cognitive mapping. Eur J Oper Res 36(1):1–13

Eden C, Ackermann F (2004) SODA. The Principles. In: Rosenhead J, Mingers J (eds) Rational Analysis for a Problematic World Revisited. Second Edition, Chichester: John Wiley & Sons Ltd.

Edwards W, Barron FH (1994) SMARTS and SMARTER: Improved Simple Methods for Multiattribute Utility Measurement. Organ Behav Hum Decis Process 60(3):306–325

Edwards W, Miles Jr RF, Von Winterfeldt D (2007) Advances in decision analysis: from foundations to applications. Cambridge University Press

Ehrgott M (2006) Multicriteria optimization. Springer Science & Business Media, Berlin

Erickson M, Mayer A, Horn J (2001) The Niched Pareto Genetic Algorithm 2 Applied to the Design of Groundwater Remediation Systems. In: Zitzler E, Thiele L, Deb K, et al (eds) Evol. Multi-Criterion Optim. SE - 48. Springer Berlin Heidelberg, pp 681–695

Figueira J, Greco S, Ehrgott M (eds) (2005) Multiple Criteria Decision Analysis: State of the Art Surveys. Springer Verlag, Boston, Dordrecht, London

Fishburn PC (1976) Noncompensatory preferences. Synthese 33:393–403

Fonseca CM, Fleming PJ (1993) Genetic Algorithms for Multiobjective Optimization: Formulation, Discussion and Generalization. In: Forrest S, (ed) Proceedings of the Fifth International Conference on Genetic Algorithms, San Mateo, California. University of Illinois at Urbana-Champaign, Morgan Kaufmann Publishers

Franco LA, Cushman M, Rosenhead J (2004) Project review and learning in the construction industry: Embedding a problem structuring method within a partnership context. Eur J Oper Res 152(3):586–601

Fraser NM, Hipel KW (1984) Conflict Analysis: Models and Resolutions. North-Holland, New York

Goodwin P, Wright G (2004) Decision analysis for management judgment. Wiley London

Greco S, Matarazzo B, Slowinski R (2001) Rough sets theory for multicriteria decision analysis. Eur J Oper Res 129:1-47

Greco S, Slowinski R, Matarazzo B (2002) Rough sets methodology for sorting problems in presence of multiple attributes and criteria. Eur J Oper Res 138:247-259

Haimes YY, Hall WA, Freedman HT (1975) Multiobjective optimization in water resources systems: the surrogate worth trade-off method. Elsevier

Hammond JS, Keeney RL, Raiffa H (1998a) Even swaps: A rational method for making trade-offs. Harv Bus Rev 76(2):137–150.

Hammond JS, Keeney RL, Raiffa H (1998b) The hidden traps in decision making. Harv Bus Rev 76:47–58.

Hammond JS, Keeney RL, Raiffa H (1999) Smart choices: A practical guide to making better decisions. Harvard Business Press

Horn J, Nafpliotis N, Goldberg DE (1994) A niched Pareto genetic algorithm for multiobjective optimization. Evol. Comput. 1994. IEEE World Congr. Comput. Intell. Proc. First IEEE Conf. IEEE, Orlando, FL, pp 82–87 vol.1

Howard RA (1992) Heathens, Heretics, and Cults: The Religious Spectrum of Decision Aiding. Interfaces (Providence) 22:15–27

Jacquet-Lagréze E, Siskos J (1982) Assessing a set of additive utility functions for multicriteria decision making, the UTA method. Eur J Oper Res 10(2):151-164

Keeney RL (1976) A Group Preference Axiomatization with Cardinal Utility. Manage Sci 23(2):140-145

Keeney RL (1992) Value-focused thinking: a path to creative decisionmaking. Harvard University Press, London

Keeney RL (2002) Common Mistakes in Making Value Trade-Offs. Oper Res 50(6):935–945

Keeney RL, Raiffa H (1976) Decisions with multiple objectives: Preferences and Value Trade-Offs. Wiley Series in Probability and Mathematical Statistics. Wiley and Sons, New York

Keisler JM, Noonan PS (2012) Communicating analytic results: A tutorial for decision consultants. Decis Anal 9:274–292

Keith WH, Radford KJ, Fang L (1993) Multiple participant multiple criteria decision making. IEEE Sys Man Cybern 23(4):1184-1189

Kersten GE (2001) Modeling Distributive and Integrative Negotiations – Review and Revised Characterization. Group Decis Negot 10(6)493-514

Kersten GE, Noronha SJ (1999) WWW-based negotiation support: design, implementation and use. Decis Support Syst 25:135–154

Kilgour DM, Eden C. (eds) (2010) Handbook of Group Decision and Negotiation, Advances in Group Decision and Negotiation 4. Springer Science.

Kilgour DM, Keith WH (2005) The graph model for conflict resolution: past, present, and future. Group Decis Negot 14(6):441-460

Kim SH, Ahn BS (1999) Interactive group decision making procedure under incomplete information. Eur J Oper Res 116:498-507

Kirkwood CW, Corner JL (1993) The effectiveness of partial information about attribute weights for ranking alternatives in multiattribute decision making. Organ Behav Hum Dec 54:456-476

Kirkwood CW, Sarin RK (1985) Ranking with Partial Information: A Method and an Application. Oper Res 33:38-48
Knowles JD, Corne DW (2000) Approximating the nondominated front using the Pareto Archived Evolution Strategy. Evol Comput 8:149–172
Korhonen P (2005) Interactive Methods. Mult. Criteria Decis. Anal. State Art Surv. SE - 16. Springer New York, pp 641–661
Korhonen P (2009) Multiple objective programming support Multiple Objective Programming Support. In: Floudas CA, Pardalos PM (eds) Encycl. Optim. SE - 431. Springer US, pp 2503–2511
Korhonen P, Wallenius J (2010) Interactive Multiple Objective Programming Methods. In: Zopounidis C, Pardalos PM (eds) Handb. Multicriteria Anal. SE - 9. Springer Berlin Heidelberg, pp 263–286
Leyva-Lopez JC, Fernandez-Gonzalez E (2003) A new method for group decision support based on ELECTRE III methodology. Eur J Oper Res 148(1):14-27
Likert R (1932) A technique for the measurement of attitudes. Arch Psychol 22(140):1–55
Miettinen K (1999) Nonlinear multiobjective optimization. Springer Science & Business Media
Morais DC, de Almeida AT (2012) Group Decision Making on Water Resources based on Analysis of Individual Rankings. Omega 40:42-45
Munda G (2008) Social multi-criteria evaluation for a sustainable economy. Springer, Berlin
Mustajoki J, Hämäläinen RP (2005) Decision Support by Interval SMART/SWING - Incorporating Imprecision in the SMART and SWING Methods. Decision Sci 36(2):317-339
Nurmi H (2002) Voting Procedures under Uncertainty. Springer Verlag, Berlin-Heidelberg, New York
Nurmi H (1987) Comparing Voting Systems. Dordrecht: D. Reidel Publishing Company
Osyczka A (1984) Multicriterion optimisation in engineering. Halsted Press
Pardalos PM, Siskos Y, Zopounidis C (eds) (1995) Advances in Multicriteria Analysis. Kluwer Academic Publishers
Partnoy, F (2012) Wait: The Art and Science of Delay. Perseus Group books
Pawlak Z, Slowinski R (1994) Rough set approach to multiattribute decision-analysis. Eur J Oper Res 72:443-459
Pedrycz W, Ekel P, Parreiras R (2011) Fuzzy Multicriteria Decision-Making: Models, Methods, and Applications. John Wiley & Sons, Chichester
Polmerol J-C, Barba-Romero S (2000) Multicriterion Decision in Management: Principles and Practice. Kluwer
Raiffa H (1968) Decision analysis: introductory lectures on choices under uncertainty. Addison-Wesley, London
Rao S (1984) Multiobjective optimization in structural design with uncertain parameters and stochastic processes. AIAA J 22:1670–1678
Rauschmayer F, Kavathatzopoulos I, Kunsch PL, Le Menestrel M (2009) Why good practice of OR is not enough—Ethical challenges for the OR practitioner. Omega 37(6):1089–1099
Rosenhead J, Mingers J (eds) (2004) Rational Analysis for a Problematic World Revisited. Second Edition, John Wiley & Sons Ltd.
Roy B (1996) Multicriteria Methodology for Decision Aiding. Springer US
Roy B, Słowiński R (2013) Questions guiding the choice of a multicriteria decision aiding method. EURO J Decis Process 1(1-2):69–97
Roy B, Vanderpooten D (1996) The European school of MCDA: Emergence, basic features and current works. J Multi-Criteria Decis Anal 5(1):22–38
Saaty, TL (1980) The Analytic Hierarchy Process. McGraw-Hill
Salo A, Hämäläinen RP (2001). Preference ratios in multiattribute evaluation (PRIME) - elicitation and decision procedures under incomplete information. IEEE Sys Man Cybern 31(6):533-545
Salo A, Punkka A (2005) Rank inclusion in criteria hierarchies. Eur J Oper Res 163(2):338-356

Salo AA, Hämäläinen RP (1992) Preference assessment by imprecise ratio statements. Oper Res 40:1053-1061
Schaffer JD (1984) Multiple Objective Optimization with Vector Evaluated Genetic Algorithms. PhD thesis, Vanderbilt University
Shakun MF (1988) Evolutionary Systems Design: Policy Making Under Complexity and Group Decision Support Systems. Holden-Day, Oakland, CA
Shakun MF (2006) ESD: A Formal Consciousness Model for International Negotiation. Group Decis Negot 15:491–510
Shakun MF (2010) Doing Right: Connectedness Problem Solving and Negotiation. In: Kilgour DM, Eden C (Eds.) Handbook of Group Decision and Negotiation, Advances in Group Decision and Negotiation 4. Springer Science
Simon HA (1955) A Behavioral Model of Rational Choice. Q J Econ 69(1):99–118
Simon HA (1960) The New Science of Management Decision. Harper & Row Publishers, Inc, New York
Simon, HA (1982) Models of Bounded Rationality. MIT Press
Slack N, Chambers S, Harland C, Harrison A, Johnson R (1995) Operations Management, Pitman Publishing, London
Slowinski R, Greco S, Matarazzo B (2012) Rough set and rule-based multicriteria decision aiding. Pesq Oper 32:213-269
Sprague Jr RH, Watson HJ (eds) (1989) Decision Support Systems - Putting Theory into Practice, Prentice-Hall
Srinivas N, Deb K (1994) Multiobjective Optimization Using Nondominated Sorting in Genetic Algorithms. Evol Comput 2:221–248
Steuer RE (1986) Multiple Criteria Optimization: Theory, Computation, and Application. Wiley, New York
Stewart TJ (2005) Dealing with uncertainties in MCDA. In: Figueira J, Greco S, Ehrgott M (eds) Multiple Criteria Decision Analysis: State of the Art Surveys. Springer Verlag, Boston, Dordrećht, London, 445-470
Thierauf, RJ (1982) Decision support systems for effective planning and control - A case study approach. Prentice-Hall, Inc., Englewood Cliffs, New Jersey
Vetschera R, de Almeida AT (2012) A PROMETHEE-based approach to portfolio selection problems. Comput Oper Res 39(5):1010–1020
Vincke P (1992) Multicriteria Decision-Aid. John Wiley & Sons, New York
Von Neumann J, Morgenstern O (1944) Theory of games and economic behavior. Princeton: Princeton University Press
Wallenius J (1975) Comparative Evaluation of Some Interactive Approaches to Multicriterion Optimization. Manage Sci 21(12):1387–1396
Weber M, Borcherding K (1993) Behavioral influences on weight judgments in multiattribute decision making. Eur J Oper Res 67(1):1–12
Zitzler E, Thiele L (1999) Multiobjective evolutionary algorithms: a comparative case study and the strength Pareto approach. Evol Comput IEEE Trans 3(4):257–271

Chapter 3
Basic Concepts on Risk Analysis, Reliability and Maintenance

Abstract: Man's level of dependence on equipment is increasing. This degree of dependence requires high levels of availability, which has been changing the impact that disruption of these systems causes. For many systems, an interruption has consequences that go beyond the dimension of financial loss, thus justifying a multidimensional consequence approach by using multicriteria (MCDM/A) models. Thus, understanding the relationships between and among risk, reliability and maintenance (RRM) is essential in order to offer more comprehensive solutions to the various problems often treated in isolation from each other, and which are the most important problems of the competitive market. This chapter discusses fundamental topics about RRM, including tools for risk analysis and hazard identification, concepts of reliability, maintenance techniques such as RCM and TPM and eliciting expert's knowledge. These topics are presented in order to provide a basis for structuring different MCDM/A problems that are addressed in several chapters. Some fundamental aspects could be used as input to decision models in different forms such as attributes, objectives, criteria, and problem context.

3.1 Basic Concepts on Risk Analysis

There are many concepts on risk found in the literature and also different perceptions to it. However, if a decision is being made and risk is involved, then, the risk concept should combine consequences and probabilities, incorporating the DM's preferences over that, as seen in Chap. 2.

Actually, a 'decision process' with no DM's preference has no decision being made, as discussed at the end of Chap. 1. Instead of that, that process either: a) has some preference structure incorporated within the model, at random; b) is just arbitrary following a previous decision of someone else.

Even so, in most of real cases, the consequences are multidimensional, and therefore, require an MCDM/A approach for building a decision model. The following topics are mainly based on the basic RRM literature and do not incorporate the idea of decision support, as given in Chap. 2. That is, DM's preferences are not necessarily considered in the model.

A.T. de Almeida et al., *Multicriteria and Multiobjective Models for Risk, Reliability and Maintenance Decision Analysis*, International Series in Operations Research & Management Science 231, DOI 10.1007/978-3-319-17969-8_3

3.1.1 Risk Context

In recent times, undertaking risk studies has become an increasingly complex task, making it of great importance in different spheres of society. Modern world facilitates access to information, making people more conscious of decision-making on risk and its consequences in the social and environmental context. On the other hand, organizations seek to manage appropriately all risks perceived as being the most relevant ones in the production of goods or services to ensure that their final product meets the minimum legal requirements, regulations and resolutions as well as society's expectations. However, it is of paramount importance to emphasize that in the so-called real world, despite organizations being concerned with identifying and monitoring risks, the restricted availability of resources is a crucial point, which leads to some risks receiving special attention with regard to the immediate allocation of resources, while others have to wait until resources become available.

Although in the literature there are several definitions of the term risk, the basic concept is associated with uncertainty in an environment and this is related to the likelihood of an undesirable event occurring and the impact of its consequences. This is why, according to Theodore and Dupont (2012), risk is defined as a measure of financial loss or damage to persons, in terms of the likelihood of an incident occurring and the magnitude of the loss. To Yoe (2012), risk is a measure of the likelihood and consequences of uncertain future events. It is the chance of an unwanted result where the lack of information about events that have not yet occurred is one of the factors inherent in the chance of its happening. Cox (2009) considers the preferences for consequences.

In the risk context, a change was recently observed by Aven (2012), who states that traditionally dangerous activities were designed and operated from references based on codes, standards and hardware requirements. However, what is verified today is that this trend is more directed towards a functional orientation, where the focus is associated with what it is sought to achieve. Therefore, the ability to define risk is the key element in each functional system. Identifying and categorizing risk are necessary to provide a decision support. The ability to define what may occur in the future, to evaluate risks and uncertainties and to choose among alternatives is what guides the decision-making process in the context of risk.

As seen in Chap. 2, the risk concept in the decision process combines the consequences with its probabilities, and incorporates the DM's preferences over that combination. Even so, in most real cases, the consequences are multi-dimensional, and therefore, involve an MCDM/A approach, which may involve tradeoffs as pointed out by Cox (2009), considering dimensions such as: financial, reliability and health.

Risk Management, Risk Assessment and Risk Analysis are supposed to ensure proper risk management and control, taking into account aspects such as procedures, use of tools, approaches and models. Attention should be paid to DM's participation that directly impacts the final results of a risk study. This requires communications about risks to be properly undertaken among the parties involved. Finally, detailed analysis to be carried out in risk studies directly impact the decision-making process. Different authors offer particular insights into these aspects.

To Modarres et al. (1999), Risk Analysis can be defined as a technique for hazard identification, characterization, quantification and evaluation. To Theodore and Dupont (2012), Risk Assessment is the process by which degrees of risk are estimated. Additionally, Yoe (2012) asserts that Risk Assessment is a qualitative, quantitative or semi-quantitative systematic process that describes the nature, probability and magnitude of risk associated with any substance, situation, action or event that includes uncertainties. Effective risk management requires the understanding of causes and conditions that contribute to the occurrence of an undesirable event and to the improvement of the system (Paté-Cornell and Cox 2014).

Regarding to Risk Communication, Fjeld et al. (2007) describe that this is an interaction process among stakeholders, risk assessors and risk managers. In this context, the objectives (often set by law), procedures and best practices seek to ensure that relevant aspects of risk analysis are identified by the stakeholders, thereby ensuring adequate analysis and a correct understanding of the decisions taken in relation to managing risk. In decision models, as seen in Chap. 1 and 2, some of these actors' role are related a DM.

On the topic of Risk Management, Yoe (2012) defines it as a process during which problems are identified, information is requested and risks are evaluated, and some initial definitions should be established to identify, evaluate, select, implement, monitor and modify actions taken to change the risk levels from unacceptable to the other two possible levels: acceptable or tolerable. To Aven and Vinnem (2007), the purpose of risk management is to ensure that appropriate measures are taken to protect people, the environment and assets from unintended consequences, as well as to balance different interests, especially with regard to health, safety, environment and cost. Risk management includes measures to avoid hazards occurring and to reduce the potential damage from them.

Tweeddale (2003) states that there are three main requirements for risk management: legal, commercial, moral (or ethical) requirements. Legal requirements will depend on the legal structure and the particular legislation in a specific locality. Commercial requirements are associated with a range of commercial implications such as loss of income due to production losses and costs related to damage to equipment, injuries or deaths, environmental damage, legal actions, and consequences for the company image. Moral or ethical requirements stress the value of human life, bearing in mind that people's health should not be measured monetarily. These requirements draw attention to the complexity of risk and show that risk has physical, monetary, cultural and social dimensions.

Although, different interests or requirements (criteria or objectives) are mentioned above, it does not seem to be dealing with multidimensional consequences, involving DM's preference, which would require an MCDM/A approach, as seen in Chap. 2. Really, integrating those dimensions (physical, monetary, cultural and social) may represent a risky complexity, whereas a non appropriate method is applied. This is identified by Aven and Vinnem (2007), mentioning MAUT for two attributes, costs and fatalities, although they recognize the difficulties of the elicitation process for obtaining the DM's preferences. This has to be evaluated in a case by case basis. All models have deviation, as seen in Chap. 2, however, in the purpose of making them useful, the appropriate effort should be made in the model building process. The successive refinement process proposed in Chap. 2 may support this evaluation.

3.1.2 Public Perception of Risk

Society deals with risks in everyday life so much so that risk analysis is an inherent characteristic of human beings. In daily routine activities, risk is always present e.g. when walking in the street, using public transportation to work, eating fatty foods, etc. Each person who participates in a hazard/risk analysis gives their own opinion, memory, attitude and global view of the situation under study. Moreover, these people are often affected by different types of personal biases such as their level of education, beliefs, experience, culture, etc. Even experts come to different conclusions when presented with the same data. The literature examines the issues that arise by discussing different physical situations and contexts.

van Leeuwen (2007) supports that perceptions of risk vary among individuals and the general public, business and other stakeholders, and change over time and in accordance with the prevailing culture. People continually assess situations and decide if the risks associated with a particular action can be justified. In some circumstances, dangerous effects are clearly associated with a particular course of action. However, in other cases, the impact of each effect can be uncertain and not immediately obvious.

To Modarres et al. (1999), the perception of risk often differs from the perception of objective measures, thereby distorting risk management decisions. Subjective judgments, beliefs and social bias with respect to events with low probability and high consequence may affect how the results of risk analysis are understood.

In this context, according to Crowl and Louvar (2001), the general public has great difficulty with understanding the concept of risk acceptability. The major problem is related to the involuntary nature of accepting a given degree of risk. For instance, designers of chemical plants who specify a level of acceptable risk assume that these risks are satisfactory to those living in the vicinity of the plant.

However, the neighborhood is often unwilling to accept any level whatsoever of industrial risk especially if the community is aware of there having been an accident involving a similar plant anywhere else in the world.

Additionally, Theodore and Dupont (2012) state that the lack of connection between public and experts is of fundamental importance, when addressing the question of why the public do not trust experts about these matters.

In view of these factors, it is important to pay attention to the fact that a coherent risk analysis involves people's perceptions about the risks under study, and should take into account all aspects that may negatively interfere in the process.

3.1.3 Risk Characterization

Risk characterization is another important aspect that should be undertaken. The definition of aspects that directly influence this analysis, the establishment of standards that ensure risk acceptability, tolerability and unacceptability are issues that should be considered in risk characterization.

Thus, according to Tweeddale (2003), the nature of the assessed risk will depend on the answer to two questions: (1) Will the undesirable event impact people, the environment, property or production? (2) How will the effects of the event be measured?

The MCDM/A approaches, as seen in Chap. 2, may answer these questions, which deals with the measurement of desirability by DM's preference over multidimensional consequences.

Therefore, Theodore and Dupont (2012) states that risk characterization estimates the risk associated with the process under investigation. The result of this characterization is to determine the likelihood of adverse effects which will be specified and enumerated arising from processes and/or leakages of substances derived from the process.

According to Smith and Simpson (2010) there is nothing which presents no risk. Physical assets always have failure rates and humans always make some kinds of mistake. Hence, this arises up the need of establishing values that qualify risks within a level considered acceptable by society. But, in practice, what does it mean when one speaks of a risk being tolerable, acceptable or unacceptable?

Again, the MCDM/A approaches may deal with establishing values for risk, bringing the DM to centre of the decision process, by means of incorporating preferences within the model.

For Smith (2011), the term 'acceptable' means that the likelihood of fatalities is accepted as reasonable, taking into account the circumstances and there being no efforts made to reduce them. The term 'tolerable' implies that although prepared for dealing with a risk level, an effort to tackle the causes of the risk is necessary in order to reduce them. Cost is an aspect that should be taken into account in this type of analysis. For Smith and Simpson (2010), the degree of risk considered as tolerable depends on a number of aspects such as the degree of control under the circumstances, the nature of risk analysis (intentional or unintentional), the number of persons subject to risk, etc. Finally, the concept of intolerable risk consists of not tolerating a specific risk level, thus not allowing activities to be developed at this level. Additional comments with regard to these definitions can be verified in the section dealing with ALARP concept.

To Crowl and Louvar (2001), it is impossible to eliminate any kind of risk completely. At some point in the design stage, someone needs to determine whether the risks are acceptable or not. In other words, are the risks under analysis greater than the daily risks that individuals are subject to in their daily lives?

According to Modarres et al. (1999), risk acceptability is a complex and controversial issue. However, making use of risk assessment results is a common way to rank the exposure level of risk, where the risk exposure levels that are socially acceptable should be defined based on risk acceptance thresholds.

In this context, some risk measures can be verified such as Individual Risk, Societal Risk, Population Risk and Risk Indices. Each of these measures expresses the risk, taking into account different aspects and contexts.

According to Smith (2011), Individual Risk refers to the frequency of a fatality for a hypothetical person with respect to a specific hazard scenario, while the Societal risk reflects the risk measure for a group of people, taking into account multiple fatalities. Theodore and Dupont (2012) describe Population Risk as the risk for the entire population, expressed as a certain number of deaths expressed as thousands or millions of people potentially exposed to danger. Theodore and Dupont (2012) also define Risk Indices, describing them as measures represented by a unique number associated with a facility. Some risk indices are quantitative while others are semi-quantitative, ranking risks in various categories. Risk indices can also be quantitative average or benchmarkings based on other risk measures.

In this context, Crowl and Louvar (2001) add that among these risk measures, losses and accidents based on statistical data are relevant measures. However, they should be considered with some caution, given that many of these statistics represent an average, and do not reflect the occurrence of a specific accident with potential losses. In contrast, no specific method is capable of measuring all aspects simultaneously. Some of those commonly used are the OSHA incident rate, the Fatal Accident Rate (FAR) and the Fatality Rate.

More specifically, according to Tweeddale (2003), FAR is a risk measure used to assess the risks associated with the employees of an industrial plant. FAR is defined as the number of fatalities, due to accidents at work, per 100 million hours worked.

In conclusion, the definition of risk measures is necessary so that reference values are established in risk studies and used in order that objectives are met regarding monitoring and controlling risk.

3.1.4 Hazard Identification

Nowadays, identifying hazards is a critical factor to ensure that safety requirements are satisfied, thereby attending to the need for assets, systems and subsystems to function adequately. Moreover, hazard identification provides data input for risk analysis in a particular production process (in part or in its entirety). For better performance, hazards should be identified by using structured techniques, and should involve experts and trained staff. What should always be taken into account in the planning stage are restrictions on resources (i.e. financial resources, experts, designers, operational and maintenance manpower, etc.) since the availability of these will have a direct impact in the outcome of the analysis.

Zio (2007) states that the first step of hazard identification is the output of this activity which is represented by a list of sources of potential hazards (i.e. component failures, deviations in processes, external events, operational errors, etc.) which have a non-zero probability of occurrence and can produce events with significant consequences.

The methods developed in this step are usually those associated with a qualitative analysis of systems and their functions, which will be included in a framework of systematic procedures. Among these methods, FMEA (Failure Mode and Effects Analysis) and HAZOP (Hazard and Operability Study) will be highlighted.

3.1.4.1 FMEA (Failure Mode and Effects Analysis)

According to Zio (2007), FMEA is a qualitative method with an inductive nature, which supports identifying failure modes of components that may disable the system or initiate accidents that can have considerable consequences.

For FMEA in order to obtain data that is sufficiently detailed, information must be collected from historical databases as must expert opinion. It is only by using FMEA in this way that all aspects of a project and system critical components can be verified. Further details regarding to FMEA, including FMECA (Failure Mode, Effects, and Criticality Analysis), a derived technique, are given in Sect. 3.2.6.

3.1.4.2 HAZOP (Hazard and Operability Study)

According to Andrews and Moss (2002), HAZOP is a method that was first used in the chemical industry, where industrial plants are evaluated with regard to identifying potential hazards to operators and society. These hazards may arise in a particular system and may be result of interaction among different systems of the industrial process.

According to MacDonald (2004), HAZOP presents well-defined stages, as shown in Table 3.1:

Table 3.1 HAZOP stages

HAZOP stages	Details
Stage 1:	Defining the scope and objectives;
Defining the process	Establishing responsibilities;
	Forming the team.
Stage 2:	Defining planning and implementation schedule;
Preparation	Data collection;
	Registration methodology;
Stage 3:	Systems division;
Verification	Identifying deviations;
	Establishing causes, consequences and setting protection measures;
	Reaching consensus on the actions;
	Repeating activities for each element evaluated.
Stage 4:	Defining spreadsheets registration;
Registrations and monitoring	Preparing reports;
	Monitoring actions;
	Re-assessing HAZOP periodically;
	Producing and distributing final report.

HAZOP is used to identify and assess hazards in production and maintenance operations. In addition, multidisciplinary teams and expert opinion must be used in preparatory studies associated with this methodology and the scope and objectives of projects must be well established. Moreover, people involved in the process must have a good understanding of the particular terminology. Deviations, guide words and project intent are some of the terms used.

According to Ericson (2005) some of the disadvantages of HAZOP that have been reported include: focusing on single events without considering the combination of more than one event; focusing on specific guide words can result in some dangers that are unrelated to these guide words not being valued; HAZOP analysis can be too much time and resources consuming.

According to Zio (2007), while FMEA is mainly based on the structural aspects of a system, HAZOP processes focus on the plant under analysis.

3.1.5 FTA (Fault Tree Analysis)

The Fault Tree (FT) is a tool widely used in industrial processes within the risk environment. It can be classified as a qualitative or quantitative tool depending on the availability of the likelihood values of failure events.

According to Ericson (2005), FTA is defined as a structured deductive technique which is used to analyze a system so as to identify and describe the root causes and the likelihood of the occurrence of a particular undesired event. FTA is applied to evaluate dynamic complex systems, in order to understand and prevent potential problems. The development of the tree is an iterative process that can be used preventively or reactively (in this case, after failures have occurred).

The FT is a graphical model built from a top event, also known as an unwanted event. It is structured in such a way as to identify and combat all possible relevant causes (root causes) of the event linked with the top event.

This tool can be used in both preventive manner (mitigation) and corrective manner. The elimination of all root causes produces the elimination of the top event. Similarly, the elimination of only some root causes results in reducing the probability of the top event.

According to Andrews and Moss (2002), the fault tree diagram shows two basic elements: gates and events (both represented by specific symbols depending on the context). The relations amongst FT events occur through logic gates that enable or inhibit the passage of failures along the tree, thereby showing the relations necessary for another event at a top-level of the tree to occur. For each gate there is a specific gate symbol, a gate name and valid causal relation. The gates most commonly used are *AND* and *OR* gates. For example, the existence of a gate *AND* means that the output event occurs if all input events occur simultaneously (since there are at least 2 input events). On the other hand, the existence of the gate *OR* means that the output event occurs if at least one of the input events occurs simultaneously (since there are at least 2 input events). An FTA example that shows a top event, *AND* and *OR* gates and basic causes, also known as root causes, is given in Fig. 3.1.

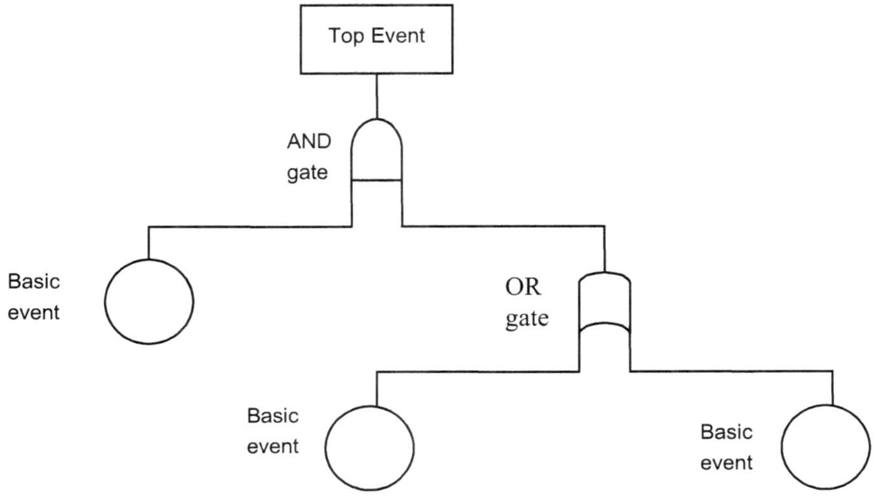

Fig. 3.1 FTA example

FTA is a technique to assist the estimation of failures likelihood (Nwaoha et al. 2013). When FTA is applied as a quantitative approach, the value of the likelihood of the occurrence of the top event is obtained based on the specific Boolean properties of the gates.

More specifically, an important matter to be noted is that a FMECA failure mode can be considered as an input to an FTA top event. Thus, each specific FMECA failure mode is a top event of a specific FT.

3.1.6 Event Tree Analysis (ETA)

According to Ericson (2005), ETA (Event Tree Analysis) is an analytical technique to identify and evaluate sequences of events in a potential accident scenario arising from the occurrence of an initiating event. ETA uses a logical tree structure known as an event tree (ET). The purpose of ETA is to determine whether the initial event will unfold in a series of unwanted events or if the event is sufficiently controlled by security systems and procedures established during the system design phase. ETA can generate several different results from one initial event, thereby allowing a specific likelihood for each outcome.

According to Bedford and Cooke (2001), the ET structure starts with an initial event *propagating* this event through the system under consideration, taking into account all the possibilities that can affect the behavior of the system/ subsystem.

ET nodes represent the possible operation (or non-operation) of a system/ subsystem. More specifically, the ET pathway that results in an accident is called an accident sequence. An example of an ET is shown in Fig. 3.2 (Brito and Almeida, 2009).

According to Ericson (2005), ETA can be used to model a system entirely, comprising subsystems, components, software, procedures, environment and human error. It can also be used at different stages such as the project design phase, and has been applied to different systems such as nuclear power, aerospace and chemical plants.

An analyst should guide the ET construction process by identifying and evaluating all possible outcomes resulting from an initial event. A positive aspect is that if applied in early stages, ETA helps to identify system security issues, thus avoiding corrective actions (Andrews and Dunnett 2000).

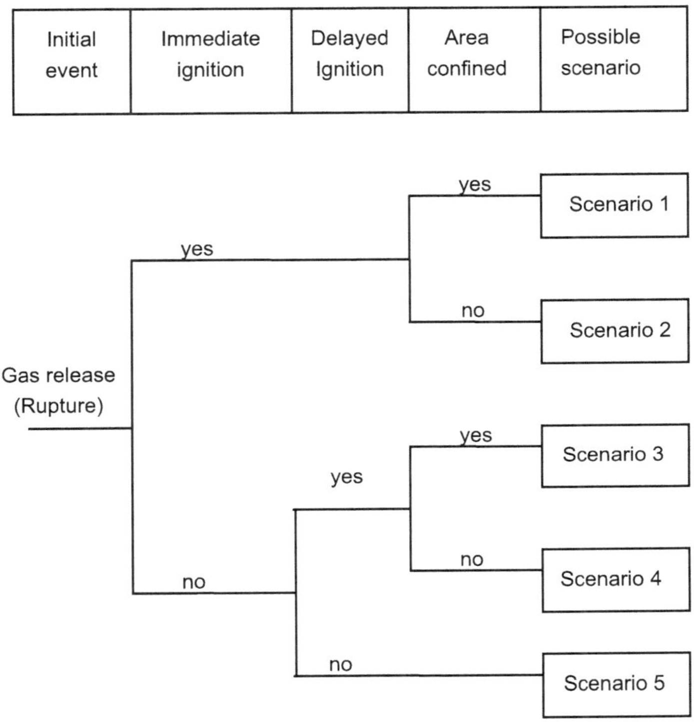

Fig. 3.2 Illustrative example of event tree applied to the risk analysis of a pipeline

Regarding the events that comprise ET accident sequences, Zio (2007) states that they are characterized by: intervention (or not) of protection systems that should come into operation (or not) to mitigate the accident (System Event Tree); the running (or not) of security functions (Functional Event Tree); and the occurrence (or not) of physical phenomena (Phenomenological Event Tree).

According to Zio (2007), these event trees types are applied in different contexts:

- System Event Tree – this is used to identify accident sequences that have developed within a plant, involving protection and security systems;
- Functional Event Tree – this is an intermediate step when constructing the System Event Tree. From the ET initial event, safety functions that need to be established are identified, and are subsequently replaced by the corresponding protection and security systems;

- Phenomenological Event Tree – this describes the evolution of a phenomenological accident that occurs outside the plant (fire, dispersion ...).

Finally, the integrated use of tools can also be checked in event trees where the Fault Tree (FT) quantitative approach is applied to obtain a value for the likelihood that a failed state will occur in any given branch of the ET. An example is shown in Fig. 3.3. Andrews and Dunnett (2000) presents a comparative analyses considering ETA and FTA.

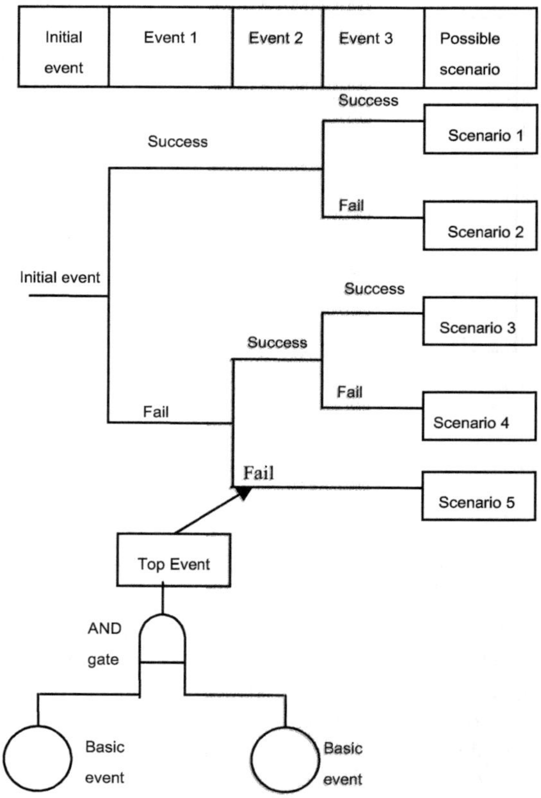

Fig. 3.3 How integrated tools (FTA and ETA) are used to determine failed states

The likelihood of an ETA failure is the same of a top event obtained from FTA, implemented for each specific failure observed in the ETA. The likelihood of success is calculated as being the complement of this failure likelihood. Otherwise, $P_{(Success)} = 1 - P_{(Fail)}$.

3.1.7 Quantitative Risk Analysis

Risk analysis techniques are devoted to supporting managerial decisions regarding risk reduction in order to achieve and maintain tolerable risk levels and therefore assuring safety.

According to Vinnem (2014) the abbreviation QRA is also used for Quantified Risk Assessment, and the context of the analysis defines which of these terms are more suitable. When an evaluation of the results is combined with the risk analysis, the term assessment should be used. This nomenclature and the term QRA are well established for offshore operations and oil and gas and chemical processes. They are also referred to as Quantitative Risk Assessment (QRA), Probabilistic Risk Assessment (PRA), Probabilistic Safety Assessment (PSA), Concept Safety Evaluation (CSE) and Total Risk Analysis (TRA), although the nuclear industry for example, adopts the terms Probabilistic Risk Assessment or Probabilistic Safety Assessment (Bedford and Cooke 2001; Vinnem 2014). Some authors consider that all these terms have almost the same meaning as the tools considered converge in order to be a scientific analysis of risk.

According to Vinnem (2014), Norway was for many years the only country that required QRA studies systematically. Norway started doing so in the 1980s. However, it took the UK almost 10 years before legislation was introduced that laid down the need for QRA studies, namely when official inquiries due to the Piper Alpha platform accident in 1988 recommended the adoption of QRA in the UK similarly to Norway which had done so ten years earlier.

When dealing with risk analysis there are many systematic techniques such as:

- Hazard and Operability Study (HAZOP);
- Safety and Operability Study (SAFOP);
- Safe Job Analysis (SJA);
- Preliminary Hazard Analysis (PHA);
- Failure Model and Effect Analysis (FMEA);
- Quantitative Risk Analysis (QRA).

Despite QRA, most of these approaches are essentially qualitative, although it is possible to incorporate quantitative information and be performed in a semi-quantitative way.

However, to perform a QRA, it is necessary initially to identify hazards and describe risks to personnel, environment and assets in a quantitative manner. Although the identification of hazards may be obtained from a qualitative study, the initiating events are evaluated in a quantitative perspective, leading to the analysis of the causes in terms of probability to estimate the probability of each scenario.

According to Vinnem (2014), for each scenario, estimates are made of consequences, effects, facility responses and associated probabilities, which enables consequences to be quantified in terms of personnel environment and assets, which represents losses in human, environmental and financial dimensions.

In the MCDM/A approaches, scenarios may be related to the state of nature (θ), which is associated to the probability $\pi(\theta)$. Also, the use of MCDM/A approaches enables multidimensional consequences to be quantified.

Vinnem (2014) describes QRA in five steps, represented by Fig. 3.4. The first two steps in Fig. 3.4 are mainly qualitative. First of all, events are identified which may also be called hazard identification (HAZID), and this requires that all possible hazards and sources of accidents should be investigated to avoid neglecting any source of accident. During this screening, levels that shall be used to classify critical and non-critical hazards are defined, providing reports that register the evaluations made to classify each hazard, in order to have a register of the reasons why and a demonstration of how a hazard was classified as non-critical, while assuring that it was safe to state that these hazards were not considered as critical.

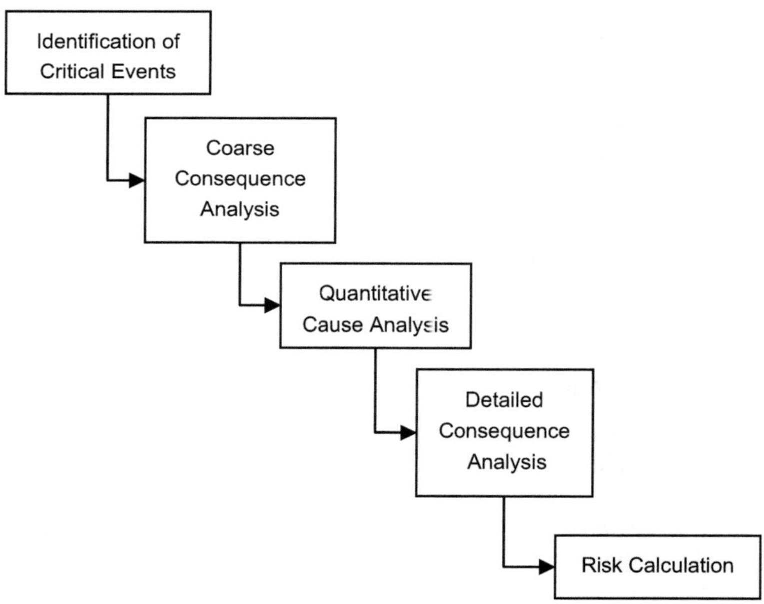

Fig. 3.4 QRA steps

Considering the tools available for hazard identification, such studies are usually supported by the use of checklists, statistics on failure, a database of accidents, HAZOP studies and similar risk analysis studies. The experiences obtained from similar projects are also an important source used to identify hazards.

After identifying the critical hazards to be considered, it is necessary to identify the causes of these hazards and which events may lead to an accident scenario occurring. Identifying the starting point for a potential accident enables the chain of events that may cause an accident to be established.

During the analysis of the cause analysis (the third step), it is determined which causes may lead to the initiating events in order to support the assessment of the probabilities of initiating events. From the cause analysis, it is possible to identify risk reducing actions that would prevent or interrupt the chain of events that may cause an accident. In the initial steps of the cause analysis, qualitative techniques are usually deployed followed by quantitative approaches if there are data available for quantification. Qualitative approaches are used to identify causes and conditions for initiating events, thereby establishing the basis for a possible subsequent quantitative analysis. With regard to the techniques used to identify the causes, there are: HAZOP, Fault Tree Analysis (FTA), Preliminary Hazard Analysis (PHA), FMEA and human error analysis techniques, which are also used in traditional reliability analysis.

Quantitative studies in cause analysis are conducted in order to establish the probability of the occurrence of initiating events, while using historical statistics to calculate the frequency of initiating events is one of the most common approaches.

The fourth step in Fig. 3.4 is related to the consequence analysis of accident scenarios. A consequence analysis considers the existence of barrier functions and elements to contain hazards and the accident sequences in order to evaluate the possible function or failure of barriers involved. According to Vinnem (2014), fire and explosions are two of the main factors evaluated, and both may be assessed by using the same calculation steps for all scenarios which may involve fire and/or an explosion. These steps depend on which conditions and sequences are related to the factors evaluated. Fire and explosions may be a result from a leakage, punctures or pinhole, any of which may expose a hazardous material which when associated with a chain of events may result in a Vapor Cloud Explosion (VCE), a Boiling Liquid Expanding Vapor Explosion (BLEVE), a Flash Fire, a Jet Fire, and so forth. Thus, the steps of the calculation are used to estimate the amount of material leaked by considering temperature and pressure conditions in the system associated with system barriers and mitigation actions. *TNO's Colored Books* present systematic procedures to assist QRA studies, especially consequence analysis, regarding the estimation of thermal radiation, ignition probabilities, conditional probabilities for fatalities, damages and other consequences. Regarding fatalities, probit functions are usually used to calculate the probability of death due to exposure to toxic substances and / or heat radiation at a given level of exposure.

The results from a QRA study arise from the risk calculation, the last step of Fig. 3.4. These results are usually compared and associated with risk tolerance level. QRA studies are usually performed until barriers and safety actions are strong enough to assure that any risk is above reference levels. A QRA study aims to provide a risk picture, which results from the hazard identification, and a cause and frequency analysis that are combined to express the risk level associated with all critical hazards.

Although the terms risk calculation, risk analysis and risk assessment can be easily misunderstood since they have the same general meaning, there are differences related to the scope of each term. Risk calculation uses information from consequence analysis and cause analysis thereby providing a risk level calculated from frequencies and the magnitudes of consequences. While risk analysis refers to the entire process described in Fig. 3.4, which includes the risk calculation, Risk assessment is the entire process of risk analysis when the results are evaluated regarding risk reference levels, which are defined by considering a notion of risk tolerance.

To ensure the reliability of the results of QRA studies, there are several factors that must be considered, such as:

- The technical description of the system (activities, operational phases);
- Purpose and target of risk analysis;
- Activity levels on the installation;
- Operation of safety systems;
- Study assumptions: how these are verified and accepted;
- Data Sources.

Thus, QRA is a systematic development of numerical estimates of the expected frequency and/or consequence of potential accidents associated with a facility or operation.

According to Arendt and Lorenzo (2000), there are two main misconceptions about QRA which are to do with the lack of adequate data on equipment failure and the cost of conducting QRA, i.e. whether it is cheap or expensive.

Regarding the availability of data, there are industry wide databases that can provide data for frequency rate estimates and regulation authorities that provide periodical reports, such as:

- The Guidelines for Process Equipment Reliability Data with Data Tables;
- IEEE Guide to the Collection and Presentation of Electrical, Electronic, Sensing Component, and Mechanical Equipment Reliability Data for Nuclear Power Generating Stations (IEEE Std 500);
- The OREDA Offshore Reliability Data Handbook;
- Non-electronic Parts Reliability Data 1991 (NPRD-91) and Failure Mode/Mechanism Distributions 1991;
- Systems Reliability Service Data Bank;

- Nuclear Plant Reliability Data System: Annual Reports of Cumulative System and Component Reliability;
- Offshore Blowouts Causes and Control;
- UK Health and Safety Executive Reports.

The accuracy of the results is a function of the resources deployed in the analysis. As the quality of the input into the model improvement, the results will become more accurate. Thus the availability of resources is the primary constraint for the quality of QRA results. There is a need to perform a cost-effective analysis, so managers (or DMs) may balance the value of QRA results compared to the cost of having such results. Thus, over the years QRA has been considered very cost-effective.

The QRA results do not show if an installation is safe or unsafe, but they give a risk picture that has to be evaluated in a risk assessment context. A DM must decide whether to seek changes and safety improvements in order to reduce risk, or even if the benefits of these safety improvements would justify the cost of making them. That is, tradeoffs regarding multiple criteria should be made, which should be based on MCDM/A methods.

Typically, QRA results report risk in terms of its consequences per year. If there is an analysis of the consequence of a human dimension, the report will contain a risk result of the expected number of fatalities and/or injuries per year or per hours of equipment operation.

If the analysis is with regard to environmental consequences, the report shall contain risk results in terms of the expected amount of chemical substances spilled and the extent/size of the affected area on the same basis as for the human consequences.

The next section tackles risk tolerance, which are used for Risk Assessment in order to evaluate if a facility or installation is safe or unsafe according to objective minimum risk targets.

3.1.8 ALARP

According to Bedford and Cooke (2001) the ALARP (As Low as Reasonably Practicable) principle has guided the setting of tolerance risk levels since the 1980s and 1990s, in order to achieve safety goals. The USNRC policy statement (NRC 1986) and the UK tolerability of risk document (HSE 1987) seek to convert the principle of setting this 'as low as reasonably practicable' (ALARP) into a numerical definition, which establishes upper levels for risk intolerance and lower levels at which risks can be considered as tolerable.

Given that criteria for risk acceptance are generally combined with risk analysis, some industries and countries have regulations that require such criteria to be defined prior to the risk analysis (Aven 2012).

In a more practical perspective, ALARP can be understood as a risk goal to be achieved in order to define investments in safety. Most of the safety standards indicate that risk evaluation should be conducted until safety improvements result in tolerable risk levels being reached. One example is the ISO/IEC: Guide 51.

Sutton (2010) describes the idea behind the concept of ALARP as being that risk should be reduced to a level that is as low as possible without requiring excessive investment, thus establishing a numerical boundary that determines whether a risk is definitely acceptable or definitely not acceptable.

Given the tradeoff inherent in considering costs and safety together, Bedford and Cooke (2001) point out that using the ALARP principle reduces the tradeoff between safety and costs so as to increase safety by implementing what is reasonably practicable. However, discussion about the value for a human life is always an issue that leads to hotly-debated argument. ALARP is usually applied to support the definition of tolerable limits for human losses.

On using the ALARP principle, it is possible to classify risks into three categories: negligible risk, tolerable risk and unacceptable risk (Macdonald 2004):

- Negligible risks are risks that fit into a category of being broadly acceptable by most people in their daily lives. This class of risk considers situations such as being struck by lightning or having a brake failure in a car;
- Tolerable risks are those risks that one person would rather not have. However they are deemed to be tolerable in view of the gains obtained by accepting this situation. For this type of risk the inconvenience in terms of burdens are balanced against the scale of risk. Thus, a compromise is accepted. An example of this situation is when a person decides to drive a car or travel by bus. Usually, in these situations, people accept that accidents can happen but try to avoid them by minimizing the chances of having an accident;
- Unacceptable risks are those that are at a level of risk that is too high to accept, and therefore are unacceptable; in other words, they have a tolerance level of zero. The losses regarding such risks are so high that they cannot be compared with any possible benefit arising from any situation where there is exposure to such risk.

The ALARP principle may be understood in the context of MCDM/A methods, considering the intra-criterion evaluation. This is similar to the constructed criterion, discussed in Chap. 2. Also, in this kind of approach may be related to a sorting problematic, in which the consequences or alternatives are classified into categories.

So this principle is used to guide hazard and risk analysis by setting tolerable risk goals to be achieved in any hazardous situation. Usually this is the first step for any assessment of a safety system.

ALARP risk regions can be illustrated as shown in Fig. 3.5, which presents each risk region according to tolerable risk levels. This is also called a carrot diagram, which is presented in most of the related literature.

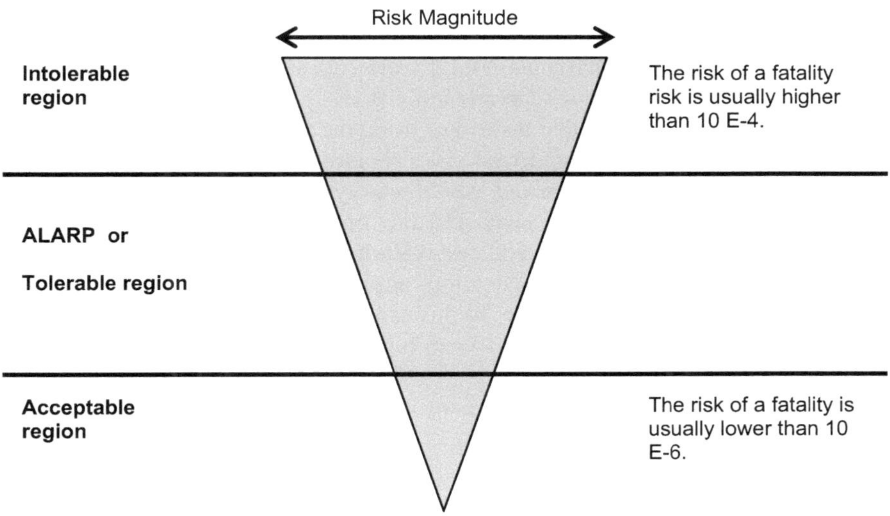

Fig. 3.5 ALARP principle: tolerance limits

The definition of ALARP regions are based on everyday risks. Thus, risks that are considered typical and commonly expected may include risks from all causes, including bad health.

To measure the risk level, the Fatal Accident Rate (FAR) is used with particular regard to the employees of some hazardous installations who are usually exposed to higher risks than those working in less hazardous workplaces.

Aven (2012) adds that in practice the value considered to reflect risk is an estimation of FAR or the probability p of a certain accident event, since the true value of FAR or p is unknown. Thus using tolerance limits means to compare an estimated value with acceptable values. This means that using a best-estimate approach may not produce clear recommendations, and thus standardized models and input data may be required. Thus, the acceptance level is a function of such models and input data.

According to Tweeddale (2003) in some cases there is a subjective opinion and a potential debate about whether ALARP standards are achieved and this can lead to such issues being questioned in court. Nevertheless, if the hazardous installation uses the best technology available and can be set up, and it also uses the best operable and maintainable management systems in order to improve safety by keeping the equipment maintained to high standards, the risk is usually an ALARP one.

There are some criticisms regarding the use of ALARP as a way to justify risk exposure. In addition, there is a problem with the term regarding 'acceptable risk'. This is because it is commonly used by those who generate the risk to excuse the fact that others will be exposed to it. And this is why some authors call the very concept of ALARP into question. Tweeddale (2003) remarks that the level of risk that an individual accepts is particular to that individual. That is, it is not an applicable standard to any individual. Moreover, what is regarded as an 'acceptable risk' may change over time. It is worthwhile to note that, in the case of MCDM/A context, since a particular DM's preference is considered, its result may not also be an applicable standard to any individual.

Tweeddale (2003) argues that instead of using the term 'acceptable risk', terms such as 'accepted risk' or 'approved risk' should be used. The former would denote that the individual involved would wisely or unwisely accept the risk, independently of whether it is considered low or high compared with everyday risks. The latter would be used to address exposure to risk that complies with rules or standards set by an appropriate statutory authority or regulator (that is, a DM) on behalf of the general community. In this case, the regulator would define what approved risks are even if those risks were higher than the everyday risks that an individual is exposed to.

Such risks may include those from many causes that can result in a fatality such as (Tweeddale 2003):

- Smoking;
- Swimming;
- Travelling by motor vehicle;
- Traveling by train;
- Accidents at home;
- Pedestrian struck by a vehicle;
- Homicide;
- Accidental poisoning;
- Fires and accidental burns;
- Electrocution (non-industrial);
- Storms and flood;
- Lightning strikes;
- Snake bite.

Some individual risks are exemplified in Table 3.2 as FAR and probability values (Macdonald 2004).

Table 3.2 Example of individual risk and FAR based in UK data

Activity	FAR per 10^8	Individual risk of death per person per year x 10^{-4}
Travel		
Air		0.02
Train	3-5	0.03
Car	50-60	2
Occupation		
Chemical industry	4	0.5
Agriculture	10	
Rock Climbing	4,000	1.4
Staying at home	1-4	

It is important to remark that these everyday risks may change from country to country; for example, even in the same country, some regions may have a significantly higher homicide rate than the rest of the country. Therefore, it is possible that a given risk level could be considered tolerable in an undeveloped country and intolerable in a developed country. Thus, the values given by Macdonald (2004), with some examples given in Table 3.2 reflect the reality of a developed country, and these might be lower than in undeveloped countries.

3.1.9 Cost-Effective Approach to Safety

After assessing the probability of hazardous events, all possible actions must be deployed in order to achieve a tolerable risk level. In fact, if the risk to life is so high that is beyond economic concern, the equipment or plant must be considered safe, otherwise it must be closed.

However, when the tolerable risk level is reached, investments in risk reduction are only justified by a cost effective evaluation. Thus, when a risk level is considered tolerable or an ALARP one, any costs to improve safety must be followed by a compatible benefit. Otherwise, it should not be implemented. Usually the cost per life saved with a previously established level is considered. Aven (2008) defines a similar measurement as the implied value of a statistical life or the implied cost of averting a fatality, by dividing the cost of the safety improvement by the number that represents the expected reduction in the number of fatalities. Therefore this ratio can also be considered by quantities other than lives saved, if for example, an environmental risk is being considered, the reference may be to tons of oil spilled.

This allows expending resources in order to improve safety by acting where one can find the greatest benefits while taking the budget allocated to improve safety into account. These costs vary according to the type of system, complexity and regulatory standards regarding the activity. Enterprises usually avoid disclosing data on levels of cost per life saved. According to Smith (2011), this value is between *£500,000* to *£4,000,000*, while if the risk has potentially multiple fatalities, then higher amounts may be considered.

Thus, the more that the number of potential fatalities increases, the more risk averse the analysis becomes, which leads to choosing a higher cost per life saved level. It is valuable to observe that utility theory provides an axiomatic structure to evaluate DM's behavior regarding risks, including risk aversion, as it seen in Chap. 2.

As examples of how these values are considered, Smith (2011) points out that for passenger road transportation, there is a voluntary aspect to the exposure and a small number of casualties per incident, so the value considered for cost per life saved is approximately *£1,000,000*. For the transportation of dangerous material, where the risk is not under an individual′s personal control which means that there is an involuntary risk, Smith (2011) presents a cost per life saved of approximately *£2,000,000* to *£4,000,000*. When considering multiple offshore fatalities, where there are a large number of fatalities and no personal control by the victims, Smith (2011) shows that the cost per life saved can be between *£5,000,000* to *£15,000,000*. Therefore, these values are quite controversial and may change when they came under scrutiny in the media or are reported as catastrophic accidents, thereby making the analysis even more risk averse.

Smith (2011) states that the maximum tolerable risk for a single fatality does not always coincide with the societal risk calculations. Thus, while societal risk measures the frequency of a fatal event, when considering individual risk, it is the frequency of individual deaths that is considered. One of the main differences between estimating individual risk and societal risk is about whether the risk is voluntary or involuntary. When considering individual risk it is important to highlight that these individuals are voluntarily exposing themselves to risk, in a specific place that sets specific conditions for the frequency and risk assessment. When considering societal risk, what is considered is the involuntary exposure to risk that may reach random individuals, and it characterizes this concept of an involuntary risk.

According to Tweeddale (2003) there is no unanimous formal agreement regarding a specific value that can be considered as denoting a tolerable level of risk. But in many countries it is typical to consider that an additional risk of 1 chance in a million per year (10^{-6} per year), due to industrial sources affecting the person most exposed to these, is a very low risk level compared to everyday risks that an ordinary person is usually exposed to without questions being raised about this. Aven (2008) points out that the probability of a fatality for a third person associated with exposure to risk in an industrial plant is required to be less than 10^{-5} per year. Therefore, the value that defines whether a risk should be considered tolerable and therefore accepted by the wider community is the Individual Risk level. Some of these everyday risks have been exemplified.

When calculating individual risk the focus must be on an event in which one specific person is seriously injured or killed. Aven (2008) defines individual risk as the frequency of death for the person or critical group of personnel most at risk from a given activity due to their location, habits or periods that make them vulnerable. Thus, individual risk is measured as the annual frequency of an accident with one or more fatalities over a homogeneous group of people who voluntarily expose themselves to risk. This is an approximation of the probability that a random person of a group who conducts a specific voluntary activity will be killed while he/she is at the industrial facility over the course of the time period considered, usually a year. This measurement is used to calculate the FAR.

As to the risk to any individual who is involuntarily exposed to some risk, consideration has to be given to the possibility that more than one person may be killed due to that risk source. Thus societal risk cannot be measured only by individual risk, but must include the possibility that there may be 1 to N fatalities. The more the number of fatalities increases the more risk averse the analysis is.

Societal risk is usually represented by F-N curves, which show the frequency of accident events with at least N fatalities.

Tweeddale (2003) recognizes the controversy regarding attempts to put a value on a human life, since human life can be considered priceless due to the emotional values that money cannot compensate for. Nevertheless, there is a need to establish a limit for the amount that can be spent per life saved, otherwise it is impossible to decide if a choice can be made on economic grounds between improving safety in order to keep an industrial plant running or closing the plant. According to Tweeddale (2003), one absolute limit to be established for cost per life saved would be obtained by dividing the annual gross national product by the annual number of births. This value represents the amount that may possibly be spent in order to extend the life expectancy of each new-born baby, if there are no other expenses in the community. As there are many other requests for financial resourcing from the wealth derived from the community, the real value of the limit to the cost per life saved would be less. Therefore the definition of this value will depend on particular priorities and other characteristics of the problem, such as those pointed out in the examples which discussed calculating risk values to do with the transportation of passengers by road, the transportation of dangerous materials and substances and multiple offshore fatalities.

3.1.10 Risk Visualization

Risk Visualization is a tool used to produce images of risk (i.e. 3D visualization and risk rich pictures) in order to illustrate and facilitate the risk perception by any actor (DMs, managers, users, etc.) in a decision making or managerial process. This subject integrates the concept of information visualization.

The risk visualization may be applied to visualization in risk management framework, considering visualization in risk identification, visualization in risk analysis, visualization in risk assessment, visualization in risk communication and visualization in risk reduction. This support can provide processed information and better control to making the more appropriate decision making for these previous modules.

Additionally, the interaction among risks (when it happens) is an important question that should be treated in risk visualization in decision making process. It allows a more complete risk appreciation (Ackermann et al. 2014).

According to Bostrom et al. (2008), understand how risk representations affect judgments and decision making is essential to comprehend the risk management and the decision-making process. Therefore, graphical representations of risk seeks to simplify some concepts and constraints related to mathematical, chemical or physical aspects, making risk management and decision making more comprehensible to the public (Ale et al. 2015).

In general, the information visualization in risk management process can aid the perception and understanding of the risk and its several aspects. Therefore, the risk visualization may be applied in several modules of the risk management framework, such as in risk identification, risk analysis, risk evaluation, risk assessment, risk communication and risk reduction. This support can provide processed information and better control to making the more appropriate decision making for these previous modules mentioned.

Eppler and Aeschimann (2009) and Horwitz (2004) highlight that the visualization in risk management is still not a frequent topic in organizations, probably because there is a difficult to describe and visualize the risk.

Some insights can arise of the answers to the following questions:

- How can the information visualization aid in the various steps of risk management? For instance, can the information visualization improve the performance in the risk identification module? Can the information visualization support the determination of the likelihood and the consequences estimation?
- How is it possible to handle differences of knowledge and skills, through the risk visualization, among the various users of the system?

Al-Kassab et al. (2014) emphasize that the way in which information is 'framed' and communicated not only helps in interactive decision process, but also provides a means of knowledge creation. Based on the literature review, they summarize the information visualization process in five steps: 1) Raw data collection; 2) Data transformation; 3) Data warehouse; 4) Visual transformation; 5) Viewer interaction.

Firstly, it is need to collect of the quantitative and/or qualitative data from different sources and store them in one place (database). Based in this set data collected, is necessary to transform and comprises these data. Then, it is necessary a visual transformation by mapping of the transformed data, and, therefore, the creation of a new 'picture' of the information that can be seen by DM, through of visual/graph structures (graphs, tables, maps, etc.). Lastly, the DM can interact with these visualization structures, allowing the transformation process at different stages of decision making. Furthermore, DM can adjust their view on the data, change the visual structure, or even affect the data transformation.

Moreover, Al-Kassab et al. (2014) identified three fundamental managerial functions of information visualization: a communication medium, a knowledge management means, and a decision-support instrument. These functions also can be contextualized within each module of the risk management framework.

The use of visualization as a communication medium function is linked with the knowledge-based processes, through of the patterns identifying, correlations, outliners, clusters data and other techniques, mainly when there is a big data. It adopts several display techniques and approaches aiming to elaborate and analyze data allowing 'transmission' of messages to be interpreted by DM and by stakeholders. Furthermore, the knowledge created by the information visualization itself should be shared and interpreted by the DM. Hence, it is essential in order to make a coherent risk evaluation, risk perceptions, preventive and mitigation actions or other strategic actions linked to risk management. This information should be communicated, understood, shared and implemented for all of the organization, or for all that suffers the impact of the risk.

The information visualization, as a function of a knowledge management means, can facilitate or obstruct the human brain's capacity to interpret information (Al-Kassab et al. 2014). Also, it is highlighted that the information visualization must take into account the context and purpose of the knowledge because this interpretation is affected by knowledge and cultural background of the DM in the risk context. It is important to note that several times in risk context, the knowledge acquired in risk management is affected harmfully by absence of information and database.

On one hand, the risk perception is linked with the past experiences of the individual, producing some biases, that can affect negatively the risk visualization and consequently the decision making process. For the other hand, if actions are taken adequately these biases can be minimized or nulled. Thus, any visualization technique presents pros and cons that need to be addressed clearly with the DM.

Finally, it is discussed the function of information visualization as a decision-support instrument. The requirement to synthesize and analyze the information in big problems can be better solved by DMs when they aided by information visualization. It can improve the process of decision making, when it properly considers the features of decision making and the characteristics of DM.

In literature, research about spatial and visual perception suggests that, generally, graphics avoid the inadequate numerical risk representations as well as countable visuals increasing the accuracy of perceived risks (Bostrom et al. 2008).

In risk map, for instance, one may use means like line thickness, textual information labels, shapes that varied in size or color and other characteristics. The color of an enclosed region may represent a 'concept' type and the size may be used to represent the magnitude this 'concept'. Thus, the reader should quickly discover the most serious undesirable incidents, since they often represent major risks.

There are some aspects to be considered when using color-coding. The number of different colors is limited by the DM's ability to remember and distinguish the colors, for the following reason: He/she can present a confused view. The use of color allows to emphasize the most serious incidents, meaning that the reader should identify them more quickly compared to using other means. An important aspect about shapes is to avoid symbols/pictures with similar shapes. When they are similar, there is an increase for the search time to differentiate them, so that this is not recommended.

A few more settings about the theme are observed in some literature studies.

Ackermann et al. (2014) present a risk map to engage multiple stakeholders and build a comprehensive view of risks. The authors use risk map as a dynamic tool to update information and create knowledge to the decision making process.

Bostrom et al. (2008) presented the foundation for designing and testing alternative ways to communicate risk and uncertainty for low-probability and high-consequence events, using the knowledge about the effects of spatial information, communication of risk, and uncertainty in spatial information and how these can be tailored effectively for earthquake risk analysis.

Fedra (1998) highlights that technological and environmental risks have an obvious spatial dimension. Floods, mudslides, and avalanches as much as toxic spills, explosions, transportation of dangerous goods, or hazardous waste management are all spatially distributed problems.

Eppler and Aeschimann (2009) present a conceptual framework for risk visualization in risk management. This framework is based in the answers for the questions of: 'why' (purposes), 'what' (contents), 'for whom' (target groups), 'when' (usage situations), and 'how' (formats).

In this context, some applications can be observed in the literature. Brito and Almeida (2009), Alencar and de Almeida (2010) and Lins and de Almeida (2012) contextualize the multidimensional risk view in the context of natural and hydrogen gas pipelines. The multidimensional risk analysis results are presented by the risk difference between pipeline sections. These risk increments provide to DMs a different interpretation with regards risk visualization, allowing that the DM allocates resources according to the risk hierarchy. It also allows the visualization of the gap size between the risks of two subsequent sections of the ranking.

Additionally, Garcez and de Almeida (2014) present a multidimensional risk assessment under an intra-criterion vision in the underground electricity distribution context. This information view allows the DM to identify the relevant consequence dimensions for each alternative and thus allocate resources to prevent and mitigate risk more effectively, prioritizing only those dimensions that impact the alternative. For example, an alternative that impacts only humans, should not receive resources that are allocated to the environmental dimension, thus preventing a misallocation of resources.

Tariq (2013) presents damage curves and maps based on estimated losses and probabilities of all floods considered. The maps illustrate the flood risk distribution over the study area, including agricultural land-use zoning and comparisons over the area before and after crop.

Finally, a specific point that should be mentioned concerns with the application of Geo Information Technology (GIT), Geo Information Systems (GIS), and software for visualization of qualitative and quantitative analyses. In this context, an overview of 3D visualization tools for quantitative analyses could be observed in Kaufmann and Haring (2014).

As an example of application, Jaedicke et al. (2014) uses a GIS (Geographic Information Systems) solution to warn avalanches in Norway. Maps are used in study showing areas susceptible to occurrence of avalanches providing an overview on the overall situation.

3.2 Basic Concepts on Reliability

First of all, to study maintenance engineering it is essential to have a thorough understanding of a key aspect of maintenance that has a strong influence on the actual effectiveness of maintenance actions. This is nothing more than the aging of the various devices that make up the system, and it is this that the dynamics of failure often reveal. Indeed, the purpose of maintenance actions is either to anticipate or remediate a failure. Thus, note that a better understanding of how failures occur serves as a starting point for developing effective plans aimed at anticipating and thus precluding the occurrence of failures.

Any piece of equipment or device that is prone to failure, prior to being regarded as piece of equipment or device, was first conceived as a design project. Accordingly, when in the design phase, several requirements are laid down and it is only after ensuring that these have been met that the final characteristics of such equipment and devices are achieved and therefore that the project can be said to have been fully completed. Among these requirements or dimensions that formed the final design there is the ability to preserve the characteristics and design features of the equipment/device over time and there is another the ease with which the device, which has developed a fault, can be returned to its operational state. These are the two most important characteristics for the process of maintenance management. The first feature is called reliability; the second concerns maintainability.

First, reliability is discussed and then the concept of maintainability. Reliability is an already well-established concept among the main ones outlined here and it makes an interesting contribution to maintenance procedures. This view leads to two main approaches towards the study of reliability. The first consists of formulating a problem in terms of relating it to the aims of a project by establishing systems and structural, technological or organizational measures to ensure that the standard of reliability required by the production system will meet the requirements set by performance issues. Such questions resonate with many problems that extend right up to the moment prior to using the system (Scarf et al. 2009).

In later chapters, some of the issues that directly affect the reliability of a project are addressed, either as a result of decisions made when choosing design requirements in general in order to achieve a certain level of design reliability, or when taking more specific actions that involve only the allocation of redundancy so as to guarantee a certain level of reliability.

It is worth mentioning that the development of reliability, as a field of study, occurred primarily in an attempt to reach a better understanding of the reasons why equipment and devices fail, and this is done by investigating aspects of design projects that ended up with products being produced (Rausand and Høyland 2004). On the other hand, it should also be noted that how pieces of equipment are operated and maintained may significantly affect the chances of their developing faults and failures.

This observation, in fact characterizes the second approach which makes use of common sense and everyday experience to highlight that the effectiveness of a functioning system depends not only on its "innate" properties, but also on the quality of its operations, maintenance, repair, or on any activity that interferes with the operational performance of the equipment. At one extreme, if all maintenance actions are limited to emergency repairs only after the system has suffered a failure, then the operational characteristics of the system are likely to be very low and the system will not operate in an efficient manner (Scarf et al. 2009). As a result, the second approach deals with numerous issues, the main concerns of which are related to the system already in operation and its nature and have regard to proposing measures that will obtain the best possible operational characteristics.

The importance of this point of view for this book is to do with the ease with which maintenance activity is seen to be related to its proper purpose. Indeed, the main objective of maintenance is to anticipate failure and consequently to reduce of the probability of its occurrence, which in turn, contributes to mitigating possible consequences associated with failure. Therefore, the way in which maintenance actions affect reliability are also discussed in this chapter and which actions can be undertaken to ensure the operating performance of equipment is good.

3.2.1 Reliability Perspectives

Despite the fact that reliability can be tackled over a wide field of study, and may cover issues not only associated with the project, but also the actions that can be performed to maximize the performance of the equipment already in place, there are other narrower views of reliability.

According to Márquez (2007) reliability, as well as risk, are in fact elements that quantify uncertainties. Thus, as the quantification of uncertainty is not in itself an end but the means by which it is possible to make better decisions, it can be said that using risk analysis methods and reliability supports the decision process under uncertainty.

Viewed from this perspective, reliability would then be a set of methods that helps in decision making regarding the performance of the system under study. Commonly reliability is deemed to have three main branches, namely:

- The reliability of hardware;
- The reliability of software;
- Human Reliability.

This chapter refers to the branch of reliability that is associated with the operation of components and equipment. On the other hand, the existence of different branches emphasized the need to study the different aspects involved in socio-technical systems: man; the machine and software (intangible elements that are used to operate these machines) (Pham 1999).

Indeed, the fact of there being different approaches to dealing, separately, with the different agents involved in the operation of production systems indicates that the procedures for using a particular approach, in fact, depict only part of the real problem. This kind of reductionism in modeling a decision problem eliminates the influences that might be present from the other actors and by doing so, this makes it feasible to find solutions.

Thus, it is important to keep always in mind that the domain of the consequences of a failure is limited to the perspective that is actually used. Consequently the awareness that reliability analysis may provide incomplete information about the actual performance of the system in its completeness warns of the need to keep an open mind and to look for complementary dimensions of decision making or other issues that were possibly left out by the reliability method adopted.

As an example, consider the behavior of a piece of equipment that had a failure. This is done by seeking to discover how the process of wear and tear due to how the equipment operated occurred which typically entails undertaking an analysis using a reliability-driven approach to machinery and equipment, in isolation, and without considering the influence of other elements (human and software). Yet, at the same time, as equipment becomes older, due to wear and tear, there is also a set of circumstances that leads operators to misusing equipment and this can lead to failure. This is not taken into account. Similarly, in an automated system, the malfunction of the system can lead to failure. Such irregular operating regimes may be linked to failures in control programs or other items of software, neither of which is taken into account under a reliability-driven approach.

Besides reducing the scope of analysis when adopting a reliability-driven approach, it is very common to limit looking at how the components of a system are impacted by changes in other parts of the system as a whole or how each of them may impact the system.

According to Jorgenson et al. (1967) this type of simplification is a way to overcome the difficulties imposed by the complexity of large systems. Moreover, reliability analysis at the component level is consistent with the actions that are carried out in practice. The most frequent failures occur in a component, so it is unnecessary to replace the complete system. Furthermore, most scheduled maintenance activities also require software components, not equipment, to be replaced.

3.2.2 Reliability as a Measure of Performance

When treated as a measure, reliability is a somewhat elusive concept. Its definition is often associated with different interpretations, such as the confidence level of operational success and the absence of failure, the durability of an item, security, etc., all of which are very abstract concepts. These are often easier to understand when the lack of reliability is considered. The failure of equipment in a production system may result in the loss of very significant amounts of money. Thus, it is easy to comprehend what reliability means, when one can visualize what might be lost in its absence.

For calculation purposes, reliability is defined in scientific texts as the probability of an item performing a predetermined function for a specific period of time and under appropriate conditions (Hotelling 1925; Lewis 1987; Barlow and Proschan 1965). As a result, reliability is a probabilistic concept that relates to the random variable T lifetime of an item, and therefore its mechanism of failure, as shown in (3.1).

$$R(t) = P(T > t)\,. \tag{3.1}$$

Because it is a probabilistic concept, one resorts to reflecting on some basic fundamentals of probability so as to be able to construct a sequence of reasoning that leads to a better understanding of reliability.

3.2.3 Reliability and the Failure Rate Function

As already explained above, reliability is frequently defined as the probability that a system will perform its specific function satisfactorily, for a determined period of time, under pre-established conditions. Within this definition, the relationship of reliability with failure is clear, namely an assessment is made of to what extent the system is far from satisfactorily performing its function. The most important variable related to reliability is that of time, and it is for this reason that most reliability phenomena are understood within the dimension of time (Carter 1986; Lewis 1987; Finkelstein 2008).

Examination of the dependence of the failure rate in relation to time adds greatly to an understanding the nature of failures; e.g. investigating whether failures occur prematurely, are random or are brought about by age. In this context, it is important to determine what the relationship between reliability and the failure rate is (Lewis 1987; Kuo and Zuo 2003; Finkelstein 2008; Kuo and Zhu 2012).

$$\lambda(t) = \frac{f(t)}{R(t)}. \tag{3.2}$$

From (3.2), (3.3) may be derived:

$$\lambda(t) = -\frac{1}{R(t)} \frac{d}{dt} R(t). \tag{3.3}$$

On solving (3.3), it follows (3.4):

$$R(t) = \exp\left[-\int_0^t \lambda(t) dt\right]. \tag{3.4}$$

Even though every care is taken over a set of items, either in the design phase, or in the phase when the item is already in use, it is observed that failures still occur. Such failures are characterized into different types depending on what the predominant mechanisms were that worked most effectively in bringing them about.

First, there are the failures that occur quite early in the life of a component. The most likely cause of this type of failure is that equipment parts were defective due to their having been improperly manufactured or constructed. It is this which leads to high rates of early failures of engineering devices. Loss of parts, substandard materials, components that are out of tolerance, and defects caused during transportation are among the causes of failures. This is indicative of inefficient quality control and results in excessive failure rates near the beginning of the life-cycle of the project (Lewis 1987).

The middle part of the bathtub curve contains the lowest levels of failure rate and shows little variation, behaving approximately as a constant. It is referred to as useful life. Failures during this period of time are often ratified as chance failures i.e., they happen irregularly and unexpectedly. They probably arise due to unavoidable loads. External loads above equipment design capacity can lead to an increase in failure rate, for example, due to the equipment material fatigue (Guedes Soares and Garbatov 1996; Garbatov and Guedes Soares 2001).

The part on the right of the bathtub curve is a region in which the failure rate increases. During this period, failures due to the aging are prevalent and the cumulative effects of such matters fatigue and as corrosion tend to be the dominant causes of these. Wear out failures are symptomatic of components aging (Lewis 1987; Bazovsky 2004). These failures happen only if the item is not appropriately maintained. In practice, it is the calculation of when the failure rate will start to increase rapidly that usually forms the basis for determining not only when parts should be replaced but also for specifying the design life of the component.

It is important to understand that different devices have different bathtub curves. This difference is present in the predominance of one of the three failure mechanisms mentioned above as well as in the different moments that most emphatically characterize the thresholds of each phase.

In practice, more than one factor or mechanism contributes to a failure (Brissaud et al. 2010). Therefore, a failure rate curve can be seen as a superposition of curves for different failure modes, as shown in Fig. 3.6.

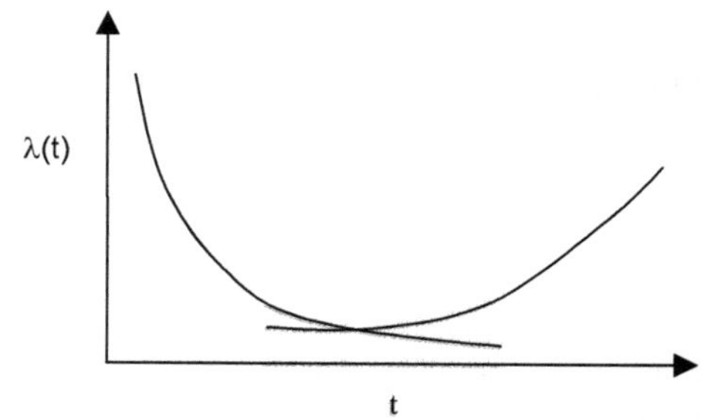

Fig. 3.6 Bathtub curve

Each failure mode and the consequent behavior of the failure rate can be represented by an analytical expression, which is associated with the distribution of the density of probability over time with respect to faults.

3.2.4 Modeling Random Failure

Random failure models are among the most widely used models to describe reliability phenomena worldwide.

For a device that needs to be free of failure, the magnitude of the effect of early failures may be limited by controlling product quality and narrowing the production process, plus a later stage of wear control before its operating life begins (burn in and debugging). Wear out failures should be limited if there is careful preventive maintenance with periodic replacement of parts or components in areas of the production system where the effect of wear is concentrated. Thus, attention is mainly focused on failures and the chances of preventing, reducing or completely eliminating consequences.

In order to do so, it is important to model this kind of failure. The lifetime distribution that describes failures, which occur at random intervals, where the number of failures is the same for equally long operating periods, is the exponential distribution. This is given by (3.5) (Bazovsky 2004)

$$f(t) = \lambda e^{-\lambda t}. \tag{3.5}$$

where λ is a constant called the chance failure rate. Its cumulative distribution is given by (3.6).

$$F(t) = 1 - e^{-\lambda t}. \tag{3.6}$$

From (3.3), the reliability function is given by (3.7).

$$R(t) = e^{-\lambda t}. \tag{3.7}$$

This reliability formula could be used on devices which are not subject to early failures, and which have not yet suffered from aging. In others words, the time where this formula is valid is the useful life of the device. This interval of time varies widely for different devices. One of the most important aspects of this kind of distribution is the fact that the reliability of a device is approximately the same for operating times of equal length. Thus, the time t in (3.6) measures the operating hours in an arbitrarily chosen operating period of a device, regardless of for how many hours the device has already been in operation before this specific operating period. During its useful life, the device is always as good as new. This is because its failure rate remains the same.

From (3.3) it follows (3.8).

$$\lambda(t) = \lambda = 1/\theta. \tag{3.8}$$

where, θ is the expected time *E(t)* for *t*, given by (3.9).

$$E(t) = \int_0^\infty t \frac{1}{\theta} e^{(-\frac{t}{\theta})} = \theta. \tag{3.9}$$

3.2.5 Models of Failure Rate Function Dependent on the Time

For early failures as well as failures due to the cumulative effect of wear and tear, also called failures due to age, it is necessary to define the most appropriate distributions which model the failure time, the context in which time influences the failure process. Although the log-normal distribution and the standard distribution are often used to represent the model that demonstrates the effect of age, the Weibull distribution is the most universally employed. The following shows some other distributions used to model the behavior of failures due to wear and tear as well as early failures related to design problems (O'Connor and Kleyner 2012).

3.2.5.1 The Weibull Distribution

The Weibull distribution is widely used, can assume a very wide variety of forms and is therefore very flexible and it can be used for various types of data (Nelson 2004; Jiang et al. 2001).

The Weibull Distribution can have two parameters, and the probability density function is given in (3.10):

$$f(t) = \frac{\beta}{\eta}\left[\frac{t}{\eta}\right]^{\beta-1} e^{\left[-\left(\frac{t}{\eta}\right)^{\beta}\right]}. \tag{3.10}$$

where:
β - the shape parameter
η -the scale parameter
One can observe a very important role related to β:
β = 1, the failure rate is constant, in which the exponential function is a special case.
β > 1, the failure rate is increasing. In this case, the corresponding failure phase caused by wear and tear can be modeled using the bathtub curve.
β <1, the failure rate is decreasing. In this case, the early failure phase can be modeled using the bathtub curve.

Fig. 3.7 displays the graph for this function

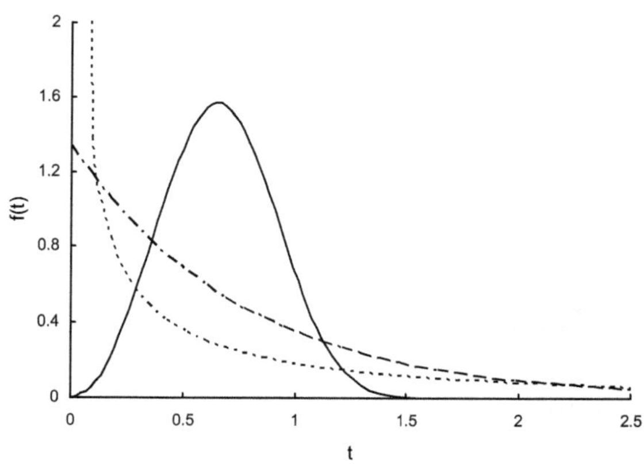

Fig. 3.7 Weibull probability density function *f(t)* for *β*=3 (—); *β*=0.5 (·······); *β*=1 (-·-)

The reliability function is given by (3.11):

$$R(t) = e^{\left[-\left(\frac{t}{\eta}\right)^{\beta}\right]}. \tag{3.11}$$

The graph for this function is shown in Fig. 3.8.

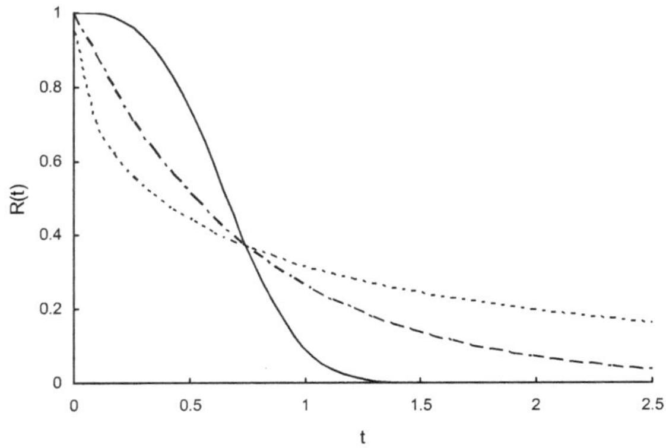

Fig. 3.8 Reliability for the Weibull Distribution *R(t)* for β=3 (—); β=0.5 (······); β=1 (-·-)

The function concerning the failure rate is a Weibull density which is given by (3.12).

$$\lambda(t) = \frac{\beta}{\eta}\left[\frac{t}{\eta}\right]^{\beta-1}. \tag{3.12}$$

The graph for (3.12) is shown in Fig. 3.9.

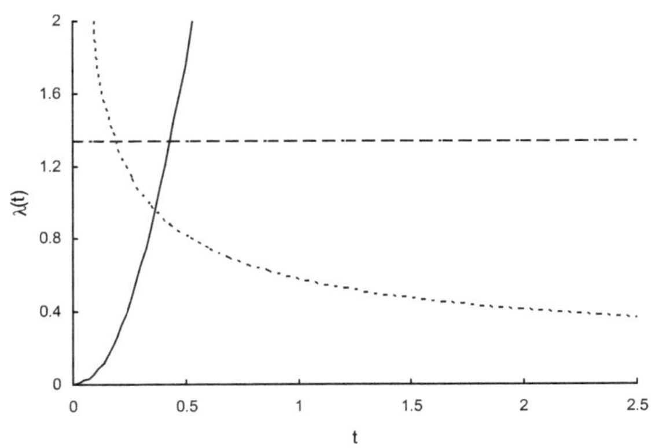

Fig. 3.9 Failure rate function for the Weibull distribution $\lambda(t)$ for β=3 (—); β=0.5 (···); β=1 (-·-)

3.2.5.2 Log-Normal Distribution

Using a Log-Normal distribution curve is appropriate for the situation where it is early failures which predominantly occur i.e. they conform to the bathtub curve law. This is known as the period of infant mortality (Martz and Waller 1982). However, it can model many types of data, due to its ability to assume several formats.

This distribution, commonly used in modeling certain types of life data, is also widely known in modeling equipment repair times.

Its density distribution can be given by (3.13):

$$f(t) = \frac{1}{\sigma t\sqrt{2\pi}}\exp[\frac{-(\ln t - \xi)^2}{2\sigma^2}], \quad 0 < t < \infty \tag{3.13}$$

where $\xi = E(ln\ T)$ and $\sigma^2 = Var\ (ln\ T)$.

The graph for (3.13) is shown in Fig. 3.10.

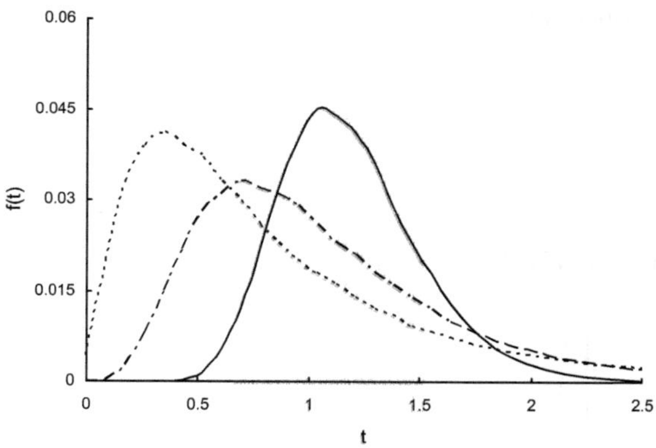

Fig. 3.10 Density function for Log-Normal Distribution for μ=3; σ=0.5 (—);σ =1 (-˙-); σ =1.5 (·······)

Equation (3.14) is the pdf, based in the standardized normal distribution.

$$f(t) = \phi\left(\frac{\ln t - \xi}{\sigma}\right)\frac{1}{\sigma t}, \quad 0 < t < \infty . \tag{3.14}$$

Given the logarithmic relationship with Normal distribution, then the Reliability measure can be obtained as (3.15):

$$R(t) = 1 - \Phi\left(\frac{\ln t - \xi}{\sigma}\right). \tag{3.15}$$

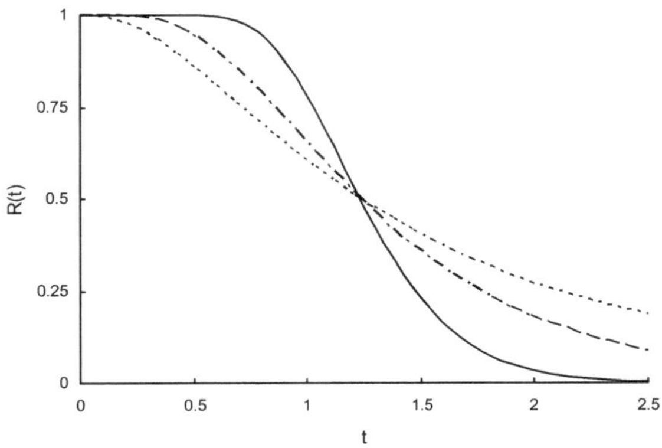

Fig. 3.11 Reliability for Log-Normal distribution for μ=3; σ=0.5 (—);σ =1 (- · -); σ =1.5 (·······)

The failure rate function is given (3.16).

$$\lambda(t) = \frac{\phi\left(\frac{\ln t - \xi}{\sigma}\right)}{\sigma t - \sigma t \Phi\left(\frac{\ln t - \xi}{\sigma}\right)}. \tag{3.16}$$

3.2.6 Influence of Reliability in Maintenance Activities

With regard to maintenance actions, it is interesting to explain how reliability can guide the process for planning maintenance.

In drawing up a maintenance plan there are different actions that must be previously defined and analyzed in order to compile it. It is easy to see that for a production system, the more complex the system under study is, the more diverse the set of actions is. Moreover, despite the great diversity of actions that make up maintenance plans, there are some similarities, especially with regard to its purpose.

For preventive maintenance actions, there are two main objectives that can commonly be identified: (1) actions that are performed to ensure the functioning of the system within the design conditions, and (2) actions that are undertaken to restore the operational condition of the project.

For the first category of preventive maintenance actions, there are routine actions, such as: cleaning, lubricating, adjusting, retightening and any others that may contribute to the permanence of the design conditions. The implementation of such actions is very important, considering that during the use of a particular device, it is possible to identify periods in which the equipment was used outside the design conditions. Should such times be long, or even if they are short but frequent, the chances are that the aging process will change. Thus, the distribution of failure times can be modified.

For the second category of preventive maintenance actions, the main goal is to control the level of wear and tear that arises from using the equipment, whether or not this is due to the conditions of use set in the design project being met. Thus, by replacing a part or component, it is expected that the condition of the device gets close to that of the original design and consequently the probability of failure is reduced.

In practice it is common to say that a device that undergoes routine intervention type 1 is only undergoing corrective actions, when no preventive action type 2 is performed. This is because a device, even when its design conditions are assured, it still degrades and ages.

Moreover, the implementation of actions, aimed only to ensure operation within the design conditions, is really the basic assumption, in order to enable estimates of reliability to be made, given that the vast majority of reliability models assume that such conditions are indeed ensured.

Thus, as a result of not taking actions in advance to avoid failure due to wear, the equipment will be doomed to fail sooner or later. Understanding these issues is essential for effective maintenance planning.

3.2.7 FMEA

FMEA (Failure Mode and Effects Analysis) emerged in the 40s, and was derived from standards set for U.S. military systems. It is a qualitative method, used to identify potential failure modes and their effects and to make recommendations regarding measures to be taken to mitigate risks which can impact the reliability of a system.

The FMEA is structured in tabular form, usually on spreadsheets, where knowledge and experience of those involved are considered as input for a historical database. Information can be extracted, for example, from drawings, process specifications, technical manuals, flow and operational procedures. The aim of applying FMEA is to identify design issues, the critical process and maintenance components.

Ericsson (2005) argues that a more detailed version of FMEA is known as: FMECA (Failure Mode, Effects and Criticality Analysis), where three criteria are usually defined order to calculate the RPN (Risk Priority Number): severity (S), occurrence (O) and detectability (D). These three criteria define the RPN (Risk Priority Number) given by (3.17):

$$RPN = SxOxD\,. \tag{3.17}$$

Despite many critical questions related to RPN being raised in the literature (Zammori and Gabbrielli 2012; Yang et al. 2008; Dong 2007; Braglia et al. 2003; Puente et al. 2002; Braglia 2000; Chang et al. 1999), the RPN value is used in many studies as a comparison measure for analysis and investigation.

Although many authors emphasize the importance of differentiating between FMEA and FMECA, Rausand (2011) states that where the frontier between them lies is rather vague and there is no reason to distinguish between them.

However, some negative aspects of FMEA deserve attention: a significant time period is required if they are to be applied effectively and they do not take human factors into account (Stephans 2004); and FMEA is not useful in the process for identifying combined failures (Nolan 2011). According to Assael and Kakosimos (2010), each individual failure is considered as an independent event which is not related to other system failures, except with regard to subsequent effects that may arise. However, it can be applied in conjunction with other techniques such as HAZOP (Hazard and Operability Study) when special investigations into complex systems are made, for example.

On the other hand, in the literature, many of its positive points are highlighted: FMEA principles are easy to understand (Stephans 2004); FMEA's description of failures provides analysts with a basis for making changes to improve a system (Assael and Kakosimos 2010); it is a useful tool for making analyses and recording recommendations for design changes (Ericsson 2005).

3.2.8 Reliability Management

According to Birolini (2014), reliability is a characteristic of an item expressed by the probability that the item will perform the function required at a stated time interval. From a qualitative standpoint, reliability can be understood as the ability of an item to remain functional. Quantitatively, reliability specifies the probability that no operational interruptions will occur during a given time interval.

In this context, how best to use reliability engineering management is a crucial issue for organizations. This should be incorporated into the strategic level to ensure that, by using appropriate methodologies and procedures, equipment/system reliability levels are maintained within the standards laid down.

Calixto (2013) points out that the success of reliability management depends primarily on four factors: organizational culture, organizational structure, the availability of resources and work routines. With respect to organizational culture, two aspects are important: satisfactory financial results and making decisions based on quantitative data. In other words, one of the minimum requirements for effective management is the availability of a reliable historical database of faults and repairs. Reliability management when considering age-dependent models are subjected to the maintenance and working conditions (Martorell et al. 1999).

Taking these aspects into account, it is observed that the wording of a suitable model definition to address questions relating to the reliability of equipment and/or system depends on both the context and the problem type being analyzed.

It is important to note that the objectives of reliability studies can affect the modeling in different ways. Different goals require different approaches and methods for modeling and analysis. Furthermore, goals can also directly impact the choice of the computational approach to be used in the analysis (Aven and Jensen 2013).

Considering the multiple factors and objectives above mentioned, as dealt with in Chap. 2, this kind of decision making process and analysis combines the multiple factors or objective and may incorporate the DM's preferences over those factors, therefore, an MCDM/A approach shall be applied.

3.2.9 Simulation

An important aspect in reliability management concerns simulation, which is used to investigate which can occur in uncertain environments, depending on the problem type being evaluated. A simulation can be used in complex environments where a more detailed analysis of a particular parameter can provide very valuable information for a particular test model.

Additionally, Yoe (2012) points out that quantitative and probabilistic methods are divided into analytical and numerical methods. Analytical methods are used when explicit equations are solved, while numerical methods have wide applicability and the flexibility to categorize the effects of natural variation and knowledge uncertainty. Among the numerical methods, the Monte Carlo simulation stands out. This basically consists of two steps: generating artificial random numbers and transforming random numbers into useful values using a frequency distribution of the variable under study.

Andrews and Moss (2002) emphasize that simulation seeks to analyze the interaction among components. The result is usually presented in terms of selected measures of system performance. The simulation should be regarded as a statistical experiment where each run of the model is an observation. In this case, the experiment is conducted entirely on a computer.

Wang and Pham (2006) emphasize the importance of simulation in assessing the reliability, availability and optimal maintenance of complex large-scale networks. The Reliability Monte Carlo Simulation generates random failure times from the failure distribution of each component.

However, Smith (2011) points out that some complicating factors can be observed in the evaluated environment, thus making it a complex one. As an example, it is noted that there are complex failure and repair scenarios where the effects of failures and redundancy depend on aspects such as the number of repair teams. Furthermore, there is the possibility of failure rates and downtimes occurring that are not constant.

The appropriate use of the Monte Carlo simulation in environments that involve an uncertainty context, such as maintenance management, is of great importance, since this enables the modeling of important events and a more accurate analysis to be made of possible outcomes of the parameters evaluated.

3.2.10 Redundant Systems

Redundant systems are used in different industrial plants so that the systems continue operating for longer, even if a failure in a system unit occurs when it is in operational mode. According to Calixto (2013), two redundancy types can be observed: passive and active redundancy. In passive redundancy, the redundant equipment (in standby mode) is for most of the time in a passive state. In other words, these passive devices operate only when the active equipment fails.

Modarres et al. (1999) add that passive redundant systems are also called standby redundant systems. The units of this system type remain out of operation until activated by a sensing and switching device. This process continues to be carried out until all standby units have been brought into operation and failed. In this last case, the system is considered failed. Calixto (2013) states that systems can provide active redundancy, in addition to passive redundancy. This occurs when similar pieces of equipment perform in conjunction the same function in a system, in an environment where there is a condition that defines production losses when several pieces of equipment fail. In some cases the charge distribution effect may occur, where some items of equipment fail and other pieces of equipment maintain the same level of production in the system, despite their degrading faster than usual due to this overload. Since in active redundancy the components operate constantly, it is expected that the mean time between failures will be lower than in the case of passive redundancy.

According to Modarres et al (1999), a reliability function for a redundant system with a standby unit is defined by the following mathematical equation:

$$R_p(t) = R_I(t) + \int_0^{\infty} f_I(t_I)dt_I R_{pp}(t_I) R'_{II}(t_I) R_{II}(t - t_I)\,. \tag{3.18}$$

where:

$f_I(t)_I$ = pdf for failure time (unit I);

$R_{pp}\,(t)_I$ = sensing and switching device reliability;

$R'_{II}\,(t_I)$ = unit II reliability in standby mode operation;

$R_{II}(t- t_I)$ = unit II reliability after coming into operation in time t_I.

Calixto (2013) states that the most part of the redundancy increases projects and maintenance costs, introducing in many types of organizations risk systems, such as pipelines and tanks.

However, redundancies are often essential systems designed according to the requirements and specifications of industrial plant systems (Kuo and Zuo 2003; Tian and Zuo 2006; Kuo and Zhu 2012). There is an extensive literature regarding redundant systems (Kuo and Prasad 2000).

3.2.11 Repairable and Non-Repairable Systems

Under reliability management, understanding the definition of repairable and non-repairable systems is crucial for a proper analysis of systems reliability, since they have very different characteristics with respect to the lifetime of the device/system and the number of possible failures.

According to O'Connor and Kleyner (2012), the reliability of non-repairable systems is defined as the survival probability over the expected life of the item or asset, or a range of its expected lifetime when only one single failure may occur. The non-repairable items can be either individual item or systems composed of several parts. Calixto (2013) adds that the availability of non-repairable equipment is defined by the same equation of reliability, where the term repair means replacement. The faulty piece of equipment in this particular case is replaced by another one. The repair time of non-repairable equipment is similar to the repair time of repairable equipment. System unavailability can be caused in both cases, with associated losses.

O'Connor and Kleyner (2012) state that after a failure occurs, the reliability of repairable items is defined as the probability that failure will not occur within the time period of interest, taking into account in this case the possibility that more than one failure can occur. Additionally, the availability of repairable items is affected by the rate at which failures occur and the maintenance period, with corrective or preventive actions.

Moreover, according to O'Connor and Kleyner (2012), in some specific situations, an item under review can be seen at different times as repairable and non-repairable. Guided projectiles, such as missiles, used in military missions are considered in the first instance as belonging to a repairable system (while stored and subjected to planned tests), and are regarded as belonging to a non-repairable system when launched towards a real target. In this case, the reliability analysis must take into account these two states in different moments.

Consequently, a system reliability study should always consider whether a system is repairable (or non-repairable) so as to ensure that appropriate actions are taken effectively at the appropriate time interval.

3.3 Basic Concepts on Maintenance

Today it is almost impossible to think of living without the conveniences brought by the technological development. In fact, our dependence on social infrastructures is so deep that the absence of them is unimaginable living for an interval of time without those technological artifacts. Consequently, the greater the importance that these devices have in our lives, the higher is the relevance of the production system responsible for producing them. The maintenance of them is equally relevant.

This is why maintenance does not stand apart from this process, due to the simple fact that all over the world, among all the humankind achievements, there is nothing that is indestructible. In some sense, everything comes to an end - from the simplest product to the most complex system. Therefore, the role of maintenance is twofold: 1) to postpone this outcome for as long as possible by undertaking activities, the objectives of which are to maintain the product or system in a working condition, and 2) to restore anything to the operational state, when was not possible to avoid the fault.

3.3.1 Characteristics of the Maintenance Function

The maintenance function has some very specific characteristics, which differ from those of the project function. For example, there is a clearly defined beginning and end of each project for which this function is responsible.

Maintenance function does not have a defined period during which each item or equipment of a system will be under the care of maintenance. Maintenance function seems to be timeless, since its objective is associated with the performance of a specific system, so whether the system is supposed to be working or not, there is no time at or during which maintenance activities can be left aside.

The fact that there is a strong demand for maintenance activities does not mean that there is a right time at which they should take place. Any time can be the time to do maintenance and these ranges from corrective maintenance that could happen randomly due to a failure, to preventive maintenance. Even when a system is not working, it is possible that some maintenance is being undertaken. Indeed, for some kinds of systems to which accessibility is very difficult, maintenance activities can only be done when the system stops operation.

Another interesting feature of the maintenance function that further enhances the first feature of timelessness is the fact that maintenance is used to cope with and counteract some natural processes, which never cease interfering in operations. For example, a set of actions has to be frequently undertaken in order to reduce the consequences of the aging process, the influence of which, most of the time, is reflected in bringing about changes in failure behavior. Therefore, if any device is under the influence of one or more natural processes, such as, damage, corrosion, and wear and tear, it is not possible to stop doing maintenance without increasing the risk of serious consequences. It is necessary to conduct maintenance actions continuously to achieve system's desirable performance.

Currently, the maintenance function has received the importance that it deserves. The point to be made about this is that this recognition of its importance was not given immediate. Maintenance departments spread around the world faced critical battles until it was recognized that maintenance is a strategic ally in the struggle to remain competitive. Indeed, even today, there are, right now, some companies where these battles are still being waged, and where the maintenance function is viewed only as an unwelcome and burdensome source of costs. In fact, even although it is no longer acceptable not to acknowledge the importance of maintenance, since the impact of poor maintenance on a company's production targets is quickly evident, there are still a great many companies that only do the minimum in terms of maintenance, i.e. they correct what has failed.

The approach of maintenance and the level of importance given to maintenance activities are related to many different issues. Therefore, the way in which the maintenance is approached may be specific for each company due to distinct characteristics of the productive system, and the different levels of development of the maintenance function. Thus, despite the fact that problems related to maintenance are similar, the particular features of each company's production system, and a set of different matters, such as company culture, make the problem unique for each company.

3.3.2 Production System and Maintenance /Basic Concepts on Maintenance

Although maintenance is a supporting function, depending on the type of system that calls for maintenance, its role can range from simple support to a central role within the plant. The truer this is, the greater the effect of the results of the maintenance action on the company's revenue and operating costs.

Thus, whether maintenance has a central or supporting role, the challenge is to carry out maintenance actions in order to make sure that they will have a positive effect on the system. In other words, the challenge is to guarantee the effectiveness of maintenance management. The problem is that between the desired and achieved outcomes, there is often a wide gap and there are many alternative routes that could be taken to narrow it. Most of them do not close it or get near to doing so. This is why the effectiveness of maintenance is not a trivial matter. The amount spent on maintenance is not directly associated with improving production performance (Scarf 1997). This finding should be one of the most important guidelines for managing maintenance. It warns that maintenance activities require structuring and planning in order to enable the system to attain levels of operational availability at the lowest possible cost by reducing the inappropriate use of resources, and by the "inappropriate use of resources" is meant using them excessively or insufficiently. This discussion raises some important questions that make think more structurally about maintenance:

1. What is maintenance management?
2. Do the functions of maintenance activities depend on the system?
3. What are, in fact, the objectives of maintenance?
4. What are the aspects that highlight the importance of maintenance?

A discussion of these points is subsequently presented.

3.3.3 What is Maintenance Management?

Maintenance can be defined as the set of activities that aims to ensure the levels of performance necessary to guarantee the achievement of production targets. This could be by avoiding failure or by restoring the operating condition when the failure has already occurred. In the first case, this is only possible by means of planned maintenance actions; in the second case, corrective actions are addressed and their purpose is to change the state of failure so as to restore the operational status of equipment, fast enough in order to ensure losses will not be high.

By a closer examination of the maintenance problems, there is a key event, around which a number of different actions must be performed in order to safeguard the competitive existence of different production systems. A failure is, in fact, the non-operating condition of a device or a condition of productive disability and this often reveals itself with different consequences, which can often be summarized as monetary losses but at other times result in very negative outcomes that are difficult to convert into objective monetary values, such as the loss of a human life and serious damage to the environment.

This diversity in the nature of consequences, coupled with the behavior of equipment failure, in which the uncertainties governing different fault events, are major complicating aspects when attempting to establish systematic maintenance actions. This makes it very difficult to adopt standard procedures, with regard to dealing with failures when they have already occurred, or anticipating such failures. They demand a more rigorous treatment, with an emphasis on using mathematical maintenance models (Dekker 1996).

3.3.4 Do the Functions of Maintenance Activities Depend on the System?

Sufficient differences which justify different levels of maintenance being carried out in different processes can be observed. What degree of rigor, for example, should be set for maintenance standards related to aircraft and aircraft engines or turbines? It can be said that these standards must be much higher than the standards set out in, for example, a small plant. Indeed, although the actions of maintenance engineering are different in each plant, being influenced by its size, type, company policy, and many other factors; it is essential to know the scope of activities of the maintenance engineering department (Corder 1976).

In general, they can be grouped into two general categories: primary and secondary functions. Primary functions are often very similar, regardless of where they are put into practice. The intention is to ensure the proper performance that is demanded of equipment. In fact, it is these functions which justify the existence of the maintenance engineering department. With regard to secondary functions, these differ greatly from company to company and are carried out by this department because of the convenience, or because of whatever other reasons that are different from those associated with the primary functions.

3.3.5 What are, in Fact, the Objectives of Maintenance?

The definition of maintenance strategies must be aligned with business goals and therefore the characteristics of the production system (Pinjala et al. 2006). Thus, although it is possible for there to be variations in the objectives of maintenance due to the peculiarities of the system, the main and common objectives in various sectors in which maintenance is conducted can be identified, such as (Corder 1976):

1. To extend the useful life of assets;
2. To ensure satisfactory levels of availability;
3. To ensure operational readiness of systems, and;
4. To safeguard people who use the facilities.

The first three objectives are, in fact, directly associated with the way, whether this is good or bad, that the maintenance activities are being performed. On the other hand, the last objective is rather indirect. Actually, maintenance is not in charge of safety. But, obviously, each time that a failure with dangerous consequences for humans and for the integrity of the system is avoided as a result of maintenance activities, these have contributed to making the plant safer.

There is no doubt that these four aspects are in fact the most important objectives of maintenance. But it may be observed in the literature some variations in these common objectives of maintenance. For example, Dekker (1996) summarizes maintenance objectives under four headings, namely: ensuring that the system functions (availability, efficiency and production quality); ensuring the system's life (asset management); ensuring safety; and, ensuring human well-being.

Despite there being some slight differences in the main objectives of maintenance, the simultaneous achievement of these objectives is not a trivial job. In fact, due to objectives conflicting with each other, it is very common to adopt only one objective under the supposition that the one chosen is the one that is the most closely associated with the strategic objective of the business (Rosqvist et al. 2009; Khazraei and Deuse 2011).

The problem with this approach is the fact that by reducing maintenance objectives to only the main one, the DM's view of the problems in the maintenance field is considerably restricted.

This restriction is even more serious if it is observed the contemporary aspects that invite the maintenance managers to think more broadly about maintenance problems. Some of these aspects are listed above (Levitt 2003; Newbrough and Ramond 1967).

3.3.6 The Aspects that Highlight the Importance of Maintenance

Following aspects of the importance of maintenance are highlighted (Newbrough and Ramond 1967):

1. the increase in mechanization. This has reduced the direct cost of manual labor, but has increased the importance of giving due regard to the maintenance of equipment;
2. the increase in the complexity of equipment. This affects the demand for highly specialized skills when conducting maintenance activities;
3. the growth of the parts and supplies inventory. In fact, this is a direct consequence of the first two factors;
4. Stricter control of production;
5. Programming stricter deliveries. This has reduced the inventory of finished products and has improved customer service. On the other hand, it has also increased the effects of disruptions in the production process;
6. Increasing quality requirements. While providing an increase in the sales potential by increasing the attractiveness of products, this also emphasizes the need for a more immediate response to any abnormality of the product or of the production process.
7. The increase in concern about environmental damage and risk of human deaths associated with failures of devices;

8. The widening of the consequence domain, and the diversification of its nature. It is quite impossible, at a failure event of some kinds of systems, to track all the agents affected by one failure, or to determine the nature of this affect;
9. The dangers that arise from managerial mistakes with regard to maintenance activities that are being currently emphasized (Levitin 2000; Wang and Pham 2006);
10. Finally the fact of the business scenario being so competitive that the aim is to avoid all failures;

All these aspects emphasize the importance of decision-making in the maintenance context. To achieve the maintenance objective, the decision maker develops or follows maintenance policies that are the most appropriate for his/ her objectives and the characteristics of the production system.

The main challenge for the maintenance manager is to structure the maintenance procedures and activities to be undertaken in such a way that the strategic objectives associated with them are achieved. This means that the manager has to plan the maintenance actions with these objectives in mind.

According to Márquez (2007), a maintenance plan is a structured set of tasks that includes activities, procedures, resources and time required to perform maintenance tasks. The implementation of maintenance planning in practice leads to establishing maintenance policies. Maintenance policy is the process of coordinating maintenance activities with the particular characteristics of each system, as well as with the goals that the decision-makers wish to reach, which reflect the company's strategic objectives.

To define a maintenance policy a mathematical model is associated with it in order to make sure that the policy achieves best results. The model, by using a performance function, defines the levels of each action and these should be used to optimize this function. The most appropriate action can be defined before defining its frequency. However, there are models in which not only the activity, but also the frequency, is defined simultaneously (Scarf et al. 2009).

The next section presents a more structured discussion about how mathematical models contribute to maintenance management, emphasizing the contribution of the maintenance policy to the maintenance management process.

3.3.7 Maintenance Policies

For a better understanding of the mathematical models on maintenance, and changes made to them over time, one can go back to the past and describe some important aspects that had to be taken into consideration at the time that these models were proposed.

According to Jorgenson et al. (1967), two distinct classes of problems are involved in asset management: inventory management and management of durable equipment. Inventories provide items for the production process; durable equipment provides services. The management of durable equipment, however, imposes two additional problems, choosing appropriate levels of service for the equipment and keeping up these services.

Years ago, choosing an appropriate level of service was discussed, while maintenance costs were assumed to be constant or nonexistent. In the 1960s, however, an entire theory on the maintenance of equipment started to be constructed. Optimal maintenance policies were proposed and characterized for a wide variety of situations.

The first studies on maintenance policies treated the problems as being deterministic (Taylor 1923; Hotelling 1925), i.e., problems in which the result of each maintenance action is non-random. Some years later, however, different studies properly faced up to the stochastic aspects of maintenance problems (Barlow and Proschan 1965; Barlow and Hunter 1961; Barlow and Proschan 1975; Barlow and Hunter 1960; Glasser 1969), for which the consequences of the maintenance actions would be random. What most motivated developing maintenance policies emerged largely and importantly from tackling practical problems of maintaining complex electronic equipment, i.e., aircraft, missiles, spacecraft, communications equipment, computers, and so on.

The methodology and the theoretical development related to stochastic maintenance have a striking resemblance to the stochastic theory of inventory management. Both have their roots in simple deterministic models. Stochastic inventory theory models usually assume that for a particular item, demand per unit time and delivery time are random variables. The corresponding stochastic elements in the theory of maintenance are the time to failure of equipment as well as repair time.

Indeed, from a broader perspective, Jorgenson et al. (1967) state that no distinction is made between the principles of management of inventory control and of durable goods. For durable equipment, the outputs of conservation activities are services rather than individual items. The level of conservation activity depends not only on acquiring a spare part of productive assets, but also various other inputs of materials and services that represent the maintenance activity. The output of service equipment can be fed into other activities.

Terborgh (1949) refers to the same issue. According to him, the hand of time lies heavily on the work or the deeds of men. He also notes that it is a fact that more practical consequences confront the owner of the item with two problems: the first is to distinguish the speed of death, or, in other words, to say if an asset that has not been exhausted physically still has a life of economic usefulness, either generally or for the particular function it performs. The second is to make the financial provision, in order to be possible to prevent wear of durable goods over its service life. Jardine (1973) classifies maintenance policies into two general classes: probabilistic models of maintenance policies and deterministic models. A striking difference between these classes of problem, besides the existence of a stochastic process that governs the events processed within the policies, is to check whether or not there has been a complete failure on an item. For example, for complex systems where the probability of its main function being stopped is very low. However, since with the passage of time there is a considerable increase in operating costs, it is quite appropriate to use a deterministic approach which consists of observing a cost function and has features very similar to inventory control models.

Therefore, Jardine (1973) establishes different subgroups in each class, which consist of models with similar features:

1. A class of Deterministic Models: models for replacing equipment when operating costs and use increase; models for replacing equipment when operating costs use, for use with finite time horizons; models for replacing equipment considered capital investment, taking into account the discounted net benefit; models for replacing equipment considered capital investment, taking into account the technological improvement.
2. Class of Probabilistic Models: Age Replacement based models, taking into account the time of repair and replacement, and finally, the block replacement model.

Sherif (1982), discusses different maintenance policies in an article that summarizes several studies on the subject. He also makes the same classification as Jardine (1973), although he makes different divisions existing within these subgroups of two broad classes.

McCall (1965) presents a survey of maintenance policies which is in a quite different format from those of Jardine (1973) and Sherif (1982)

McCall (1965) does not consider any deterministic model and makes an in-depth analysis of maintenance policies for systems with stochastic failure. According to McCall (1965), the development of such policies is based on a variety of mathematical techniques. This foundation, along with a variety of applications, sometimes obscures the underlying structure common to all policies. The first author's purpose is to identify this common structure, and thus clarify the relationships between the various maintenance policies.

McCall (1965) classifies models into two categories. The first corresponds to the class of models called (preparedness) preparation or readiness, when equipment fails stochastically. Its state, in fact, is not known with certainty. Alternative maintenance actions for such equipment include inspection and replacement. Preventive maintenance models constitute the second class of maintenance models. In these models, the machine is subject to stochastic failures, and machine status is always known with certainty. If the equipment displays an increasing rate of failures and, moreover, if it is more costly to repair the fault when the system is in operation than to replace the equipment before it fails, then it may be advantageous to replace the equipment before it fails. The problem is to determine a suitable replacement plan (Nakagawa 1984; Nakagawa 1989).Currently the first class of models has gained considerable development. Investigating the state of equipment is now supported by a large number of technological tools, in addition to which the research field has increased and diversified in recent years. This class has been referred to as condition-based maintenance and has been declared as a new milestone of a new generation of approaches in the practice of production and maintenance management (Ahmad and Kamaruddin 2012; Wang 2012; Baker and Christer 1994).

In contrast with this, for the second class, the current preventive maintenance models are almost the same as the earliest ones. Basically, these models deal with the process of failure, by observing the lapse of time since the last preventive maintenance activity (Chang 2014). Despite the small evolution of preventive maintenance models per se, one of the most important contributions, currently, is the combination of the two distinct strategies, for instance, checking the state and replacing preventively after some time. These kinds of combination sometimes are known as hybrid policies, where both actions could be taken following different rules. For some examples of this policy (Scarf and Cavalcante 2012; Scarf and Cavalcante 2010).

Another very important class of policies that is very useful for systems comprising more than one part or multi-component is known as an opportunistic policy. The main distinguishing feature of these policies is that maintenance actions for a piece of equipment depend on the state of the rest of the equipment. The connection established between the states of the components allows more favorable outcomes to occur when compared with those that could arise from individual policies for each component. By restricting the vision to one device at a time, the opportunity to observe actions for multiple components simultaneously is lost, as are the savings that would be made by dealing with opportunities in the most intelligent way.

Just as in the preventive maintenance policies, in which the most recent major contributions are associated with the combination of different actions (Drapella and Kosznik 2002; Jiang and Jardine 2007; Thangaraj and Rizwam 2001), the combination of actions was also used in order to improve the opportunistic maintenance policies. The advantages arising from the combined use of the activities inherent to the two most typically known groups: preventive and corrective maintenance.

This combination was systematically studied by representatives of the RAND Corporation in the early sixties (Radner and Jorgenson 1963; McCall 1965; Jorgenson and McCall 1963). As a result, opportunistic maintenance basically refers to the situation where preventive maintenance is performed arising from opportunities related to the choice of a date or constraints due to the impossibility of postponement, given a failure event. In many cases, it is assumed that the process of generating opportunity is completely independent of the fault (Dekker 1996). On the other hand, it is common to consider the opportunities that coincide with the time of failure of individual components. Due to economies of scale in the cost maintenance function, the undesirable event of a fault in a component is also considered as an opportunity for preventive maintenance of other components.

One must note that in many situations, a combination of preventive and corrective maintenance repairs is not realistic. The need for corrective maintenance arises unexpectedly, while preventive maintenance can be planned. Thus, if there is a combination of both types of activity, the character of schedulable preventive maintenance is lost, or is forced to ignore that a device is flawed for some period of time. Nevertheless, there are situations in which this loss and non-action are acceptable, particularly when the corrective repair of a single component requires the entire system to be disassembled. Thus, combining a corrective repair of a component with the pre-emptive repair of its neighboring components can be lucrative.

As previously mentioned, there are two options for making a combination. On the one hand, preventive maintenance can be brought forward when a failure occurs, and thus when repairs cannot be postponed. On the other hand, when faulty components can be kept idle for a limited period of time, one can opt to delay corrective action until to the next preventive maintenance instant. There are several studies that further develop and refine this type of policy (McCall 1965; Radner and Jorgenson 1963; Woodman 1967; Jorgenson and McCall 1963; Zheng and Fard 1991; Zheng 1995).

Even at different times, several authors such as (Barlow and Proschan 1965; McCall 1965; Dekker 1995; Dekker 1996) reported on the growing interest in developing and implementing maintenance policies for systems with stochastic failure. Undoubtedly, this interest was caused by the high costs and the extra-ordinary demands arising from more complex equipment such as jet aircraft, electronics, computers, etc.. It was also observed that, unlike the consideration of maintenance as an expensive nonsense - a concept that has prevailed for a long time - its real importance has been identified in the face of operational require-ments that are achieved as a result of implementing relatively sophisticated maintenance policies.

3.3.8 Structure of a Decision Problem in Maintenance

Specific literature on maintenance try to give a vision regarding to the structure of a decision problem in maintenance, although it do not use the basic principles of decision making area, particularly concerning to the MCDM/A methods.

According to McCall (1965), the general structure of these problems has elements that are characteristic of decision theory models. While in operation, the equipment in question may take one of several states, with the two extreme states being as good as new and the faulty state. Between these two state-limits there is a set of intermediate states, which denote different degrees of deterioration (Grall et al. 2002; Bérenguer et al. 2003; Fouladirad and Grall 2014). The move from state to state is governed by a stochastic mechanism the behavior of which could be unknown, partially known or completely known by the DM. A neglected piece of equipment moves stochastically from one state to another in a natural way, to reach the state of absorption that corresponds to failure. The behavior of the device can, however, be regulated by choosing a particular action at each decision point. These actions include doing nothing, conducting an inspection, carrying out repairs and replacing different types, or performing a complete overhaul thereby renewing the equipment. The sequence of actions chosen by a DM reflects the maintenance policy and the difference between the controlled and non-inhibited degradation process of the equipment. It is a measure of the influence of policies. The performance of the policy can be measured in terms of costs, by associating an occupancy cost to each state and a cost of intervention with each action. The goal of the DM is to choose maintenance actions such that the cost per unit time of operation of the equipment is minimized (McCall 1965; Jorgenson and McCall 1963; Jardine 1973; Radner and Jorgenson 1963; Dekker and Scarf 1998).

Regarding to the approach of decision maintenance problems using MCDM/A methods, there is an extensive work found on the literature, which is given in subsequent chapters. The following sub-sections give an idea of the kind of maintenance problems to be approached in this vision.

3.3.8.1 Decision Problems on Maintenance Planning

The ever increasing need for higher productivity, in the face of growing competition, has demanded of the various sectors of the economy a constant search for tools that will enable them to acquire competitive advantages.

In order for these organizations to meet these requirements, it is essential that their production systems are able to operate under normal conditions; in other words they must be reliable and available. It is the maintenance function that is in charge of ensuring the normal operation of these systems. To be successful in this objective, paying due attention to the maintenance structure is the best way to deal with common problems related to the management of maintenance.

According to Kelly (1983), maintenance planning is a traditional practice, recommended for the maintenance of machinery, equipment and tools, and should be conducted by preparing work plans and setting norms and standards for their conduct. Márquez (2007) states that what is needed for any level of maintenance is a structured set of tasks that includes activities, procedures, resources and defines the time required to perform maintenance tasks. These definitions explain the scope of maintenance planning:

- What should be done?
- When should it be done?
- Which resources should be employed?

The more correct the answers are to these three points, the more efficient the planning of maintenance resulting from these issues is. In this sense, effective maintenance planning enables managers to take actions using the correct equipment, at the right time, and with the proper tools. The successful implementation of maintenance activities is directly related to precise preplanning.

The answers to these questions usually follow a hierarchical order. So, initially, it is essential to specify which activity or activities to conduct on each device; subsequently establish the frequency with which each of the activities should be undertaken on each of the items of equipment, and finally define the set of resources that will be used.

3.3.9 Main Techniques for Maintenance Management

Maintenance used to be defined as a simple task of restoring the original condition of equipment and systems and it is currently conceived, in a broad and modern way, as a process that ensures reliability and the availability of the function of the equipment and facilities for a production process or the provision of services, with security, while preserving the environment and being conducted at appropriate cost.

In accordance with British Standard BS EN 13306 (2010), maintenance is the combination of all technical, administrative and managerial actions during the life cycle of an item, which are intended to retain it or restore it to a state in which it can perform the required function.

Maintenance is the term used to address the way in which organizations try to avoid the failures of their assets. It is an important part of production systems, particularly when it is critical to the company's business. For example, this applies to power plants, airlines, refineries and petrochemical plants.

Although the paradigm of the past dictated that maintenance professionals should perform a good repair service when prompted, now maintenance work is being given more recognition. Skills and technologies have been developed to prevent failures instead of correct them.

Maintenance professionals are increasingly required to have several core competencies such as:

- Sizing and integrating physical, human and financial resources in maintaining systems, and doing so efficiently and at least cost, while considering the possibility of continuous improvement;
- Using management methodologies, mathematical and statistical tools to support the planning and control of maintenance systems and thus to aid decision making;
- Incorporating quality concepts and techniques into maintaining production systems, in technological and organizational aspects, improving processes, and producing standards and procedures for control and audit;
- Using performance indicators, costing systems, and assessing the economic and financial viability of projects;
- Information management in companies using appropriate technologies;

Significant research work has been conducted in various subfields of maintenance, based on more specific aspects of maintenance. Such research includes issues such as data analysis and fault repair, preventive maintenance models, reliability models, asset management, human reliability, accelerated testing, diagnosis and prognosis models in predictive maintenance, performance evaluation of maintenance policies. This specificity and focus is essential for developing and validating contributions to scientific research.

On the other hand, there is a set of management approaches and a more systemic and generalist view of maintenance management as a process that involves resources such as human, material and financial resources to develop better performance and thereby greater plant availability. Among this set of approaches there are TPM (Total Productive Maintenance) and RCM (Reliability Centered Maintenance).

Many organizations have adopted managerial maintenance approaches such as TPM and RCM, since these approaches are committed to the long-term improvement of maintenance management. Several authors have reported maintenance management as a strategic management activity that can contribute significantly to the success of business (Reis et al. 2009). In following sections, an overview of some managerial techniques used in the field of maintenance management is given.

3.3.9.1 Total Productive Maintenance (TPM)

Total Productive Maintenance (TPM) is defined as the productive maintenance performed by all employees through small group activities in which productive maintenance is the form of maintenance management that recognizes the importance of reliability, maintenance and economic efficiency in the design of plants (Nakajima 1988).

The term productivity in TPM is related to the goal of the maximum overall efficiency of equipment, which is a measurement of the capacity of machines versus the amount actually produced in time. Availability, quality and labor saving because of plant modifications are essential aspects of TPM. This maximum efficiency can be achieved through quality management, which has the function of controlling the possible defects that may occur during the process. TPM seeks to eliminate losses and achieve zero defects, zero breakdowns and zero accidents, so that the length of time that the production line is available is longer and therefore it can produce at maximum capacity. TPM is a management philosophy that promotes change in the organizational culture towards greater quality and productivity at all levels in the company. TPM tries to eliminate the different losses that adversely impact the effective operation of the system (Pintelon and Gelders 1992).

Tajiri and Gotoh (1992) and Shirose (1992) state that a definition of TPM contains the following five points:

1. It aims at getting the most efficient use of equipment.
2. It establishes a total (company-wide) planned maintenance system, (preventive maintenance, and improvement related maintenance).
3. It motivates the participation of department workers, equipment operators, and equipment designers.
4. It involves everyone from top management down.
5. It promotes and implements planned maintenance based on autonomous, small group activities.

In other words, the goal of TPM is to redesign the system of the company, by seeking to improve the performance of people and equipment. Improving staff performance is based on training employees (operators and maintenance workers) so that they can maintain the machines working as per their specifications, and when an abnormality occurs, the operator himself is able to identify it and solve the problem, whenever possible.

Improving equipment consists of structural modifications that represent some kind of benefit to the yield of machine and operator. Another relevant point is to reduce future maintenance costs when evaluating the purchase of new machines. Companies want to increase their productivity and reduce losses. TPM is one of the tools used to eliminate such losses, which can be classified, according to Shirose (1992), as:

1. Breakdown losses – these can be failures because of stoppage in the operation, which is caused suddenly, or by deterioration in function, which is a partial reduction in the capacity and function of the equipment compared to the original state. This loss is related to the loss of function of the equipment, and leads to both chronic failures and sporadic faults, resulting in the loss of time and productivity;
2. Setup and adjustment losses – these happen when one device produces different products, so it may take excessively long to adjust the equipment so that it is able to produce another product with the desired quality;
3. Idling and minor stoppage losses – occur when short breaks are not taken into consideration, but when added together can result in a high loss of time, and empty operations;
4. Reduced speed losses – occur when the machine operates for any reason at a slower speed than normal;
5. Quality defects and rework – occur when there are defects that can lead to disposing of the product and so time and materials for production are lost, while defects need to be corrected and to fix them an additional amount of operating and labor time will be necessary which entails sustaining a loss;

6. Startup / yield losses (reduced yield between machine startup and stable production) – correspond to the period in which the performance conditions of the equipment after it has been triggered do not reach stable production.

So that TPM may develop in organizations, it is necessary that the foundations, called pillars, are constructed in teams and coordinated by managers or leaders of each team. The eight pillars of TPM philosophy form a support system that targets ensuring productive efficiency for the entire organization. Nakajima (1988) lists the eight pillars of TPM as being:

1. Autonomous Maintenance - places responsibility for routine maintenance, such as cleaning, lubricating, and inspection, in the hands of operators;
2. Planned Maintenance - schedules maintenance tasks based on predicted and/or measured failure rates;
3. Quality Maintenance – detects design errors and prevents them from entering into production processes. It applies root cause analysis to eliminate recurring sources of quality defects;
4. Focused Improvement - Small groups of employees work together pro-actively to achieve regular, incremental improvements in the operation of equipment;
5. Early Equipment Management - directs practical knowledge and understanding of manufacturing equipment gained through TPM towards improving the design of new equipment;
6. Training and Education – fill in gaps in knowledge which is required to achieve TPM goals. Training and educational opportunities are given to operators, maintenance personnel and managers;
7. Safety, Health, Environment – These are about maintaining a safe and healthy working environment;
8. TPM in Administration - Applying TPM techniques to administrative functions.

The concept of overall equipment effectiveness (OEE) is an important TPM topic. It is calculated by multiplying the availability of equipment by its performance efficiency and by its quality rating. OEE gives a useful measure for tracking the progress and improvements from the TPM program; but it does not give enough detail to determine why the equipment is better or worse (Mobley et al. 2008).

According to Nakajima (1989), the results from TPM are an increase in the machine availability index by decreasing the number of breakdowns; a decrease in the number of failures in the process and thus a decrease in the number of customer complaints; a reduction in production costs; and a decrease in the number of workplace accidents. All this is possible by preparing and developing people, combined with greater integration between man and machine that operates so as to improve productivity and increase the competitiveness of the entire organization.

There is a trend for many companies to adopt TPM as a tool since they are interested in the potential success of this methodology. It is also true that many of the targets are quite challenging, which is why it is important to motivate people to seek continuous improvement in order to achieve zero losses in the production environment and equipment. Some companies have not been successful in implementing the TPM, and this is due to several causes:

- No support is given from upper management and implementation does not follow the "top down" direction recommended. This is a key point since it is necessary to change the culture of staff so they will adopt new practices, and to invest in improvements in equipment, since, without the support of top management, this challenge becomes more difficult;
- The internalization required for autonomous maintenance is missing, in which the minimum requirements are often not guaranteed and the pillar performs tasks that are more aesthetic than about implementing techniques;
- Without there being an effective program of planned maintenance, there is a change of attitude in the maintenance sector and the environment remains the same as it was before implementing TPM
- Without systematic measurements and the monitoring of losses that compromise the performance of the equipment, it becomes difficult to manage the improvement process
- Without changing the practices of how new systems and spare parts are acquired, maintenance performance may not be effective.

There are implementation procedures for TPM in the literature (Manzini et al. 2009). TPM recommends deployment steps to be followed, and indicates that the maintenance plan must choose which of the various types of policies will be more profitable, but does not explain how to do so in detail. Thus, TPM leaves a gap in supporting decision making about the best maintenance policy and there are different interpretations as to how to implement TPM.

3.3.9.2 Reliability Centered Maintenance (RCM)

RCM is a methodology for identifying maintenance needs in physical or industrial processes. It came from the aeronautics industry in the 1970s, and was adopted by the American defense industry. Then it was extended to the nuclear energy area, and several industrial sectors. RCM is widely used in various industries. (Nowlan and Heap 1978). The process involves the assessment of a structured set of questions that sequentially identify some aspects of the equipment: Main functions, functional failures; Failure modes; effects of failures and consequence of failures.

RCM is a program that integrates various engineering techniques that aim to ensure the functioning of industrial equipment. This program has been recognized as a very efficient way of addressing maintenance issues, since it uses a rational and systematic approach to solve problems (Moubray 1997). Moreover, according to Ben-Daya (2000), RCM is an approach used to optimize a preventive maintenance strategy and its main focus is on maintaining the function of a system rather than wanting to restore it to its optimal condition.

To be at its most effective, RCM needs to be based on certain factors such as:

- The involvement of engineers, operators and maintenance technicians;
- Due importance being given to the study of the consequences of failures that drive maintenance tasks;
- The scope of the analysis, as this should include safety issues, the environment, and operation costs;
- Suitable importance given to proactive activities that involve predictive and preventive tasks;
- Avoiding hidden failures that reduce system reliability;

According to Moubray (1997) there are seven basic questions that should be used by an RCM program:

1. What are the functions and associated performance standards of the asset in its present operating context?
2. In what ways does it fail to fulfill its functions?
3. What causes each functional failure?
4. What happens when each failure occurs?
5. In what way does each failure matter?
6. What can be done to predict or prevent each failure?
7. What should be done if a suitable proactive task cannot be found?

In RCM, the four most important terms are: system, subsystem, functional failure and mode of failure:

- System: This is the plant as a whole or a subdivision thereof which is identified in the RCM analysis;
- Subsystem: This is a group of items of equipment and/or components which together perform one or more functions and can be considered as a separate functional unit within the system;
- Functional failure: Every subsystem performs a certain function. A functional failure describes how each subsystem failure occurs;
- Failure Mode - Identifies each specific condition related to a specific piece of equipment which causes loss of function of a subsystem.

RCM provides functional requirements and standards for the desirable performance of equipment; For each function, functional failures are defined, and the failure modes and effects of failures analyzed using FMEA (Failure Modes and Effects Analysis). Consequences of each failure are analyzed for impacts arising from the effects of the failure modes. Errors fit into one of four categories: hidden; associated with safety or the environment; operational; and non-operational.

According to Rausand and Vatn (2008), the RCM analysis process can be carried out over the following 12 steps:

1. Study preparation;
2. System selection and definition;
3. Functional failure analysis (FFA);

4. Critical item selection;
5. Data collection and analysis;
6. Failure modes, effects, and criticality analysis (FMECA);
7. Selection of maintenance actions;
8. Determination of maintenance intervals;
9. Preventive maintenance comparison analysis;
10. Treatment of non-critical items;
11. Implementation;
12. In-service data collection and updating.

RCM training involves the basic concepts, functional failures, failure patterns, block diagram, concepts of reliability, redundancy, FMEA, predictive, corrective and preventive maintenance, an RCM decision diagram and deployment steps. In the selection phase of relevant maintenance activities, more effort is devoted to the most critical components. Maintenance tasks can be predictive, which are based on wear; preventive when time-based; and reactive, when equipment runs until failure.

In terms of documenting maintenance activities, RCM suggests a worksheet that contains diagrams of the system, subsystem, components, description of activity, its frequency and person responsible for. When you have quantitative data you can base the study on reliability, or when such data are scarce, the work team must define the periodicity of maintenance. It is important that activities are documented, and many of them are carried out by maintenance staff, but can also be performed by staff from operations, by engineers or by a third party.

In the implementation of RCM, the establishment of targets and indicators is fundamental for a successful application. Initially, indicators are defined and then the current situation is set, to then be able to develop goals that are coherent and feasible to achieve, yet while challenging, they are not impossible. The indicators need to be monitored so that feedback is given to work teams.

The review of the RCM program should be performed regularly because implementation is an evolutionary process. The conditions of the equipment and the resources of maintenance change so often that it is necessary to review maintenance procedures in order to be up-to-date. Furthermore, it is important to note that work teams′ knowledge is always increasing and if used in a good way, this can contribute to the continued development of the RCM program.

Finally, it may be added that according to Ben-Daya (2000), if RCM is implemented in combination with TPM, better results can be achieved. He states that RCM offers a framework for optimizing the maintenance effort and getting the maximum out of the resources committed to the planned maintenance program and he argues that RCM can help achieve better results from implementing TPM. Moubray (1997) states that RCM can achieve greater safety and environmental integrity, improve operating performance further (output, product quality and customer service) and lead to maintenance being even more cost-effective, to prolonging the useful life of expensive items, to making the database more comprehensive, to motivating individuals more and to better teamwork.

3.4 Prior Knowledge of Experts in Risk, Reliability and maintenance

In reliability, risk analysis and maintenance models, it is essential to incorporate uncertainty into the modeling. These uncertainties are usually derived from natural variation (random pattern), lack of knowledge or lack of understanding of cause-effect relationships in the present or future condition. Therefore, uncertainty may arise from the uncertain knowledge of some aspect, such as the inaccuracy of the measurement techniques, lack of data, lack of detail, and other factors that directly affect the measurement of uncertainty.

Furthermore, the variability associated with estimates and uncertainties may come from the lack of a clear specification of what is required; the lack of experience in certain activities; complexity in terms of the factors of influence and interdependence of variables; limited analysis of the processes involved in the activities; and, the possibility of particular and rare events or conditions occurring that may affect the activity under analysis.

Besides, under the aspect of the uncertainty of the data, in the context of decision-making, there are uncertainties related to the objectives, priorities and acceptable tradeoffs that decision-makers have to deal with. There must be a complete understanding between the parties involved (clarifying the goals and the reasons for them). Therefore, the various parties involved introduce uncertainties due to ambiguities with respect to: specifying responsibilities; their perception of roles; communication interfaces; contractual conditions and their effects; and with respect to mechanisms for coordination and control.

According to Berger (1985), an important element of many decision problems is the prior information concerning the state of nature θ. A convenient way to quantify each medium of information is by using probability distribution ($\pi(\theta)$), also known as prior probability distribution. Therefore, the experience acquired by experts about a variable can be used in the form of a probability distribution (Martz and Waller 1982).

The use of measures of probability/possibility/occurrence in risk management, maintenance and reliability models and decision analysis models is a very strong requirement. To estimate these measures, information is needed about the several events (failure mode, incidents, accidents, etc.). It happens that in many situations, there are items of information that are unthinkable or extremely unlikely to occur, i.e., they form a set of rare events. For example, rare events are those that might be included in an analysis of systems judged to be highly reliable (e.g. nuclear systems, aircraft systems, space systems, etc.), or also of the occurrence of rare events (catastrophic events by natural disasters, nuclear accidents, accidents in new technologies, certain conjunctions of causes and effects, etc.).

Hence, on those occasions it becomes quite hard to determine the precise values of probability of failure or outcome of an accident. Yet, even in ordinary circumstances, in an industrial system (in which events occur with a higher frequency), determining likelihood is affected by the lack of a comprehensive database of failures, failure modes and accidental events and their consequences.

Therefore, managers require alternative means of acquiring knowledge about the context. With a view to minimizing this obstacle, in the literature there are methodologies aimed at eliciting experts' prior knowledge, all of whom should be familiar with the theoretical approach and have experience of the context analyzed. Thus, this section will focus on one brief context: addressing the main characteristics of the use of expert's knowledge, and ways to aggregate them.

According to Walley (2002), theories of statistical inference can be divided into two broad classes: those that satisfy the principles of probability and those in which the inferences are grounded on interpreting what would happen if historical events were to be repeated and on data sampling (the frequentist approach) .

This approach is very useful for solving decision problems. As its name suggests, it uses historical data or data obtained from trials on which to base its claims.

In risk analysis methodologies, a frequentist approach may be used to tackle failure modes analysis. However, as previously mentioned, a purely frequentist concept of probability cannot always be applied due to the fact that there are some rare events, the repetition of which are almost impossible (very hard) to predict, especially when considering the operations of a small production system or unique system, or when the accumulated amount of historical data is small (insufficient). Thus, it becomes impractical to establish a probability, based on the past experience of the company, due to the absence of such data (Garcez et al. 2010).

According to Garcez et al. (2010), one way to overcome the lack of an internal database is to use an external database (e.g. the database of other local companies or international organizations). However, the simple use of an external statistical database as a benchmark may be mistaken because some characteristics that directly influence this probability represented in the external database, such as regulations, operational structures, levels of technology employed, safety, societal culture, etc. may not reflect the environment of the system that will be analyzed, thus generating differences in statistics.

Therefore, it is necessary to correlate the factors influencing the probabilities with the technical characteristics of the system being analyzed and its nearby systems. To do this, all the experience gained by experts in the field is applied (by the Bayesian Approach), using their expertise and also knowledge about the operating system analyzed, thereby providing valuable information to the decision process (Clemen and Winkler 1999).

The analysis of the data from the database allows a better view of the historical statistical relationships and their relationship to accidents, while the Bayesian approach enables a realistic representation of the expert's knowledge about the dynamics of operation and failure modes in the systems analyzed (O'Hagan 1998).

As an alternative way to determine the rate at which accidents are caused by failure, the methodology defined in Raiffa (1968) can be used. This calls for prior knowledge (using the Bayesian hypothesis) to be elicited along with an analysis of historical data on accidents and failures, coming from external (local or international) databases, internal data of the company itself or similar companies, and thus may enjoy the advantages of each approach.

3.4.1 Elicitation of Expert's Knowledge

According to Kadane and Wolfson (1998), the purpose of eliciting prior knowledge is to capture the main characteristics of an expert's opinion, and thereby to integrate their experience and their academic knowledge. For O'Hagan and Oakley (2004), frequentist inference only enables probability to be interpreted, while Bayesian statistical methods are based on a personal (or subjective) interpretation of probability.

Subjective probability is the degree of belief of the expert in the chance of a particular event occurring, i.e., there is not a correct (accurate) probability, but there is a probability distribution that can be assigned to an event, following all the basic postulates of probability theory (Berger 1985).

For Keeney and von Winterfeldt (1991), formal elicitation of an expert's view of probability elicitation consists of the following steps:

- Identifying and selecting problems;
- Identifying and selecting experts;
- Discussing and refining the problematic;
- Training experts on why and how knowledge is elicited;
- Elicitation process;
- Analyzing, aggregating (outcomes) and resolving disagreements;
- Documenting and reporting results.

According to Garthwaite et al. (2005), the procedure for eliciting the prior knowledge of the expert can be separated into four stages:

1. Arranging for (setup), selecting and training experts and identifying aspects of the problem to be elicited;
2. Eliciting, interaction with experts;
3. This relates to adjusting the probability distribution of the result of the elicitation;
4. The last step is linked to assessing the adequacy of the elicitation process.

In order to elicit an expert's prior knowledge properly, Kadane and Wolfson (1998) list some important points, namely: there must be consensus on the elicitation procedures: it is only expert opinion that should be elicited; experts should be questioned only on observable quantities; experts should not be asked to estimate moments of distribution (in the first instance), they should be asked to review quantiles or probabilities of predictive distribution; frequent feedback should be given to the experts during the elicitation procedure; and, experts should be asked to evaluate hypothetical data, unconditionally and conditionally.

3.4.2 Equiprobable Intervals Method

This section discusses a methodology for eliciting an expert's prior knowledge, given by Raiffa (1968), who uses the method of equiprobable intervals. Subjective probability refers to the degree of belief in a proposition. At one extreme, there is $P(A)=1$ if event A is trusted to be completely true; and at the other, there is $P(A)=0$ if event A is trusted to be completely false, so the points in the interval [0,1] express beliefs that lie between $P(A)=1$ and $P(A)=0$.

Therefore, this method is based on successive subdivisions of equiprobable intervals (intervals with equal probability), i.e., percentiles, about which the interview with the expert takes place. This methodology is structured as follows:

1. Explain the process to the expert in general terms, warning him/her of the fact that the goal is to estimate the most likely value for θ and not its exact real value;
2. Establish a range of possible values of θ. Define the minimum expected value of $\theta_{0.001}$ (the minimum value of the event that is unlikely to occur – a false event) , and the maximum expected value of $\theta_{0.999}$ (the maximum value of the entire event that is likely to occur – a true event);
3. Start subdivision into equiprobable intervals, initially obtaining the value $\theta_{0.5}$, for which $P(\theta_{0.001}) = 0.5$;
4. Divide the interval between $\theta_{0.001}$ and $\theta_{0.05}$, thus obtaining $\theta_{0.25}$ where $P(\theta_{0.25}) = 0.25$;
5. Divide the interval between $\theta_{0.5}$ and $\theta_{0.999}$, thus obtaining $\theta_{0.75}$, where $P(\theta_{0.75}) = 0.75$;
6. Repeat the procedure for the division of other percentiles that need analysis ($\theta_{0.001}, \ldots, \theta_{0.125}, \ldots, \theta_{0.375}, \ldots, \theta_{0.625}, \ldots, \theta_{0.875}, \ldots, \theta_{0.999}$);
7. In the final step, apply a consistency test on the expert, by asking him/her: What is the range in which θ is most likely to fall? Is it within or outside the range $\theta_{0.25}$ and $\theta_{0.75}$? For this question the expert may only give only one of three answers: within, outside, or indifferent. In this case, the correct answer would be indifferent, because, if there is consistency in the elicited values, the probability of being within or outside the range is *0.5*. Should the expert answer either within or outside, one must reevaluate the points with the expert because either answer appears to be inconsistent, i.e., there was probably some inconsistency.

After having determined the percentiles and checked the consistency thereof, a statistical analysis will be undertaken in order to fit the points to a given probability distribution function.

3.4.3 Experts' Knowledge Aggregation

In the context of decision-making and risk assessment, the required information is not always complete or available (Zio 1996), or when there is a need to consider the uncertainty, experts must quantify their knowledge and generate a distribution of subjective probability.

Should the DMs require as much more information as possible, they can consult other experts who have more information or knowledge, and preferably those who have skills in and knowledge of the area of interest, and thus several experts can be used.

However, the absence of any knowledge based on data, models, analogies, theories, physical principles, etc. to assist the experts, can result in judgments that are mere "assumptions" (Garcez et al. 2011).

For Fischer (1981), assessments of subjective probabilities can improve substantially when the opinions of a group of experts are aggregated, so that more than just a probability distribution is considered. However, the expert must be rational when evaluating the uncertainty of the results, and the expert's views must be internally consistent with the theory of probability.

Winkler et al. (1992) list several reasons why the knowledge of multiple experts should be combined:

1. The combined probability distribution produces a better overview than a single probability distribution, both from the perspective of a psychological standpoint (as in the idiomatic expression: two heads are better than one) or a statistical standpoint (when representation by the average of several samples is better than the average of a single sample);
2. The set of probability distributions may be considered as a form of agreement between the various expert's knowledge;
3. It is more reasonable and practical to use a single probability distribution than several distributions. Therefore the analysis is more complete.

When the probability distributions represent the judgments of several experts, a distribution can be obtained that will represent the consensus between them. Thus, the problem of determining this distribution may be treated as a probability distribution agreement/aggregate/combined problem (Winkler and Cummings 1972; Hampton et al. 1973; Ekel et al. 2009). This probability distribution must fully reflect the information provided by these experts (Winkler 1981; Kaplan 1992).

To justify using an aggregate of expert's knowledge, Fischer (1981) argues that the general individual probability forecasts tend to be too radical, i.e., events that are considered highly likely to occur are much less frequent than expected; and events that are considered extremely unlikely to occur, occur much more frequently than expected. Thus, the evaluation of the opinions of multiple experts enables a less radical view of the probability of the event to be reached.

In contrast, Clemen and Winkler (1999) argue that a group of experts can defend a course of action that is more risky than that by an individual or a group of experts reached without discussion. This is probably because experts rely on information provided by others, or there is a sharing of responsibilities among experts.

In order to choose which procedure (method) should be used to aggregate experts' knowledge, it is necessary to consider pragmatic issues such as cost and acceptance. Cost considerations generally favor using simpler procedures, such as, the statistical average. However, when there are considerations that affect acceptability, it is likely that more complex aggregation procedures for interaction between experts are more favorable, such as face-to-face procedures or using the Delphi methodology, for example (Fischer 1981).

Clemen and Winkler (1999) list some general guidelines to determine what approach to aggregating knowledge from experts should be considered:

- What information is provided by experts? Is the probability distribution complete? It is not, if there is only partial information on some of these distributions (e.g., means, variances, etc.);
- Who is involved? A single or a group of experts?;
- What degree of modeling should be performed?;
- What type of aggregation rules are to be used?;
- What parameters necessary for the aggregation method? (e.g., setting weights), and;
- What is the level of complexity of the aggregation process to be adopted?

In the literature, there are two main approaches to aggregating experts' knowledge (opinions), when it is represented by a probability distribution: the mathematical approach and behavioral approach.

The mathematical aggregation procedures consist of analytical models that work on each individual probability distribution so as to produce a combined probability distribution. Aggregation in the behavioral approach tries to generate associations between experts by their interacting with each other and reaching agreement. This can be face-to-face or may involve the exchange of information without direct contact. This approach considers the quality of individual information and dependence between these (Garcez et al. 2011).

Instead of probabilities aggregation, there are other approaches related to fuzzy logic, which may be found in the literature Ekel et al. (2009) specify two main approaches to reach a consensus: first, expert's opinions are combined into a collective opinion, using weighted aggregation. The disadvantage of this approach is when there is an expert who has a deep knowledge of the problem and there is a discrepancy between her/him and other experts. Also, there is a disadvantage when an expert can be neglected due to reducing the weight to his/her opinions, and also defining the set of weights may require significant computational effort.

The second approach, described by Ekel et al. (2009), is to maintain the weights of each expert constant. To reach consensus, the weight given to the expert who most disagrees with the rest of the group is reevaluated. A disadvantage of this approach is that an expert who disagrees may have to change his/her opinion drastically (perhaps unjustifiably), or this expert may be repeatedly asked to revise the opinions of his/her initial position, which requires greater intellectual effort.

Furthermore, decisions typically require the multiple views of different experts, as a single person may not have sufficient knowledge about the problem and therefore cannot solve it alone (Ekel et al. 2009; Parreiras et al. 2010).

As to aggregating experts' knowledge, even though experts can agree what the relevant variables to be analyzed are, this does not mean that they have a consensus on the probability distribution. If they do not disagree on any point, there is no need to consult more than one expert and therefore there would be no need to make expert aggregation (Clemen and Winkler 1999).

In other words, the members of the group would have a uniform opinion and therefore their knowledge would be the same as the one that would have been made, if there had been only one expert. Although such a situation rarely occurs, in cases where the consequences of taking wrong decisions are potentially very serious, when experts are selected, it may be valuable to make a preliminary effort to determine whether they do disagree over the probability distribution of the variables.

To increase the overall satisfaction level of the solution (collective opinion), experts should have the chance to influence the consensus by providing information about their individual knowledge.

References

Ackermann F, Howick S, Quigley J, et al. (2014) Systemic risk elicitation: Using causal maps to engage stakeholders and build a comprehensive view of risks. Eur J Oper Res 238:290–299

Ahmad R, Kamaruddin S (2012) An overview of time-based and condition-based maintenance in industrial application. Comput Ind Eng 63(1):135–149

Ale B, Burnap P, Slater D (2015) On the origin of PCDS – (Probability consequence diagrams). Saf Sci 72:229–239

Alencar MH, de Almeida AT (2010) Assigning priorities to actions in a pipeline transporting hydrogen based on a multicriteria decision model. Int J Hydrogen Energy 35(8):3610–3619

Al-Kassab J, Ouertani ZM, Schiuma G, Neely A (2014) Information visualization to support management decisions. Int J Inf Technol Decis Mak 13(2):407–428

Andrews JD, Dunnett SJ (2000) Event-tree analysis using binary decision diagrams. Reliab IEEE Trans 49(2):230–238

Andrews JD, Moss TR (2002) Reliability and risk assessment. Wiley-Blackwell, Suffolk

Arendt JS, Lorenzo DK (2000) Evaluating process safety in the chemical industry: a user's guide to quantitative risk analysis. American Chemistry Council

Assael MJ, Kakosimos KE (2010) Fires, explosions, and toxic gas dispersions: Effects calculation and risk analysis. CRC Press Taylor & Francis Group, Florida

Aven T (2008) Risk analysis: assessing uncertainties beyond expected values and probabilities. John Wiley & Sons, Chichester

Aven T (2012) Foundations of Risk Analysis. 2nd ed. John Wiley & Sons, p 198, Chichester
Aven T, Jensen U (2013) Stochastic Models in Reliability. Stochastic Modelling and Applied Probability. Springer, New York
Aven T, Vinnem JE (2007) Risk Management with applications from the offshore petroleum industry. Springer Series in Reliability Engineering. Springer-Verlag, London
Baker RD, Christer AH (1994) Review of delay-time OR modelling of engineering aspects of maintenance. Eur J Oper Res 73(3):407–422
Barlow R, Hunter L (1960) Optimum Preventive Maintenance Policies. Oper Res 8:90–100
Barlow RE, Hunter LC (1961) Reliability Analysis of a One-Unit System. Oper Res 9:200–208
Barlow RE, Proschan F (1965) Mathematical theory of reliability. John Wiley & Sons, New York
Barlow RE, Proschan F (1975) Statistical theory of reliability and life testing: probability models. DTIC Document
Bazovsky I (2004) Reliability theory and practice. Dover Publications, Mineola
Bedford T, Cooke R (2001) Probabilistic Risk Analysis: Foundations and Methods. Cambridge University Press, New York
Ben - Daya M (2000) You may need RCM to enhance TPM implementation. J Qual Maint Eng 6(2):82–85
Bérenguer C, Grall A, Dieulle L, Roussignol M (2003) Maintenance policy for a continuously monitored deteriorating system. Probab Eng Informational Sci 17(2):235–250
Berger JO (1985) Statistical decision theory and Bayesian analysis. Springer Science & Business Media, New York
Birolini A (2014) Reliability Engineering. Theory and Practice. Springer Berlin Heidelberg
Bostrom A, Anselin L, Farris J (2008) Visualizing Seismic Risk and Uncertainty. Ann. N. Y. Acad. Sci. Blackwell Publishing Inc, pp 29–40
Braglia M (2000) MAFMA: multi - attribute failure mode analysis. Int J Qual Reliab Manag 17(2):1017–1033
Braglia M, Frosolini M, Montanari R (2003) Fuzzy TOPSIS approach for failure mode, effects and criticality analysis. Qual Reliab Eng Int 19(5):425–443
Brissaud F, Charpentier D, Fouladirad M, et al. (2010) Failure rate evaluation with influencing factors. J Loss Prev Process Ind 23(2):187–193
Brito AJ, de Almeida AT (2009) Multi-attribute risk assessment for risk ranking of natural gas pipelines. Reliab Eng Syst Saf 94(2):187–198
BS EN 60706-2 (2010) Maintenance. Maintenance terminology, British Standards Institution
Calixto E (2013) Gas and Oil Reliability Engineering: Modeling and Analysis. Gulf Professional Publishing, Oxford
Carter ADS (1986) Mechanical reliability. Macmillan London
Chang C, Wei C, Lee Y (1999) Failure mode and effects analysis using fuzzy method and grey theory. Kybernetes 28(9):1072–1080
Chang C-C (2014) Optimum preventive maintenance policies for systems subject to random working times, replacement, and minimal repair. Comput Ind Eng 67:185–194
Clemen RT, Winkler RL (1999) Combining Probability Distributions from Experts in Risk Analysis. Risk Anal 19:187–203
Corder AS (1976) Maintenance management techniques. McGraw-Hill
Cox LA Jr (2009) Risk analysis of complex and uncertain systems. Springer Science & Business Media, New York
Crowl DA, Louvar JF (2001) Chemical Process Safety: Fundamentals with applications. Prentice Hall, Boston
Dekker R (1995) On the use of operations research models for maintenance decision making. Microelectron Reliab 35(9):1321–1331
Dekker R (1996) Applications of maintenance optimization models: a review and analysis. Reliab Eng Syst Saf 51(3):229–240

Dekker R, Scarf PA (1998) On the impact of optimisation models in maintenance decision making: the state of the art. Reliab Eng Syst Saf 60(2):111–119

Dong C (2007) Failure mode and effects analysis based on fuzzy utility cost estimation. Int J Qual Reliab Manag 24(9):958–971

Drapella A, Kosznik S (2002) Combining preventive replacement and burn-in procedures. Qual Reliab Eng Int 18(5):423–427

Ekel P, Queiroz J, Parreiras R, Palhares R (2009) Fuzzy set based models and methods of multicriteria group decision making. Nonlinear Anal Theory, Methods Appl 71:e409–e419

Eppler MJ, Aeschimann M (2009) A systematic framework for risk visualization in risk management and communication. Risk Manag 11(2):67–89

Ericson CA (2005) Hazard analysis techniques for system safety. John Wiley & Sons

Fedra K (1998) Integrated risk assessment and management: overview and state of the art. J Hazard Mater 61:5–22

Finkelstein M (2008) Failure rate modelling for reliability and risk. Springer Science & Business Media, London

Fischer GW (1981) When oracles fail – A comparison of four procedures for aggregating subjective probability forecasts. Organ Behav Hum Perform 28:96–110

Fjeld RA, Eisenberg NA, Compton KL (2007) Quantitative environmental risk analysis for human health. John Wiley & Sons

Fouladirad M, Grall A (2014) On-line change detection and condition-based maintenance for systems with unknown deterioration parameters. IMA J Manag Math 25(2):139–158

Garbatov Y, Guedes Soares C (2001) Cost and reliability based strategies for fatigue maintenance planning of floating structures. Reliab Eng Syst Saf 73(3):293–301

Garcez TV, Almeida-Filho AT de, de Almeida AT (2011) Procedures for aggregating experts' knowledge and group decision model approaches. In: Bérenguer C, Grall A, Soares CG (eds) 20th European Safety and Reliability (ESREL 2011) annual conference, Troyes, September 2011. Safety, Reliability and Risk Management. 2012. Taylor and Francis, London, p 3076

Garcez TV, Almeida-Filho AT de, de Almeida AT, Alencar MH (2010) Experts' elicitation of prior knowledge on accidental releases in a natural gas pipeline. In: Bris R, Soares CG, Martorell S (eds) European safety and reliability conference, Prague, September 2009. Reliability, Risk, and Safety: Theory and Applications, Vol. 1-3. 2009. Taylor and Francis, London, UK, p 2480

Garcez TV, de Almeida AT (2014) Multidimensional Risk Assessment of Manhole Events as a Decision Tool for Ranking the Vaults of an Underground Electricity Distribution System. Power Deliv IEEE Trans 29(2):624–632

Garthwaite PH, Kadane JB, O'Hagan A (2005) Statistical Methods for Eliciting Probability Distributions. J Am Stat Assoc 100:680–701

Gertsbakh IB (1977) Models of preventive maintenance. North-Holland, New York

Glasser GJ (1969) Planned replacement- Some theory and its application (Probability theory applied to age and block replacement models in preventive maintenance of parts, noting inspection cost distribution). J Qual Technol 1:110–119

Grall A, Bérenguer C, Dieulle L (2002) A condition-based maintenance policy for stochastically deteriorating systems. Reliab Eng Syst Saf 76(2):167–180

Guedes Soares C, Garbatov Y (1996) Fatigue reliability of the ship hull girder accounting for inspection and repair. Reliab Eng Syst Saf 51(3):341–351

Hampton JM, Moore PG, Thomas H (1973) Subjective probability and its measurement. J R Stat Soc Ser A 136:21–42

Horwitz R (2004) Hedge Fund Risk Fundamentals: Solving the risk management and transparency challenge. John Wiley & Sons

Hotelling H (1925) A General Mathematical Theory of Depreciation. J Am Stat Assoc 20(151):340–353

HSE (1987) The Tolerability of Risk from Nuclear Power Stations. HMSO - Health and Safety Executive, London

Jaedicke C, Syre E, Sverdrup-Thygeson K (2014) GIS-aided avalanche warning in Norway. Comput Geosci 66:31–39
Jardine AKS (1973) Maintenance, Replacement and Reliability. John Wiley, New York
Jiang R, Jardine AKS (2007) An optimal burn-in preventive-replacement model associated with a mixture distribution. Qual Reliab Eng Int 23:83–93
Jiang R, Murthy DNP, Ji P (2001) Models involving two inverse Weibull distributions. Reliab Eng Syst Saf 73(1):73–81
Jorgenson DW, McCall JJ (1963) Optimal Replacement Policies for a Ballistic Missile. Manage Sci 9(3):358–379
Jorgenson DW, McCall JJ, Radner R (1967) Optimal Replacement Policies. Rand McNally
Kadane JB, Wolfson LJ (1998) Experiences in Elicitation. J R Stat Soc Ser D The Stat 47:3–19.
Kaplan S (1992) "Expert information" versus "expert opinions". Another approach to the problem of eliciting/combining/using expert knowledge in PRA. Reliab Eng Syst Saf 35:61–72
Kaufmann R, Häring I (2014) Comparison of 3D visualization options for quantitative risk analyses. In: Steenbergen RDJM, VanGelder PHAJM, Miraglia S, Vrouwenvelder ACWMT (eds) 22nd Annual Conference on European Safety and Reliability (ESREL), Amsterdam, 2013. Safety, Reliability and Risk Analysis: Beyond the Horizon. Taylor & Francis Group, London, UK, p 758
Keeney RL, von Winterfeldt D (1991) Eliciting probabilities from experts in complex technical problems. IEEE Trans Eng Manag 38:191–201
Kelly A (1983) Maintenance planning and control. Butterworths, London
Khazraei K, Deuse J (2011) A strategic standpoint on maintenance taxonomy. J Facil Manag 9(2):96–113
Kuo W, Prasad VR (2000) An annotated overview of system-reliability optimization. Reliab IEEE Trans 49(2):176–187
Kuo W, Zhu X (2012) Importance Measures in Reliability, Risk, and Optimization: Principles and Applications. John Wiley & Sons, Chichester
Kuo W, Zuo MJ (2003) Optimal Reliability Modeling: Principles and Applications. John Wiley & Sons, New Jersey
Levitin G, Lisnianski A (2000) Optimization of imperfect preventive maintenance for multi-state systems. Reliab Eng Syst Saf 67(2):193–203
Levitt J (2003) Complete guide to preventive and predictive maintenance. Industrial Press New York
Lewis EE (1987) Introduction to reliability engineering. Wiley, New York
Lins PHC, de Almeida AT (2012) Multidimensional risk analysis of hydrogen pipelines. Int J Hydrogen Energy 37:13545–13554
Macdonald D (2004) Practical – Hazops, Trips and Alarms. Newnes – Elsevier, Oxford
Manzini R, Regattieri A, Pham H, Ferrari E (2009) Maintenance for Industrial Systems. Springer London
Márquez AC (2007) The maintenance management framework: models and methods for complex systems maintenance. Springer Science & Business Media
Martorell S, Sanchez A, Serradell V (1999) Age-dependent reliability model considering effects of maintenance and working conditions. Reliab Eng Syst Saf 64(1):19–31
Martz HF, Waller RA (1982) Bayesian Reliability Analysis. John Wiley & Sons, New York
McCall JJ (1965) Maintenance Policies for Stochastically Failing Equipment: A Survey. Manage Sci 11(5):493–524
Mobley RK, Higgins LR, Wikoff DJ (2008) Maintenance engineering handbook. McGraw-Hill
Modarres M, Kaminskiy M, Krivtsov V (1999) Reliability Engineering and Risk Analysis: A Practical Guide. CRC Press
Moubray J (1997) Reliability-centered maintenance. Industrial Press Inc., New York
Nakagawa T (1984) A summary of discrete replacement policies. Eur J Oper Res 17(3):382–392
Nakagawa T (1989) A replacement policy maximizing MTTF of a system with several spare units. Reliab IEEE Trans 38:210–211

Nakajima S (1988) Introduction to TPM: total productive maintenance. Productivity Press
Nelson WB (2004) Applied life data analysis. John Wiley & Sons
Newbrough ET, Ramond A (1967) Effective maintenance management: organization, motivation, and control in industrial maintenance. McGraw-Hill, New York
Nolan DP (2011) Handbook of fire and explosion protection engineering principles for oil, gas, chemical and related facilities. Gulf Professional Publishing, Oxford
Nowlan FS, Heap HF (1978) Reliability-centered Maintenance. Dolby Access Press
NRC (1986) Safety Goals for Nuclear Power Plants. US Nuclear Regulatory Commission NUREG-0880
Nwaoha TC, Yang Z, Wang J, Bonsall S (2013) A fuzzy genetic algorithm approach for analysing maintenance cost of high risk liquefied natural gas carrier systems under uncertainty. J Mar Eng Technol 12(2):57–73
O'Connor P, Kleyner A (2012) Practical reliability engineering. John Wiley & Sons, Chichester
O'Hagan A (1998) Eliciting expert beliefs in substantial practical applications. J R Stat Soc Ser D The Stat 47:21–35
O'Hagan A, Oakley JE (2004) Probability is perfect, but we can't elicit it perfectly. Reliab Eng Syst Saf 85:239–248
Parreiras RO, Ekel PY, Martini JSC, Palhares RM (2010) A flexible consensus scheme for multicriteria group decision making under linguistic assessments. Inf Sci (Ny) 180:1075–1089
Paté-Cornell E, Cox LA Jr (2014) Improving Risk Management: From Lame Excuses to Principled Practice. Risk Anal 34(7):1228–1239
Pham H (1999) Software reliability. John Wiley & Sons
Pinjala SK, Pintelon L, Vereecke A (2006) An empirical investigation on the relationship between business and maintenance strategies. Int J Prod Econ 104(1):214–229
Pintelon LM, Gelders LF (1992) Maintenance management decision making. Eur J Oper Res 58(3):301–317
Puente J, Pino R, Priore P, Fuente D de la (2002) A decision support system for applying failure mode and effects analysis. Int J Qual Reliab Manag 19(2):137–150
Radner R, Jorgenson DW (1963) Opportunistic Replacement of a Single Part in the Presence of Several Monitored Parts. Manage Sci 10(1):70–84
Raiffa H (1968) Decision analysis: introductory lectures on choices under uncertainty. Addison-Wesley, London
Rausand M (2011) Risk assessment. Theory, Methods, and Applications. Wiley, New Jersey
Rausand M, Høyland A (2004) System reliability theory: models, statistical methods, and applications, vol 396. John Wiley & Sons, New Jersey
Rausand M, Vatn J (2008) Reliability Centred Maintenance. Complex Syst. Maint. Handb. SE - 4. Springer London, pp 79–108
Reis ACB, Costa APCS, de Almeida AT (2009) Planning and competitiveness in maintenance management: An exploratory study in manufacturing companies. J Qual Maint Eng 15:259–270
Rosqvist T, Laakso K, Reunanen M (2009) Value-driven maintenance planning for a production plant. Reliab Eng Syst Saf 94(1):97–110
Scarf PA (1997) On the application of mathematical models in maintenance. Eur J Oper Res 99(3):493–506
Scarf PA, Cavalcante CAV (2010) Hybrid block replacement and inspection policies for a multi-component system with heterogeneous component lives. Eur J Oper Res 206(2):384–394
Scarf PA, Cavalcante CAV (2012) Modelling quality in replacement and inspection maintenance. Int J Prod Econ 135(1):372–381
Scarf PA, Cavalcante CAV, Dwight RA, Gordon P (2009) An Age-Based Inspection and Replacement Policy for Heterogeneous Components. Reliab IEEE Trans 58(4):641–648
Sherif YS (1982) Reliability analysis: Optimal inspection and maintenance schedules of failing systems. Microelectron Reliab 22:59–115

Shirose K (1992) TPM for Workshop Leaders. Productivity Press, New York

Smith DJ (2011) Reliability, Maintainability and Risk. Practical methods for engineers. BH (Elsevier), Oxford

Smith DJ, Simpson KGL (2010) Safety Critical Systems Handbook. A straightforward guide to functional safety, IEC 61508 (2010 Edition) and related standards. BH, Oxford

Stephans RA (2004) System safety for the 21st century. The update and revised edition of system safety 2000. Wiley-Interscience, New Jersey

Sutton I (2010) Process Risk and Reliability Management Operational Integrity Management, William Andrew – Elsevier, Oxford

Tajiri M, Gotoh F (1992) TPM implementation, a Japanese approach. McGraw-Hill Companies

Tariq M (2013) Risk-based flood zoning employing expected annual damages: the Chenab River case study. Stoch Environ Res Risk Assess 27:1957–1966

Taylor JS (1923) A Statistical Theory of Depreciation. J Am Stat Assoc 18(144):1010–1023

Terborgh GW (1949) Dynamic equipment policy. McGraw-Hill Book Co

Thangaraj V, Rizwam U (2001) Optimal replacement policies in burn-in process for an alternative repair model. Int J Inf Manag Sci 12(3):43–56

Theodore L, Dupont RR (2012) Environmental Health and Hazard Risk Assessment. Principles and Calculations. CRC Press, Boca Raton

Tian Z, Zuo MJ (2006) Redundancy allocation for multi-state systems using physical programming and genetic algorithms. Reliab Eng Syst Saf 91(9):1049–1056

Tweeddale M (2003) Managing Risk and Reliability of Process Plants. Gulf Professional Publishing. Burlington

Van Leeuwen CJ, Vermeire TG (2007) Risk Assessment of Chemicals. An introduction. Springer, Dordrecht

Vinnem J-E (2014) Offshore Risk Assessment. Principles, Modelling and Applications of QRA Studies Vol. 2, Springer-Verlag, London

Walley P (2002) Reconciling frequentist properties with the likelihood principle. J Stat Plan Inference 105:35–65

Wang H, Pham H (2006) Reliability and Optimal Maintenance. Springer-Verlag, London

Wang W (2012) An overview of the recent advances in delay-time-based maintenance modelling. Reliab Eng Syst Saf 106:165–178

Winkler RL (1981) Combining Probability Distributions from Dependent Information Sources. Manage Sci 27:479–488

Winkler RL, Cummings LL (1972) On the choice of a consensus distribution in Bayesian analysis. Organ Behav Hum Perform 7:63–76

Winkler RL, Hora SC, Baca RG (1992) The quality of expert judgment elicitations. San Antonio, TX: Center for Nuclear Waste Regulatory Analyses

Woodman RC (1967) Replacement policies for components that deteriorate. OR 18:267–280.

Yang Z, Bonsall S, Wang J (2008) Fuzzy Rule-Based Bayesian Reasoning Approach for Prioritization of Failures in FMEA. Reliab IEEE Trans 57:517–528

Yoe C (2012) Principles of Risk Analysis – Decision Making under uncertainty. CRC Press, Boca Raton

Zammori F, Gabbrielli R (2012) ANP/RPN: a multi criteria evaluation of the Risk Priority Number. Qual Reliab Eng Int 28:85–104

Zheng X (1995) All opportunity-triggered replacement policy for multiple-unit systems. Reliab IEEE Trans 44(4):648–652

Zheng X, Fard N (1991) A maintenance policy for repairable systems based on opportunistic failure-rate tolerance. Reliab IEEE Trans 40(2):237–244

Zio E (1996) On the use of the analytic hierarchy process in the aggregation of expert judgments. Reliab Eng Syst Saf 53:127–138

Zio E (2007) An Introduction to the basics of Reliability and Risk Analysis. Series in Quality, Reliability and Engineering Statistics vol 1. World Scientific, Singapore

Chapter 4
Multidimensional Risk Analysis

Abstract: Accidents involve critical consequences that require an appropriate and efficient form of risk management. A multidimensional risk analysis allows a broader view. MCDM/A approaches enable more consistent decision-making, taking into account the DM's rationality (compensatory or non-compensatory), DM's behavior regarding risk (prone, neutral or averse) and the uncertainties inherent in the risk context. This chapter presents numerical applications illustrating the use of multicriteria models in two different contexts: a natural gas pipeline and an underground electricity distribution system. Two different MCDM/A approaches are considered: MAUT (Multiattribute Utility Theory) and the ELECTRE TRI outranking method. In the numerical applications, MCDM/A approach steps for building decision models are presented: identifying hazard scenarios, estimating the set of payoffs, eliciting the MAU function (Multi-attribute Utility function), computing the probability function of consequences and estimating multidimensional risk. Loss functions are introduced in the models to calculate the probability distribution functions over the multiple criteria such as impact on humans, and environmental and financial losses. Therefore, Decision Theory concepts are applied to estimate risk in industrial plants and modes of transportation. Finally, other decision problems related to multidimensional risk analysis, using MCDM/A, are considered in different contexts, such as: power electricity systems, natural hazards, risk analysis on counter-terrorism, nuclear power plant.

4.1 Justifying the Use of the Multidimensional Risk

The perceived level of risk is directly linked to the perceived intensity of consequences to people and society as well as to issues related to the level of probability. These consequences are multidimensional and are associated to the objectives, represented by criteria and can be approached with an MCDM/A or a multiobjective method (see Chap. 2). Many studies show that using a single dimension of risk may not be realistic (Morgan et al. 2000; Willis et al. 2005; Apostolakis and Lemon 2005; Brito and de Almeida 2009; Garcez et al. 2010; Alencar et al. 2010; Alencar and de Almeida 2010; Brito et al. 2010; Garcez and de Almeida 2014b; Lins and de Almeida 2012; Garcez and de Almeida 2014c).

The perception of risk and its tolerability is highly affected by recent events. For example, in the maritime risk context after accidents such as Amoco Cadiz

A.T. de Almeida et al., *Multicriteria and Multiobjective Models for Risk, Reliability and Maintenance Decision Analysis*, International Series in Operations Research & Management Science 231, DOI 10.1007/978-3-319-17969-8_4

(in 1978), Derbyshire (in 1980), Herald Free Enterprise (in 1987) and Piper Alpha (in 1988) many maritime sectors started to seek for the improvement and application of risk modeling and decision making techniques (Wang 2006). Such risk evaluations and safety concerns has to be observed not only by companies involved with the specific context, as maritime transportation and offshore operations, but also other companies related with the sector, in this case, ship designers and shipbuilders in order to improve safety (Guedes Soares and Teixeira 2001).

Since the widely accepted notion of risk (see Chap. 3) is also based on consequences, there is a need to estimate consequences/loss/severity. For many authors, risk assessment deals with estimating possible losses and is an essential procedure, whose outcome is the foundation on which the DM justify his/her decisions. Thus, the consequences of an event can be represented based on some of these aspects, e.g., the number of fatalities, the number of people injured, financial loss, damage to property, environmental losses, etc. (Alencar et al. 2014; Alencar and de Almeida 2010; Brito et al 2010; Luria and Aspinall 2003).

Furthermore, both Individual Risk and Societal Risk concepts only consider the scale of human loss. Cox (2009) argues that, in the risk context, a rational decision making seeks to ensure that a risk analysis builds evaluations and comparisons of proposed risk management actions and interventions, not merely describing the current situation.

In some studies, under a more conservative view, the risk of human loss is assessed from the perspective of a result of the occurrence of injuries, and not only fatalities. Therefore, precisely how people are injured (for example, first or second degree burns) are considered as a consequence for the calculation of risk (Brito and de Almeida 2009). In this context, Cox (2009) reinforces that dominated actions should be eliminated, choosing the best option among non-dominated alternatives and guaranteeing that those alternatives are not ignored. An evaluation of the total consequences is necessary, in order to provide an effective risk management. For each alternative, the overall consequences are calculated taking into account the summing of all the impacts of proposed alternatives on humans exposures.

Although, there is a need for a multidimensional risk view, under the relevant aspects of various ways of analyzing the result of accidents in industrial plants and modes of transportation in various parts of the world, most of the studies consider only issues related to a single dimension. Doing so makes a single dimension approach inadequate or incomplete when the issues involved are complex. Furthermore, nowadays, analyzing the consequences must satisfy the expectations of society, the state (public sector) and private companies. The magnitude and severity of the consequences make it essential to develop a more appropriate and efficient form of risk management, which provides for positive outcomes. In this sense, Beaudouin and Munier (2009) present a critic on industrial risk management techniques based on procedures derived from health, safety, and environment within quality management programs, and draw attention that decision analysis techniques derived from experiments and theoretical foundations are more efficient practices for risk management. Hence, there is a need for a multidimensional assessment, which enables more consistent decision-making to be made and which takes into account the DM's preferences and the context of uncertainty.

Almeida-Filho and de Almeida (2010b) emphasize that risk has been a topic of interest for many years, however the majority of studies avoid considering multiple risk dimensions, in most of the cases there is multiple risk dimensions and it's evaluated through different indexes that are difficult to aggregate into a joint evaluation. There have been used in the literature risk evaluation frameworks from NORSOK and ISO to evaluate risk in an oil and gas context, however, these approaches do not provide a multiple dimension evaluation, only seeks to achieve tolerable risk levels, disregarding decision maker's judgment about the relation between different risk dimension levels and the level difference in each risk dimension. Thus they presented a framework based on well-established risk evaluation framework in the literature (NORSOK and ISO) considering the multidimensional risk aspects.

Still under the one-dimensional view of risk, many studies and risk analyzes take as a basis the financial aspect associated with a monetary value as a criterion of the loss to be used. This approach could appear to be broader, because it regards risk from a more managerial point of view and by analyzing costs. However, considering only the financial aspect is not always an appropriate measure. This can be verified, for example, when the monetary value is not the only measure of value or when certain considerations cannot or should not be converted into an equivalent financial value, e.g. fatalities.

Tweeddale (2003) points out that establishing which economic factors are associated with risk is a point widely discussed in the literature, in which different approaches are taken. A critical point in this context is the attempt to associate a financial value with the loss of a human life. For many, a human life is priceless. For others there is the question of the emotional value of life for friends and family that cannot be compensated by any amount of money. Moreover, according to Hobbs and Meier (2000), some value judgments of interest such as the value of a human life are made by analysts and cannot be properly dealt with in calculations.

Hobbs and Meier (2000) states that other aspects may also be considered with respect to the monetization of criteria, such as the issue that some techniques associated with monetization may be difficult to implement, or even impossible to apply in practice, thus increasing the time required to attempt to do so or this may lead to less suitable methods being used.

In this same perspective, Bedford and Cooke (2001) state that cost-benefit analysis is a well-established method where monetary values are defined for a particular unit (for example, human life). Cost-benefit analysis is used to guide the decision-making process in the area of the ALARP principle (see Chap. 3). Thus, cost-benefit analysis reflects how society prioritizes the various attributes considered, which in principle will be the dimensions of human and financial loss.

For most of the contexts, besides risk, the cost-benefit analysis is appropriate for capturing the society priorities, rather than individual preferences, which are better captured by MCDM/A methods. The former is related to Societal Risk concept, whereas the latter is related to Individual Risk concept. This issue is related to the comparison between the use of cost-benefit analysis and MCDM/A methods (Almeida-Filho and de Almeida 2010a).

Another important point to be highlighted is that financial losses cannot always be measured with complete accuracy. This occurs due to firms having to take account of pressure groups in society who are well informed of possible dangers that their companies may present to society at large. To counter this, several companies have a strategy for differentiating themselves from their competitors. This includes creating an image that the care for the environment, and that their first priority is to safeguard the safety of their employees, their customers and the community they form part of. Nevertheless, when any kind of accident occurs that bring losses to any of the "users" of the system, there is pressure from society not to consume the products of this company. These results in losses to the company that are not only brought about by the accident itself (in the monetary perspective) but also because they lose customers and suppliers; contracts may be broken; their business image is damaged. These losses cannot be "easily" or completely (precisely) measured in financial terms.

Hence, traditional approaches to risk analysis do not consider the multiple dimensional impacts (consequences) that industrial accidents may cause. However, a multidimensional risk view, in many different contexts is necessary.

Furthermore, nowadays, analyzing the consequences must satisfy the expectations of society, the state (public sector) and private companies. The magnitude and severity of the consequences make it essential to develop a more appropriate and efficient form of risk management, which provides for positive outcomes. In other words, the results must be at an acceptable level of safety and also from an economic point of view, the survival of the company is necessarily called into question in the sense that the cost of taking measures that prevent and mitigate risks has to be balanced against the likelihood of accidents happening to people or extensive damage to the environments exposed to them.

As already shown in several studies, an approach to risk assessment that uses only a single dimension of risk cannot be sufficiently comprehensive to ensure that the most realistic and efficient assessment of risk is made (Alencar and de Almeida 2010; Apostolakis and Lemon 2005; Brito and de Almeida 2009; Brito et al. 2010; Garcez and de Almeida 2014b; Lins and de Almeida 2012; Morgan et al. 2000; Willis et al. 2005). Additionally, for Brito and de Almeida (2009), even if other effects are not as important as the risks to human beings, they also require substantial attention from DMs.

Hence, there is a need for a multidimensional assessment, which enables more consistent decision-making to be made and which takes into account the context of uncertainty risk. In many decision problems, more than one factor influences the

DM's preferences with respect to possible outcomes (Montiel and Bickel 2014; Bedford and Cooke 2001).

According to Salvi et al. (2005), if environmental assessment and risk management is given more importance, all stakeholders will take part in the decision-making process. Probably, this feature results from the development of society, which has increasing access to global information, and this is combined with people's concerns related to the sustainable development of society. Society has also taken a cautious attitude due to experiences caused by industrial disasters (e.g., Flixborough, Chernobyl, Bhopal, and more recently Fukushima on 11 March 2011).

The occurrence of these and other disasters have shown that there must be public consultation with various stakeholders, and that this dialogue should not occur independently of the risk management process, the main objective of which is to ensure the long term security of populations. Therefore, the maintenance and consent of an industrial activity is strongly dependent on society accepting the risks that the activity generates.

Hence, companies are coming to recognize the need to take the different opinions and preferences of the various stakeholders into account in the risk decision-making process. MCDM/A methodologies can be extremely useful to aggregate these different opinions (criteria, preference, weights) so the most appropriate decision may be taken both at the national and local level (Roy 1996).

Therefore, according to Yoe (2012), what is observed is that a process of decision making can be simple or complex depending on a few factors that need to be considered. When the analysis is of a single dimension of the problem and there is only one DM, the process is simpler. The same is not true under the risk management process. It is considered to be a complex process due to there being a number of aspects, such as, the views of interested and involved parties, the processes of identification, analysis and risk assessment, and the analysis of consequences, not merely financial impacts.

The process of risk management involves managers and stakeholders with different values, priorities and objectives. In this process, consideration is given to such aspects as tradeoffs between risks, costs, benefits, social values and other impacts of conflicts of values as a result of many perspectives represented by stakeholders in the decision making process.

Hobbs and Meier (2000) affirm that MCDM/A methods present many positive as well as negative aspects. The positive points are:

- Emphasis on learning and understanding by the users;
- Tradeoffs more explicit as to the interests involved;
- Values obtained directly by the stakeholders;
- Reject dominated alternatives.

They argue on some the critical points:

- Large amount of information regarding alternatives and criteria (often not properly interpreted by stakeholders);
- Possible failure of priority of stakeholder groups;
- Improper application of MCDM/A methods, generating distortions of the DM's preferences, as well as inconsistencies in value judgments.

However, if care is taken in the definition, study and use of MCDM/A methods, the potential occurrence of these negative issues can be avoided. Furthermore, this issue has to be seen for each case. It should be reminded that for the purpose of making the model useful, the appropriate effort should be made in the model building process.

Hence, the justification for the use of MCDM/A approaches associated with managing risk is that is made of a set of techniques, methodologies and models, with the goal of working in a better way with aspects associated with uncertainty, understanding conflicts and the tradeoffs involved. Another point highlighted by Cailloux et al. (2013) is that a multicriteria decision aiding approach helps with the subjective part of risk assessment.

In strategic risk management, DMs usually have to consider various conflicting objectives under uncertain decision parameters (Comes et al. 2011). Since MCDM/A methods are easy to use in structuring complex problems and building consensus, they have often been used successfully to support DM in emergency management (Geldermann et al. 2009).

According to Hobbs and Meier (2000), the aim of MCDM/A methods is to improve the quality of decisions involving multiple criteria by making decisions more explicit, rational and efficient. Some aspects of this should be considered: structuring the decision problem; tradeoffs among the criteria; value judgments of the people involved in the process; helping people develop more consistent assessments with respect to risk and uncertainty; facilitating negotiation and; documenting how decisions are made.

For Linares (2002), risk analysis also presents some advantages when combined with a multicriteria decision approach: it allows including the DM's preferences in relation to risk, and may also be taken consistently with a compromise programming approach.

In the multicriteria approach, there is a multidimensional value because multiple criteria are taken into consideration. Thus, instead of considering a single dimension (aspect) such as human or financial loss, other dimensions are taken into consideration depending on the context studied and persons (entities) that are part of the decision process.

Some loss dimensions that can be considered in this context are:

- The human dimension, which can take into account the damage to people affected by the consequences of a failure event which can be estimated by the number of people affected (injuries and/or fatalities);

- The environmental dimension, which may include, e.g., areas affected as a result of the event (Alencar et al. 2014; Brito et al. 2010; Alencar et al. 2010; Brito and de Almeida 2009);
- The financial dimension where we can consider monetary losses arising from events occurring;
- The operational dimension that considers the influence of the consequences of the event and the behavior of the production system;
- Several others that somehow express the needs or preferences that DMs wish to consider.

Given the existence of the uncertainty associated with risk analysis, the use of MAUT to develop models for multicriteria decision is quite appropriate in this context of risk analysis (Keeney and Raiffa 1976; de Almeida 2007; Brito et al. 2010; de Almeida et al. 2015).

In utility theory, measures are obtained based on multiple attributes, where the DM establishes the degree of preference for possible multidimensional results (Keeney and Raiffa 1976; Berger 1985; Bedford and Cooke 2001).

MAUT is used because it presents a well-structured protocol, supported by a very solid and consistent axiomatic framework for decisions involving multiple criteria. Moreover, according to Keeney and Raiffa (1976), in the modeling step, probabilistic uncertainties are inserted within the axiomatic structure, thereby enabling a more consistent approach to the application of MAUT in multicriteria decision problems under conditions of uncertainty. Furthermore, the probabilistic modeling is a complement to modeling the DM's preferences.

In this context, the two next sections present risk evaluations and decision models built with the use of MCDM/A methods and the next section presents a procedure for building models for risk evaluation, using MCDM/A.

4.2 Multidimensional Risk Evaluation Model

This section presents an MCDM/A procedure for building risk evaluation and decision models, which is adapted from that of Chap. 2, incorporating a specific situation, in which the DM's behavior regarding to risk (prone, neutral, averse) can be approached via utility theory. According to Cox (2012), the application of utility functions rather than simple risk formulas – composed by terms such as exposure, probability and consequence - allows to take into account DM's risk attitudes, improving the effectiveness of the decision making process to reduce risks. This procedure has been applied in several contexts described in next section (Brito and de Almeida 2009; Brito et al. 2010, Alencar and de Almeida 2010; Lins and de Almeida 2012; Garcez and de Almeida 2014a; Garcez and de Almeida 2014b).

According to Geldermann et al. (2009), emergency situations caused by humans or by Nature require effective and consistent management, and always involve complex decisions. Many conflicting objectives need to be solved; priorities need to be set, while the various perspectives of different stakeholders should converge towards a consensus.

Brito et al. (2010) state that risk management is a critical activity for many processes and systems, especially for systems that transport hazardous materials. The consequences of accidents highlight the importance of developing a proper and effective risk management technique for this type of process. Additionally, the complexity inherent in the process of decision making on risks, which involves considering technical, economic, environmental, political, psychological and social issues, is an increasingly important aspect of risk management that requires to be tackled more thoroughly.

The decision model presented in this section uses a multicriteria approach based on MAUT, and incorporates the DM's behavior in the decision making process. The model enables the decision maker (DM) to define actions in priority classes in order to mitigate risks in the context under consideration. MAUT provides a well-structured protocol, supported by a solid and consistent axiomatic framework for making decisions involving multiple criteria. Moreover, in the probabilistic modeling step, uncertainties are inserted within the axiomatic structure, thereby enabling a more consistent approach to be taken to a MAUT application in multicriteria decision problems under uncertainty. This stage of probabilistic modeling can be understood as a complement to that of modeling the DM's preferences. The model to be presented takes into account aspects of Decision Theory which will be presented in more detail.

According to Berger (1985), during the decision-making process, it is of great importance to take the possible states of nature into consideration. θ is used to denote the set of all possible states. Typically, when procedures are developed to obtain information about θ, experiments are designed so that the observations are distributed according to some probability distribution that presents θ as a parameter of uncertainty.

In Decision Theory, there is an attempt to combine information from samples with other relevant aspects of the problem, thereby enabling the best decision be made. In addition to this information from samples, two other information types are relevant. The first is knowledge of the possible consequences of decisions. Commonly, this knowledge can be quantified by defining the loss (or gain) that is expected to occur for each possible decision and possible θ values. The second refers to a priori knowledge. Generally, these items of information are derived from past experiences in similar situations involving similar θ.

The model presented is a quantitative model that incorporates the DM's preferences and his/her behavior with respect to risk, thus enabling alternatives to be prioritized by making a hierarchical ranking of the risks, which allows a multidimensional view to be taken of risks from the perspective of different consequences. To illustrate the stages of the model, the structure of a decision model in the context of a natural gas pipeline is shown in Fig. 4.1.

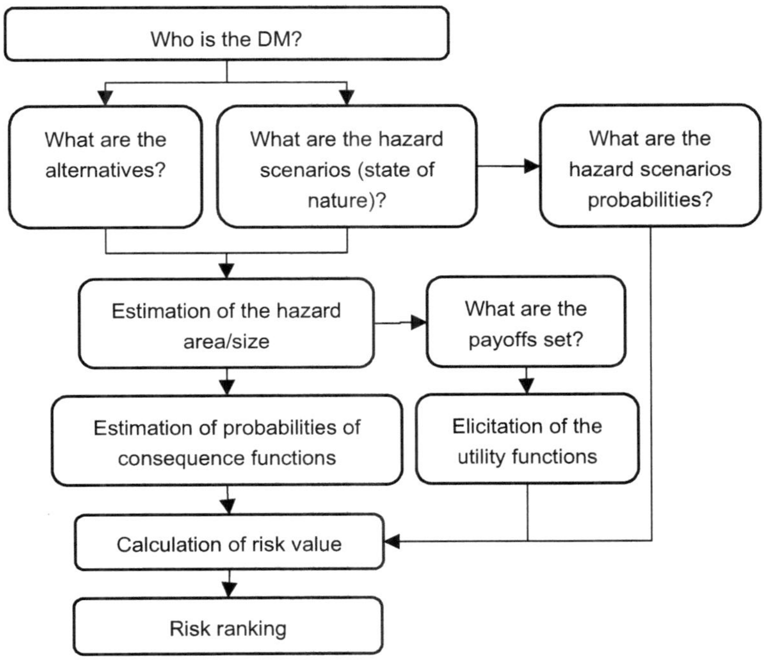

Fig. 4.1 Structure of decision model

Another important aspect that has prompted developing the model is that it uses Decision Support Systems (DSS) to assist in routing between steps in order to make the process more dynamic, thus making it possible for the DM to make a more detailed study of all stages of the steps of the risk analysis. Furthermore, use of DSS aims to support the decision making process, and takes into account both technical aspects such as its stochastic nature and the variety of the parameters which will be entered into the model as well as factors related to the decision making process on risk analysis (Lopes et al. 2010).

Finally, it is worth mentioning that the steps of the proposed methodology are not static. In other words, there is a transition between steps which allows the DM to return to the previous steps to adjust a parameter in order to make the result more dynamic and realistic. Further details of these aspects will be observed throughout the text.

4.2.1 Contextualizing the System

In this step, the system should be contextualized, since it is necessary to describe the general characteristics of the system. The reason for this is that it is only by questioning the purposes of the system and why, in overall terms, it is structured the way it is that some methodological approaches can be more fully understood. Therefore it is necessary to become familiar in overall terms with technical, environmental, social and external environment issues that impact the system and to determine the extent to which each of these, separately, or by interacting with each other affect the performance of the system. The answers to such questioning will guide the decision making process within the type of multidimensional risk analysis that will be selected and applied.

4.2.2 Identifying the Decision Maker

This is the stage used to define who will be responsible for the decision, since it is this DM's preference structure which will be adopted. It is extremely important to identify the DM correctly because decision making in complex environments (such as transport systems for hazardous products, electric power systems, nuclear systems, critical infrastructure, etc.) involves potential severely adverse impacts on society, the environment, economic losses, etc.

Therefore, it is necessary that the DM is thoroughly familiar with the context of the risk analysis. For example, he/she must be fully alert to possible accident scenarios, be fully aware of the consequence dimensions of accidents, and be able to draw up and implement protective and mitigation measures. In other words, not only must the DM be knowledgeable about the context in which decisions about risk may have to be taken but also about the needs of the various stakeholders involved in the decision making process.

It is worth mentioning that the DM's preferences should reflect the interests and goals of the organization (company) and also of the managers who are responsible for any consequences arising from the decision. In some situations it is necessary to include the preferences of various DMs. This process is characterized as a group decision, which may involve three main actors: the company representative of the system considered, the government representative (regulators) and the representative of the community in which the system is located.

In this model, it is assumed that there is a single DM who fully meets the requirements of having the necessary experience, the required level of responsibility and thorough knowledge of the system. This DM is responsible for seeing to it that public safety (regulatory body) standards are met, and as DM assumes appropriate responsibilities to society.

Additionally, is worth noting that the information from the risk management should serve as input that should be passed on to other managers with a view to guiding them on how to perform their functions more adequately. This applies to such managers as those in charge of maintenance, health, environment and safety, or even the production manager. The DM can also be the planning or project manager, where there is an already established system or new systems are being implemented. Thus, the proposed model can be applied to systems that are not yet in operation or those that will be developed. It will determine which alternatives will require most attention in the project design or project execution stage. This should then lead to preventive and mitigation measures being drawn up and taken so as to minimize risks at the project level.

Apart from the DM, another person who has an important role to play in the decision-making process is the expert. Experts provide technical and theoretical support to assist the DM with any questions or issues that may influence the decision-making process. Since this model is intended as a tool that assists risk management. Some experts with relevant knowledge who perform important functions in the organization can be included.

On some specific occasions, the DM plays the role of DM and an expert at the same time, due to his/her having technical knowledge regarding some related to such matters as likelihood, repair times, failure rates, and the characteristics of the system. The DM's preference structure is also incorporated into the problem since it reflects the preference structure of the company, represented by managers' decisions. However, this is not necessarily a requirement of the model. The model allows preference aggregation, when the DM is aided by several specialists. This occurs when the DM does not have the necessary knowledge about specific information.

4.2.3 Identifying Hazard Scenarios

This step consists of defining all the possible scenarios which have resulted from system/subsystem failure modes. These scenarios describe the set of states of nature $\Theta = \{\theta_{11}, \theta_{12}, \ldots, \theta_{21}, \theta_{22}, \ldots, \theta_{jk}\}$ related to the failure mode j and the resulting hazard scenario k.

Hazard scenarios do not define the causes of the failure mode or accidents, but rather the phenomena or accidents associated with the failure mode, which are influenced by the type of failure mode and by the existence of other interacting factors (e.g. there is immediate or delayed ignition and a confined space).

In this context, Crowl and Jo (2007) state that accidents originate from incidents. An incident can be defined as a loss of control over a material or form of energy. Many incidents are followed by a series of events which propagate accidents. This can include fire, explosions and toxic gas leaks. According to the authors, a single section of equipment may have dozens of scenarios, each of which must be identified.

A widely used technique to determine possible accident scenarios is Event Tree Analysis. This technique enables the sequencing of initial events to be analyzed as well as their interactions with the factors that affect the evolution of the event to its final result. This analysis is conducted based on a failure mode.

Once every possible hazard scenario $\Theta = \{\theta_{11}, \theta_{12}, \ldots, \theta_{21}, \theta_{22}, \ldots, \theta_{jk}\}$ is known, the DM must indicate which scenarios the model will consider.

4.2.4 Defining and Selecting Alternatives

At this stage of the model, the alternatives are defined for the DM. The multicriteria decision model produces a risk hierarchy related to the company's systems or subsystems, and it is these which are the alternatives.

In an alternative, the features must be homogeneous, and take into consideration both technical and social issues as well as aspects that influence the probability of a hazard scenario occurring. Expert opinion is important, because it is the expert who has prior knowledge about the behavior of the system. For example, for technical issues related to a natural gas pipeline system, extremely important characteristics include the diameter of the pipe, gas pressure, age of the pipe, characteristics of the soil, composition of the pipe material, the corrosion protection used, etc. These factors along the sections (alternatives) impact on the variation in failure rates and the consequences of accidental releases of natural gas from the pipeline (Jo and Ahn 2002; Jo and Ahn 2005; Sklavounos and Rigas 2006; Jo and Crowl 2008; Brito and de Almeida 2009; Garcez et al. 2010; Alencar et al. 2010; Brito et al. 2010)

Regarding environmental dimension, characteristics that could be considered include the type of the surrounding vegetation, the presence of wildlife exposed to risk, the importance degree of the environment, environmental impact, etc. As to the human dimension, characteristics that should be considered include land use, population density and community type.

Returning to the context of natural gas pipelines, Henselwood and Phillips (2006) assert that these factors may influence the likelihood of an accidental ignition of a natural gas leak. As an example, in an industrial region, the ignition, due to the presence of large numbers of ignition sources, of leaking gas is more likely than in a rural area, where the population density and the presence of ignition sources are low. More details about these aspects are given in Brito and de Almeida (2009), Alencar and de Almeida (2010) and Lins and de Almeida (2012).

Finally, it is important to emphasize that the uniformity of the characteristics listed above in each system/subsystem comprises a distinct discrete set ($A = \{a_1, a_2, \ldots, a_n\}$), where the final system is the sum of all the subsystems analyzed.

4.2.5 Estimating the Probability of Accident Scenarios

Risk analysis enables system failures to be anticipated, thereby helping to identify potential causes and possible consequences. They can be anticipated by analyzing accidents that have previously occurred in similar facilities and which have been recorded in the specialized literature or databases. This analysis allows a statistical evaluation to be made of the most common causes and local conditions which favored the occurrence of claims (Garcez et al. 2010).

In this step a priori probabilities ($\pi_{ai}(\theta_{jk})$) of accidental scenarios defined in the previous step are estimated for each alternative i established. According to Raiffa (1968), the Bayesian approach has become important in situations where there are few or even no data. In these situations, it does not make sense discard a priori knowledge that a specialist has about a variable (or variables) in question. A priori knowledge is a result of variables interacting with the structure, conditioning factors and intervening aspects of the problem and its details, and it is these which make it possible to explain this knowledge using a probability distribution. These probabilities can be obtained from different procedures. One of the best-known is that of eliciting an expert's prior knowledge (Bayesian hypothesis).

4.2.6 Analysis of Objects Exposed to Impacts

At this stage objects that are exposed to impacts due to an accident scenario having occurred θ_{jk} will be analyzed in a particular alternative i, and in the different consequence dimensions ($C = \{c_1, c_2, \ldots, c_r, \ldots, c_m\}$) considered. As mentioned earlier, these consequence dimensions may consider impacts on human health, environmental impacts, financial loss, company image losses, operating loss, etc.

For each hazard scenario and alternative, mathematical models are used and numerical applications made on several features of the objects in the surroundings exposed to hazard. Through this mathematical study, possible impacts are estimated on the different consequence dimensions considered.

However, in the first place, it is necessary to determine what the area or danger zone (S_i) is that results from each scenario and each specific alternative. Having done so, estimates can be made of the impacts and consequences in the dimensions considered in a particular alternative. The danger zone, according to Dziubiński et al. (2006), is a region where impacts exceed critical limits, causing injury to persons, property and environment losses.

4.2.7 Estimating the Set of Payoffs

During this stage possible impacts (consequences) or payoffs that arise from accident scenarios (θ_{jk}), are verified, in a danger zone (S_i) which has been defined in the previous step.

The model consists of a set of multidimensional consequences involving risks. For each consequence dimension considered, the maximum impacts (losses) resulting from an accident should be defined.

4.2.8 Eliciting the MAU Function

According to Brito and de Almeida (2009), the traditional representation of risk considers probabilities or the multiplication of probabilities and consequences that do not reflect people's aversion to harmful events with low-probability and high (often catastrophic) consequences. An approach that considers the DM' preferences is required. The consequence utility function is a way to incorporate a DM's preference in the context of risk where consideration is given to losses due to accidents.

MAUT can be used to aggregate preference values and consequences with respect to multiple dimensions taking into account the DM's preferences and behavior, considering cases with uncertainty (Brito and de Almeida 2009; Alencar and de Almeida 2010).

In MAUT, compensation between criteria implies the use of a synthetic function that aims to aggregate all criteria in a single analytic function. Thus, the structure of the DM's preferences should be based on a compensatory notion. Moreover, MAUT incorporates utility theory axioms. The basic idea of utility theory is to quantify the DM's desire, by assigning values to assets such that these values represent a rule of choice for the DM.

Keeney and Raiffa (1976) break the MAU function elicitation procedure down into five stages that should be used when modeling a problem:

- Introduction to terminology and ideas;
- Identifying the independence assumptions;
- Evaluating the conditional utility functions;
- Evaluating the scale constant;
- Checking and validating consistency.

The first step consists of ensuring that the DM understands the purpose of the utility function and the consequence space. Therefore, one of the most important insights the DM can have is the issue that there is no great preference to be defined, but rather a set of consequences in which the DM demonstrates his/her preferences. As preferences are the DM's subjective representations, there is not a correct choice.

Before engaging with the utility elicitation procedures, it is essential to familiarize the DM with concepts such as: decision analysis, utility functions, and lotteries. Details of these concepts can be found in Keeney and Raiffa (1976), Roy (1996) and Vincke (1992).

Another relevant aspect, according to Keeney and Raiffa (1976), concerns the Von Neumann-Morgenstern expected utility that can be used to characterize an individual risk attitude through simple lotteries.

The concept of a simple lottery can be seen in the following example where the DM maker has, certainly, an amount of money to gamble (e.g. \$*t*.00) and needs to set the probability value p that makes him indifferent towards two situations: keeping the money or making the lottery bet. In other words, the DM remains indifferent between having \$*t*.00 with certainty and risks in a lottery with two possible outcomes: receiving an X amount with probability p or losing the game with probability $1-p$. Graphically, this may be represented by Fig. 4.2.

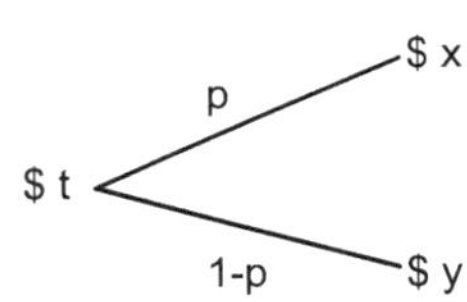

$$\$t \sim \langle \$x, p, \$y \rangle,\ x > t > y,\ p = [0,1]$$

Fig. 4.2 Graphical representation of a payoff lottery

When the DM understands the concepts, the structure of the decision problem and the consequence space are established. To reach a better understanding of this, an example with three consequence dimensions (c_1, c_2, c_3) will be presented (Brito and de Almeida 2009; Alencar et al 2010; Garcez et al. 2010; Brito et al 2010), where c_1 represents losses in the human dimension (e.g.: the number of people exposed to fatality), c_2 represents losses in the environmental dimension (e.g.: a vegetation area exposed to fire) and c_3 represents the losses in the financial dimension (e.g.: the maximum monetary amount disbursed). A graphical representation of this is given in Fig. 4.3.

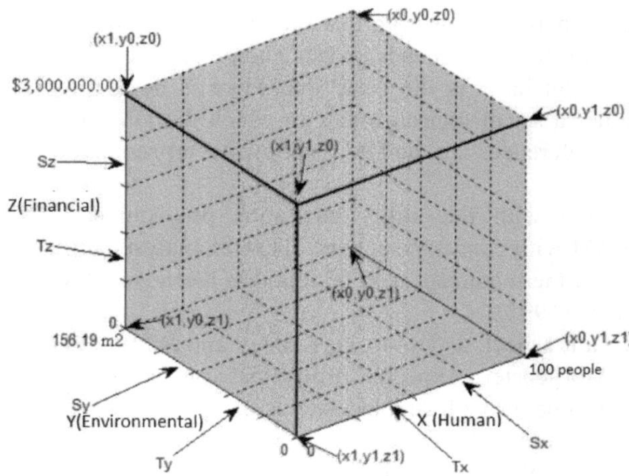

Fig. 4.3 Graphical representation of the consequences space of the MAU function

Eliciting the utility function occurs over a closed interval of consequences, where the maximum value is limited to a null result (no impact). In other words, the most desirable utility is $u\left(c_1^1, c_2^1, c_3^1\right) = 1$. The minimum utility value is linked to the scenario of the worst consequences estimated by the alternatives. Thus, $u\left(c_1^0, c_2^0, c_3^0\right) = 0$ is the least desirable consequence, since we are dealing with losses.

It is worth mentioning that, although it is possible verify discrete and quantifiable consequence values (e.g. the number of people injured), the consequence sets in each dimension can be considered continuous for the purposes of evaluating the utility function.

Therefore, the following values of the consequences space are observed:

- $c_1^0 \le c_1 \le c_1^1$ (e.g.: 100 dead people $\ge x \ge 0$ dead people);
- $c_2^0 \le c_2 \le c_2^1$ (e.g.: $156m^2$ burnt vegetation $\ge y \ge 0m^2$ burnt vegetation);
- $c_3^0 \le c_3 \le c_3^1$ (e.g.: loss of $\$3{,}000{,}000 \ge z \ge \$0.00$).

To confirm the DM's understanding with respect to the limits of the consequence space and his/her preferences, he/she is asked regarding to define his/her preferences with respect to the points S_{c1} and T_{c1}, S_{c2} and T_{c2} and finally S_{c3} and T_{c3} defined in Fig. 4.3. What consequence points does the DM prefer:

- S_{c1} or T_{c1}?
- S_{c2} or T_{c2}?
- S_{c3} or T_{c3}?

If there is any inconsistency in the DM's answers (the DM must state his/her highest preference for one of these points: T_{c1}, T_{c2} or T_{c3}), the DM must be given a new explanation that will lead him/her to a correct understanding of the limits of the consequence space and the conceptual basis of utility theory.

According to Keeney and Raiffa (1976), some independence utility assumptions should be verified after defining the limit values of the utility functions and checking that the DM understands them correctly.

According to Alencar and de Almeida (2010), an attribute c_1 is additively independent of an attribute c_2 if two lotteries are equally preferable for all (c_1 and c_2) and for a 'c_1 and c_2' arbitrarily chosen, as presented in Fig. 4.4.

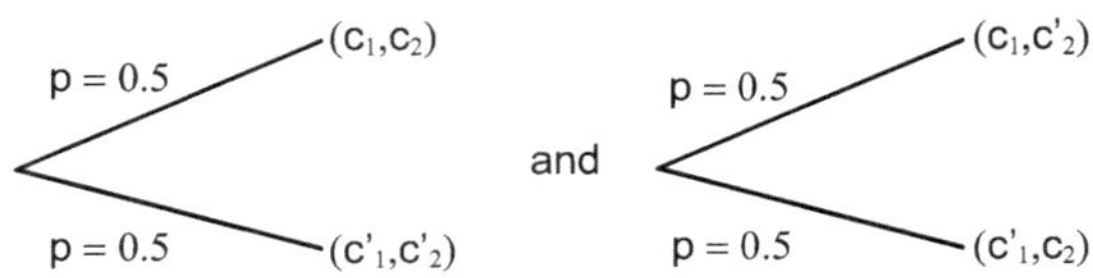

Fig. 4.4 Lotteries to check the additive independence

According Figueira et al. (2005) when attributes (from the perspective of the Von Neumann-Morgenstern utility model) and the DM's preferences are consistent with the conditions of utility independence, then $u(c_1, c_2, \ldots, c_r, , \ldots, c_m)$ can be decomposed into additive, multiplicative or another well-defined structure in order to simplify the evaluation of these relations.

The MAUT can be expressed in an additive form, if and only if, c_r attributes are mutually independent in utility and the additive independence between the attributes is observed. Then:

$$u = \sum_{r=1}^{m} k_r u(c_r) \tag{4.1}$$

where u_r represents the one-dimensional utility functions $[0,1]$; and, k_r represents the scale constants estimated by the elicitation process based on the comparison of lottery payoffs. The sum of the scale constants must be equal to one $\left(\sum_{r=1}^{m} k_r = 1\right)$.

On continuing with the utility function elicitation process, it is necessary to estimate the functions that depict one-dimensional utility functions on the m consequence sets analyzed by the model. The procedures for eliciting the one-dimensional utility function are also described in Keeney and Raiffa (1976).

According to Keeney and Raiffa (1976), to evaluate the scale constants, a structured set of questions should be applied in which the DM makes probabilistic choices of lotteries involving payoffs in the dimensions analyzed.

Returning to the three-dimensional example, the DM is asked to find the p value where the DM is indifferent between the certainty of having consequence $\left(c_1^1, c_w^0\right)$ (in this case $p = 1$) or playing the lottery $\left\langle\left(c_3^0, c_w^0\right), p, \left(c_3^1, c_w^1\right)\right\rangle$ where the value of c_w^0 corresponds to the consequence $\left(c_1^0, c_2^0\right)$ and the value c_w^1 is equivalent to the consequence $\left(c_1^1, c_2^1\right)$.

Once the p value is defined, the DM is asked about the value of q in which he/she is indifferent between the certainty of having consequence $\left(c_1^0, c_2^0\right)$ or playing the lottery $\left\langle\left(c_1^0, c_2^0\right), q, \left(c_1^1, c_2^1\right)\right\rangle$. Having obtained the p and q estimated values and the condition $\sum_{r=1}^{m} k_r = 1$, the following may be defined: $k_1 = p$, $k_2 = (1 - p)q$ and $k_3 = (1 - p)(1 - q)$.

The last step consists of verifying the consistency and the variability of the results if some parameters are modified. Due to the associated uncertainty related to the parameters of the model, this phase can capture the impact of the results by using sensitivity analysis on the model.

4.2.9 Computing the Probability Functions of Consequences

Several uncertainties are present in scenarios (θ_{jk}) and estimating hazard zones (S_n), as shown in the earlier stages of the model. These uncertainties are undesirable, because it becomes impossible to define deterministically which multidimensional consequences can occur due to an accident scenario. For this, it is necessary to estimate the probability distributions of the consequences, represented by a consequence function P, defined by the probability of obtaining a consequence p, since a scenario θ_{jk} occurred in alternative a_i.

In this step of the model, there is a need to estimate the joint probability distribution over the possible values in "m" consequence dimensions $P(c_{1,\dots,m} \mid \theta_{jk}, a_i)$ for each alternative and hazard scenario adopted.

According to Brito and De Almeida (2009), in some contexts it may be considered that different consequence dimensions can have small or even negligible correlations between them. This is because the hazard radius covers several dozen meters. The combination of these consequence dimensions occur randomly and independently, depending on the specific characteristics of each alternative, so that the probabilities $P(c_1 \mid \theta_{jk}, a_i), \dots, P(c_r \mid \theta_{jk}, a_i), \dots, P(c_m \mid \theta_{jk}, a_i)$ can be estimated independently.

However, in some risk analysis contexts, the probability distributions of these consequences are not treated independently, as is the case of risk analysis regarding petroleum extraction platforms, nuclear power plants, etc. where the danger zones usually extend over a wide area and the size of impact interferes non-randomly in various consequence dimensions.

In the case of probability distributions independent of consequences it is possible to define mathematical formulations to model the consequence functions for each loss independently.

There are several models in the context of natural gas pipelines considering consequence functions for estimating the human, environmental and financial risk dimensions (Brito and de Almeida 2009; Garcez et al. 2010; Alencar et al. 2009; Brito et al. 2010). Similarly, for the context of hydrogen gas pipelines with required adaptations, same approach is considered for estimating risk dimensions (Alencar and de Almeida 2010; Lins and de Almeida 2012). A model for risk evaluation in underground vaults of an electricity distribution system considers the same decision analysis principles for assessing risk dimensions of human impacts, financial losses, operating losses and disturbance on the local transit vehicles (Garcez and de Almeida 2014b).

4.2.10 Estimating Multidimensional Risk Measures

In the context of decision making, the DM must choose an action in order to ensure that the consequences are those that are the most favorable ones possible for him.

Decision Theory is a mathematical formalization of this paradigm. It allows rational decisions under uncertainty. According to Berger (1985), Decision Theory involves the following aspects:

- Analyzing past and current information of the system under study, based on the objective and / or subjective information available;
- Eliciting probability distributions to model uncertainties;
- Developing a mathematical model that describes the system and its revision level, which considers the level of accuracy required;
- Eliciting the DM's preferences and values;
- Identifying or designing alternative actions that lead to the desired goals;
- Using mathematical logic to combine alternative actions, utilities and probabilities with the mathematical model of the system in order to identify the best action course for the DM;
- Implementing the action(s) chosen in the previous step;
- Returning to the first step and restarting the process to correct errors and distortions regarding the data, probabilities, utilities and action alternatives.

According to Berger (1985), by Decision Theory, the loss function can be defined as the negative of the utility function of the expected consequence, expressed by:

$$c_r = -u(c_r \mid \theta_{jk}, a_i) \tag{4.2}$$

It can be considered that the consequences are results of the impact dimension of a given action, which can be estimated by using a probability distribution function $P(c_{1,\ldots,m} \mid \theta_{jk}, a_i)$.

Keeney and Raiffa (1976) point out that if an appropriate utility is assigned to each possible consequence and the expected utility of each alternative is calculated, what is observed as the best course of action is an alternative with the highest expected utility. Thus, the consequence utility is the expected value of the utility:

$$c_r = E[u(c_r)] = \int_{c_r} P(c_r) u(c_r) dc_r \tag{4.3}$$

Therefore, the utility function $u(c_r)$ can be calculated by:

$$u(\theta_{jk}, a_i) = u(P(c_r \mid \theta_{jk}, a_i)) = \int_{c_r} P(c_r \mid \theta_{jk}, a_i) u(c_r) dc_r \tag{4.4}$$

After having obtained knowledge about an a priori probability distribution of the states of nature $\pi_{ai}(\theta_{jk})$, which depends on the characteristics/conditions of each system (alternative) analyzed, it is possible to calculate the risk associated with each alternative, using a risk perspective such as a consequence/damage/severity added with the uncertainty, as can be seen in the following equation:

$$r(a_i) = \sum_{r=1}^{m} \left(\sum_{\theta} \pi_{a_i}(\theta) \left(- \int_{c_r} u(c_r) P(c_r \mid \theta_{jk}, a_i) dc_r \right) \right) + (-1)\pi_{a_i}(\theta_N) \tag{4.5}$$

where r represents the various dimensions (attributes) of the analysis. In other words, these are the consequence dimensions ($c_1, c_2, \ldots, c_r, \ldots, c_m$), after having considered the occurrence of all hazard scenarios $\Theta = \{\theta_{11}, \ldots, \theta_{jk}\}$ and alternatives a_i analyzed. The value of $\pi_{ai}(\theta_{jk})$ depends on the characteristics/conditions of each system analyzed.

The state of nature θ_N represents the normality scenario of the system, where the system operates under normal conditions, without any dangerous scenario occurring, thus justifying the loss function value equal to -1. The risk values could be found in the range $[-1,0]$, where the value -1 is related to the lowest risk and the value 0 to the highest risk.

Thus, the risk concept based on Decision Theory assesses the consequences (c_r) of the hazard scenarios (θ_{jk}), by combining both uncertainties associated with: *(i)* the consequences $P(c_r|\theta_{jk},a_i)$; and *(ii)* the hazard scenarios $\pi_{ai}(\theta_{jk})$.

Additionally, the risk measure used considers the DM's preference structure in the set of expected consequences, through utility functions $u(c_r)$, representing the "desirability" that the DM has about property losses (in this particular case, the consequences of an accident scenario occurring) and allowing a probabilistic evaluation of the consequences under uncertainty.

These risk measures comprise a descending risk hierarchy of several of the alternatives (a_i) evaluated. Consequently, the results of this hierarchy serve as input to the decision-making process and risk management.

4.3 Risk Decision Models

Several applications of a multidimensional risk evaluation and decision models have been conducted, based on the previous procedure, adapted from Chap. 2. These applications incorporate the situation in which the DM's behavior regarding to risk is represented by a utility function. This procedure has been applied in several contexts: natural gas pipeline (Brito and de Almeida 2009; Brito et al. 2010), hydrogen gas pipeline (Alencar and de Almeida 2010; Lins and de Almeida 2012) and electricity distribution system (Garcez and de Almeida 2014a; Garcez and de Almeida 2014b).

In this section, three applications of a multidimensional risk evaluation model are presented. The first application is made in the context of risk analysis in natural gas pipelines and is based on Brito and de Almeida (2009). The second application concerns the context of an underground electricity distribution system. This application is based on Garcez and de Almeida (2014b). The third application considers a different MCDM/A method, taking into account a non-compensatory rationality, according to the procedure presented in Chap. 2 (Brito et al. 2010).

4.3.1 Risk Evaluation in Natural Gas Pipelines Based on MAUT

Natural gas is a fossil fuel with reserves available in many parts of the world. Its use has grown over the last 30 or so years due to a number of factors, including, for example, economic and environmental aspects. The high demand for it in widely scattered different locations requires a mode of transportation to convey large amounts of gas from its source to its destination, quickly and safely. Thus, among the existing modes of transportation, pipelines stand out. Although using pipelines is considered a safe system, some accidents have occurred over the years, some of which have had critical consequences.

In this context, this subsection will present a numerical application of a Multidimensional risk evaluation, taking into account the characteristics of the model that have been presented earlier in this chapter, well as some additional points, specific to the context of natural gas pipelines.

Thus, multidimensional risk analysis in natural gas pipelines is conducted using hazard scenarios, in order to estimate the probability of the occurrence of a hazard scenario and the possible consequences that might result from pipeline failure.

Additionally, the model presents a ranking of pipeline sections in a multi-dimensional risk hierarchy, in which three dimensions of risk are considered, namely the human, financial and environmental dimensions. These dimensions are the main ones to be considered that arise from the operation of the pipeline sections under analysis. A ranking of these segments under a risk hierarchy is presented so as give insights into the process of managing pipeline risk, thereby contributing to defining mitigating actions according to the risks associated with each section analyzed. A single DM was considered.

The total length of the pipeline analyzed in this application is *18,000m* divided into 9 sections that comprise a discrete set $X = \{x_1, \ldots, x_9\}$, where each element presents specific features.

Probabilities of each scenario are obtained as per procedures presented by Brito and de Almeida (2009). These authors use a conservative risk assessment for each scenario and pipeline extension, and include the most critical danger zone for each segment associated with the worst accident scenario that may occur in that specific extension.

A conservative estimate of the radius of maximum danger *CDR* is given in (4.6), considering the operating pressure *Po*, the diameter *d* of the pipe and length of the pipeline *L* from the compressor station. More details can be found in Jo and Ahn (2002).

$$CDR \cong 1{,}512 \cdot \frac{P_o^{1/2} \cdot d^{5/4}}{L^{1/4}} \tag{4.6}$$

Settled danger areas for each section, and the human, environmental and financial consequences should be defined. This set of consequences will be included in the analysis using the model, for which the most pessimistic values in each consequence dimension will be input.

The proposed model seeks to assess risks considering three risk dimensions in natural gas pipelines: Human Risks (r_h), Financial Risks (r_f) and Environmental Risks (r_m). The reasons why it is primarily these dimensions that are considered are based on values that are normally found in both productive organizations and in other organizations or institutions involved. These will be translated into principles of social and environmental responsibility and ethical aspects of human relationships. These aspects should influence company actions that seek to secure the financial return aimed at.

As to the human dimension, Brito and de Almeida (2009) assume that human consequences are estimated by the number of people affected physically due to a particular accident scenario, and who receive at least second degree burns, and not necessarily by the number on fatalities.

With regard to the environmental dimension, the area of vegetation affected is used as a measure for the environmental consequences, taking into account the extent of environmental impacts caused by this type of accident (Alencar et al. 2010; Garcez et al. 2010; Brito et al. 2010; Alencar et al. 2014).

Finally there is the financial dimension for which disbursements on foregone income, contractual fines for supply disruptions, fines and other indemnifications for harm caused to people, environment or organizations and companies are considered. Additionally there are expenses related to maintenance and operational actions taken with a view to re-establishing the operational conditions of the pipeline.

The next step corresponds to eliciting a MAU function *U*(*h*, *f*, *m*), which it is considered an additive function. The property of additive independence implies that there is preferential Independence among the payoff sets. *U*(*h*, *f*, *m*) can be expressed by the following (4.7).

$$U(h,f,m) = k_h \int_h P(h \mid \theta, x_i) U(h) dh + k_f \int_f P(f \mid \theta, x_i) U(f) df + k_m \int_m P(m \mid \theta, x_i) U(m) dm \tag{4.7}$$

The calculation of the average radiation flux (due to a hazardous scenario of deflagration) is obtained from (4.8) (Jo and Crowl 2008).

$$I = \frac{(\eta \cdot \tau_a \cdot Q_{eff} \cdot H_c)}{4\pi (CDR)^2} \tag{4.8}$$

where I is the average radiation flux, τ_a is the atmospheric transmissivity, η is the ratio of the irradiated heat over the total heat released, H_c is the combustion heat of the natural gas, CDR is the critical danger radius and Q_{eff} is the effective rate of gas leak.

The estimate of risk is based on Decision Theory principles. According to Berger (1985), risk is considered as the expected value of the loss and can be defined by (4.9) verified in Alencar and de Almeida (2010).

$$r(x_i) = \sum \pi_i(\theta_{jk}) L(\theta_{jk}, x_i) \tag{4.9}$$

Knowing that:

$$L(\theta_{jk}, x_i) = -u(P(p \mid \theta_{jk}, x_i)) \tag{4.10}$$

In this way, losses associated with each scenario and section are summed in the three dimensions discussed, multiplied by accident scenario probabilities and added to the losses associated with a normal scenario (θ_N), as shown in (4.11).

$$r(x_i) = E_\theta[L(\theta_{jk}, x_i)] = \sum_j \sum_k L(\theta_{jk,x_i}) \pi_i(\theta_{jk}) + (-1)\pi_i(\theta_N) \tag{4.11}$$

Due to the additive independence properties of the MAU function and the independence in probability of the probability distributions over the consequences, the risk $r(x_i)$ is given by (4.12).

$$r(x_i) = \sum_j \sum_k \begin{bmatrix} k_h \int_h P(h \mid \theta, x_i) u(h) dh \\ + k_f \int_f P(f \mid \theta, x_i) u(f) df \\ + k_m \int_m P(m \mid \theta, x_i) u(m) dm \end{bmatrix} \pi_i(\theta_{jk}) + (-1)\pi_i(\theta_N) \tag{4.12}$$

Using the risk values obtained from (4.12), pipeline sections can be ordered in descending order, thereby obtaining a ranking of pipeline sections that should be used as input for risk management activities.

The MAUT interval scale allows an incremental comparison between the risk sections in line with the utility value between the alternatives. Thus, (4.13) and (4.14) are applied to analyze the relationship between alternatives, showing respectively the absolute difference between alternatives and the difference ratio between alternatives. The difference ratio *DR* is used to interpret the values in relation to the calculated risks.

$$DA = r_b(x_i) - r_{b+1}(x_i) \tag{4.13}$$

$$DR = \frac{r_b(x_i) - r_{b+1}(x_i)}{r_{b+1}(x_i) - r_{b+2}(x_i)} \tag{4.14}$$

where the index (b) represents the position in the ranking of the section and $r_b(x_i)$ represents the risk value related to a specific section. Through the analysis obtained from the results of these equations, the DM can define which sections should be included given the resources available, thus representing how much more a section adds to the risk when compared to another section placed further down in the ranking provided by the risk model.

Therefore, taking into account all the calculation steps described earlier in this section, Table 4.1 presents the sections prioritized based on comparisons of the increments of risk. The values listed in $r_b(x_i) - r_{b+1}(x_k)$ column must be multiplied by (10^{-5}).

Based on Table 4.1, some interpretations may be made. A descending ranking of values is applied for risk assessment, where S_1 shows the highest value of risk among the sections evaluated. The highest losses associated with the likely consequences of accidents are expected for S_1. Additionally, it is observed that the increment in the risk values from S_4 to S_1 is 1.3098 times greater than that from S_7 to S_4. In the same way, the increment in the risk values from S_9 to S_6 is almost 14 times greater than that from S_8 to S_9.

Table 4.1 Ranking Positions, DA and DR of the analysis

Ranking Position (β)	Section (x_i)	*DA*	*DR*
1	S_1	0.7277	1.3098
2	S_4	0.5556	0.0450
3	S_7	12.3355	0.5135
4	S_6	24.0237	13.5551
5	S_9	1.7723	1.8107
6	S_8	0.9788	1.9436
7	S_2	0.5036	1.4291
8	S_3	0.3524	-
9	S_5	-	-

According to Brito and de Almeida (2009), given financial, technical and manpower constraints, the ranking obtained helps to prioritize the most critical pipeline sections in order to allocate a greater amount of resources for mitigating actions to those sections deemed most critical in the DM's view, bearing in mind that his/her preferences were incorporated throughout the development of the model, based on different risk dimensions. The DR analysis enables the DM to analyze the sections considered more consistently, making it possible for him/her to establish better planning actions, as well to allocate resources better.

In conclusion, all these improvements observed by using a MAUT application in this multidimensional risk model provided consistent results that can support managers in planning activities. Additionally, the ranking of risk values enables managers to analyze the existing context better, leading the organization to consider these aspects of mitigating risks and to consider preventive actions linked to the risk mitigation process.

4.3.2 Multidimensional Risk Evaluation in Underground Electricity Distribution System

Typically, energy distribution systems are big and complex. These systems are considered as being among the main elements of the critical infrastructure. Several other external systems such as systems of water supply, telecommunications, traffic, public transport, health, food supply, gas distribution, and others are dependent on this system. Therefore, several impacts other systems can be caused by small faults in the power system, and to generate a chain of consequences, which is why it is a critical part of the infrastructure for society.

Greater initial investment is required by installation of the infrastructure of an underground system. In generally, it is more complex than overhead systems. There are some disadvantages by use of the underground systems, such as it incurs higher costs associated with maintenance; it is also difficult to access underground networks; to upgrade the system (physical and limited space configuration); and, to operate and maintain auxiliary ventilation systems, etc.

Though, this system have advantages: the operation of underground systems is more safer and reliable than overhead systems for the population; more immune to interference from nature (storms, winds, storms, falling trees, etc.); better accessibility of disabled people, low visual pollution in the city and presenting less impact on the occurrence of traffic accidents.

Regardless of being safer than overhead systems, many underground vaults events have occurred. Hundreds of accidents in vaults occur every year in Ney York, such as smoke, explosions, fires, etc. (Radeva et al. 2009; Rudin et al. 2010; Rudin et al. 2011; Rudin et al. 2012).

The low frequency of the occurrence of accident scenarios, its magnitude of their consequences and the complex environment surrounding the hazard zone make the risk management becoming even more complex and uncertain (Garcez and de Almeida 2014a; Garcez and de Almeida 2014b). Also, the large number of subsystems, with each having particular characteristics, and there is a lack of (or incomplete) historical data of accidents and its failure modes and past events make the decision process even more complex.

Hazard scenarios can produce various consequences, for instance, fatalities and injuries to people, blackout, disruptions to local vehicular traffic, explosions and fires in nearby locations, impact of the company image, the population being afraid (on account of the uncertainty of when and where an accident will occurs), affect the system reliability and safety and other consequences which cannot be in financial terms (Garcez and de Almeida 2014a; Garcez and de Almeida 2014b). Hence, these consequences can disturb directly or indirectly the sector of the society, the public sector and business.

According to Garcez and de Almeida (2014b), assessing the risks comprehensively and realistically is extremely important. It may generate knowledge that can be applied to assist a DM to choose and implement preventive and mitigating measures.

Furthermore, the several resources available by company, such as: money, time, work teams, technology, safety equipment, etc. are limited and scarce. For optimization the use this resources, it is necessary to use decision-making tools that assess the consequences and uncertainties. Moreover, it is necessary to evaluate risks together with the DM's preference structure, thereby solving the problem more adequately (Garcez and de Almeida 2014a; Garcez and de Almeida 2014b; Garcez and de Almeida 2014c).

Therefore, it is necessary a decision making tool to aid the DM, generating a hierarchy of the multidimensional risks from the several underground vaults. The aim is to prioritize available resources to implement actions (preventive and mitigate actions) that increase system safety.

As seen, the MCDM/A, MAUT, permits the use of multiple value judgments; thereby incorporating the uncertainty and subjectivity inherent in the problem of estimating and evaluating different dimensions of the risks involved; and aggregating the DM's preferences.

According Berger (1985), a good decision should be a logical consequence of what one wants, what one knows and what one can do, so that the DM can choose an action (or actions) in order to bring about the most favorable consequences/ results for the DM. In this context, the Decision Theory is a mathematical formalization of this paradigm. It allows for rational decision-making under uncertainty, where the loss function is established as the negative of the utility function of the expected consequence.

The consequences are the result of the impact of the accident, which can be estimated using a probability distribution function $P(c|\theta, V_q)$, where θ are the states of nature (hazard scenarios); c is the consequences; and, V_q is the underground vault analyzed.

By MAUT concepts, Decision Theory and probabilistic independence, the risk measure can be expressed by (4.15).

$$r\left(V_q\right) = \sum_i \left(\sum_\theta \left(\pi(\theta) \left(-\int_c u(c) P\left(c \middle| \theta, V_q \right) dc \right) \right) \right) + (-1)\pi\left(\theta_N\right) \tag{4.15}$$

where i represents different dimensions of consequences and the state of nature θ_N is the normal setting of the system (there are no consequences – justifying that the value of the loss function is *-1*, the operation of company is normal without any accident occurrence). $\pi(\theta)$ is the probability of the hazard scenario. These risk values $r(V_q)$ are in the range [*-1,0*], where the value *-1* is related to the lowest risk and the value 0 to the greatest risk.

This section presents a numerical application based on the study realized by (Garcez and de Almeida 2014b). The hazard scenario, internal explosion caused by an arc flash, was considered. It is regarded as having the greatest impact and causes the manhole cover to be blown off and projected. The study evaluated the consequences (c) from four dimensions: operational impacts (c_O), financial impacts (c_F), disruptions to vehicular traffic (c_T) and human impacts (c_H).

The c_O corresponds to the impact on the supply operation of the electricity distribution company (downtime). The c_T is evaluated by the process of how traffic jams form on the streets around the accident area. The c_H deals with injuries caused by the projection of manhole covers and burns of at least the second degree due to exposure to incident energy from an arc flash. Lastly, the c_F is about any kind of monetary compensation related to an accident occurring.

Equiprobable Intervals method (Keeney and Raiffa 1976), based on results of Walsh and Black (2005), were used to estimate the distance projection of the manhole cover. Other hazard zone, calculated by IEEE Standard 1584 (IEEE1584 2002), also known as the Flash Protection Boundary, can be calculated as the minimum distance from the arc flash at which people could be safely exposed to incident energy without suffering second-degree burns. Estimates of the risk measures are made from the perspective of DM by Eq. (4.15).

As it is supposed that the DM's preference structure is additive independent between the criteria, the utility functions from the perspective of a one-dimensional utility ($U(c_O), U(c_T), U(c_H), U(c_F)$), can be elicited separately. To do so, the procedures described in Keeney and Raiffa (1976) were followed. It was considered that the DM is risk averse in the human dimension and risk prone in the remaining dimensions. The values of the scale constants obtained were: $k_{cO} = 0.12$; $k_{cT} = 0.16$; $k_{cH} = 0.29$; and $k_{cF} = 0.43$.

Hence, the multidimensional risk measure is calculated (4.15). The ranking of the multidimensional risk assessment is shown in Table 4.2.

The risk difference is calculated by (4.16).

$$r_i(V_q) - r_{i+1}(V_q) \tag{4.16}$$

The risk ratio is calculated by (4.17).

$$\left(r_i(V_q) - r_{i+1}(V_q)\right) / \left(r_{1^{st}}(V_q) - r_{i_n}(V_q)\right) \tag{4.17}$$

As the conclusion, V_{q3} is ranked as first underground vault and V_{q2} as second. Furthermore, it is observed that the difference between these risk values corresponds to approximately 44% of the total range of risk. Therefore, it is evident that is necessary to allocate more resources as a priority to preventive and mitigating actions on the first vault.

After the risk of the first alternatives (V_{q3} and V_{q2}) has been attend, there is another gap between the alternative ranked second V_{q2} and the one in third place V_{q5} (14% of the total range of the risk). Again, one prioritizes additional actions to prevent and mitigate the risk addressed in the first two vaults.

Another relevant information, it is that there is a homogeneous group of alternatives with similar risk values (V_{q5}, V_{q1}, V_{q6}). This information is important to the DM, because the DM can direct different and additional resources to preventive and mitigation actions to these alternatives, since they have very similar risk values.

Table 4.2 Results of ranking the risk

Rank	Vq	Risk Difference	Risk Ratio
1st	V_{q3}	1.66E-03	44%
2nd	V_{q2}	5.64E-04	15%
3rd	V_{q5}	5.64E-05	1.5%
4th	V_{q1}	2.12E-04	5.6%
5th	V_{q6}	5.74E-05	1.5%
6th	V_{q4}	1.24E-03	32.8%
7th	V_{q7}	-	-

Other issues (criteria) can be considered by the DM to choose which underground vault. DM will tackle first within this homogeneous group of risk. Another aspects can be considered by DMs when a decision making is taken: which actions and what alternatives will generate benefits more earlier? Additionally, in what alternative could be more efficiently? Finally, in another view that could be taken into account is decision-making for policy issues.

Under an inter-criteria approach, as shown in Fig. 4.5, on analyzing the risk values, it is concluded that: the first alternative shows that the traffic impact is the major one, while in the last-placed alternative the human impact is nonexistent. Furthermore, all alternatives have a financial impact and the only major value of the impact of these last-placed alternatives is on the financial dimension.

The comparison among the increments in risk, in inter-criteria analysis, is a different strategic information (Garcez and de Almeida 2012). This analysis allows identify the criterion that contributes to the greatest difference in risk between alternatives. By analysis, as shown in Fig. 4.6, the comparison pair-to-pair of the alternatives V_{q2} and V_{q5} can conclude that there are major impacts between the consequences of the financial, operational and human loss dimension. Therefore, the DM can conclude that preventive and mitigating actions that direction on disturbances to traffic loss dimension will not produce any impact in the difference in risk between these two alternatives. However, focusing on preventive and mitigating actions in the operational or human loss dimension of alternative V_{q2}, would result in reducing the amount of global risk compared to alternative V_{q5}. Thus, resources of the company can be reallocated to a manage risk more effectively.

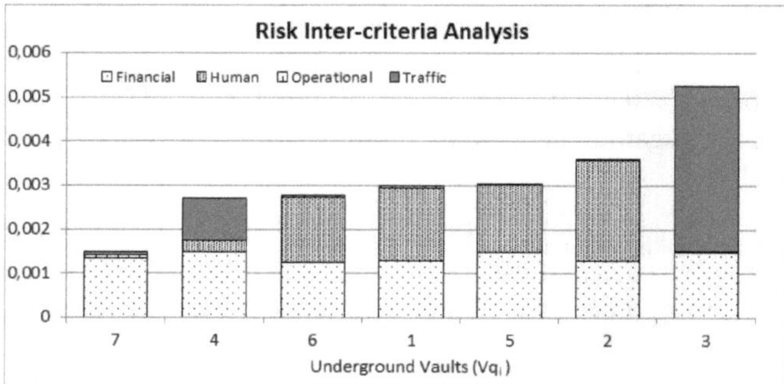

Fig. 4.5 Analysis of the measures of inter-criteria risk

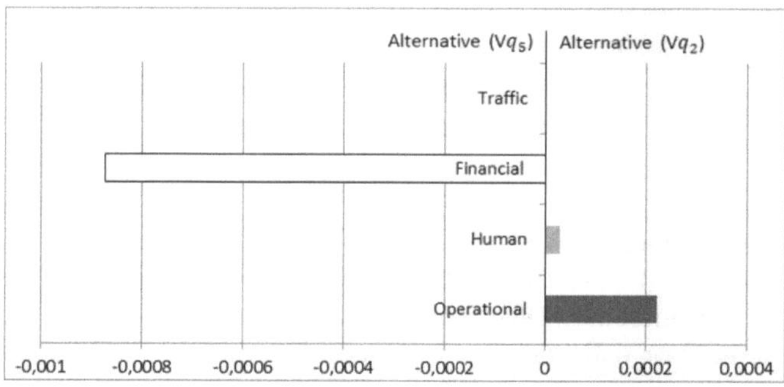

Fig. 4.6 Analysis of the intra-criteria of the risk differences of alternatives V_{q_2} and V_{q_5}

4.3.3 Risk Evaluation in Natural Gas Pipelines Based on ELECTRE Method and Utility Function

This section presents the application (Brito et al. 2010) of a different MCDM/A method, which is integrated with utility function, the ELECTRE TRI method. Three main issues should be highlighted, when compared with the two previous models. First, it is a non-compensatory approach, taking into account a specific kind of DM's rationality. Second, the problem consists of a sorting problematic, since the managerial issues in this application are distinct from the two previous. Third, it integrates the ELECTRE method with utility theory, in order to incorporate the DM's behavior regarding to risk (prone, neutral or averse) into ELECTRE.

As details given subsequently, this application illustrates the step 6 in the decision process given in Chap. 2, which involves the identification of DM's rationality (compensatory or non-compensatory).

In several situations, it is quite difficult (or even incoherent) for the DM to confront directly or indirectly monetary losses on non-monetary losses such as loss of life, injury to people, environmental damage, company image losses (Faber and Stewart, 2003) and social impacts. Therefore, it is considered that the DM feels more comfortable using a non-compensatory rationality approach, due this kind of procedure does not demand the condition of full comparability as must be done in the compensatory approach.

Specifically, in the risk management context, according to the DM, a low risk in a given criterion (with higher weight) does not compensate directly a high risk in another criterion, as should happen in an aggregation procedure with compensation. Therefore, for these cases, a non-compensatory approach for inter-criterion evaluation is more appropriate for representing the DM's structure of preferences.

Several gas pipeline problems, including new projects and concessions might be related to other DMs linked to other private or public institutions. Thus, one can be admitted that the DM wishes indirectly to consider his perception regarding the opinion of other actors (stakeholders, including population, government authorities and regulatory agency) in the decision process and this may change his final structure of preferences.

Moreover, one can consider some incomparability that may arise in the process of inter-criteria evaluation, due to a particular context (Brito et al. 2010).

As specified in step 6, in the decision process shown in Chap. 2, the decision model assumes a DM's non-compensatory structure of preferences for inter-criterion evaluation (among each risk dimensions). Hence, the outranking approach, including methods of ELECTRE's family, is more appropriated in the inter-criterion evaluation of risks to natural gas pipelines.

Another important point, as it was highlighted at the beginning of this section, is related to the problematic applied. In the two previous models, the ranking problematic was applied, based on MAUT. These models provide a comparison of alternatives with information on how large the difference in risk evaluation is between two alternatives. Differently, in this model under discussion, the DM faced different challenges related to maintenance and risk management, where for some situations a sorting problematic may be more appropriate (Brito et al. 2010). The classification (sorting) of the natural gas pipeline sections into categories allows the DM to organize particular management approaches for each risk category.

The ELECTRE TRI method, more detailed in the Chap. 2, deals with a sorting problematic, assigning each alternative s_i from a set S to a category or class C_k. For the context of this model, s_i represents sections of natural gas pipeline to be sorted, and the profiles b are comparison sections for the categories of risk.

The model application makes an evaluation of several sections of pipeline according to their multiple risk dimensions, which allows a comparison of these sections with the risk profiles in order to classify the sections into risk categories defined by the natural gas transportation/distribution company's management.

In this context, the profiles b that define the particular risk categories, depend essentially on the perception that the DM has on different risk levels related to his system, the availability of resources, the occurrence of previous accidents, society pressures, as well as being dependent on the number of different strategies, policies, and measures that the company possesses to deploy among the categories.

The highest risk category contains an alternative with higher probabilities of occurrence of financial, environmental and human consequences. This category demands relatively urgent actions that often require changes in some aspects of the project, and that demand a major financial investment in order to obtain significant reductions of these risks. Similarly, a lower class of risk presents sections of pipeline with lower levels of risk, thus allowing a little longer planning time to find effective solutions and at satisfactory costs (Brito et al. 2010).

Brito et al. (2010) highlight that the manner of determining the reference profiles b for the each risk categories must be carried out very carefully by the DM, since the sorting process is fundamentally guided by comparisons with these profiles.

A procedure to aid DM to infer theses profiles is proposed by Mousseau and Slowinski (1998). It enables the inference by means of a sample of alternatives directly sorted by the DM.

The third point highlighted on beginning of this section is the integration between the ELECTRE TRI method and utility theory, in order to incorporate the DM's behavior regarding to risk (prone, neutral, averse). The utility theory presents an axiomatic approach that can assess the DM's behavior with regard to the risk (Keeney and Raiffa 1976) when there are accidents consequences.

Let D be the set of all outcomes in a given accident impact dimension. Uncertainties are related to the states of nature θ, the resulting accidental scenarios of a pipeline accident, and to its impacts under a given dimension of outcomes. For dealing with uncertainties on D, it is necessary to use a probabilistic approach, represented by a probability distribution over the deterministic consequences and by the elicitation of the utility functions for these consequences (Brito et al. 2010).

This procedure is applied in the intra-criterion assessment process (for each risk dimension) with the aim of risk evaluation for human, environmental and financial dimensions posed by each section of pipeline.

As defined in (4.11), the risk is assessed as the expected loss, which is estimated for each section of pipeline. The loss is given by combining the probability over the deterministic consequences p in D, named by $P(p|\theta,s_i)$, and the utility function ($U(p)$, where $p \in D$) over these consequences, as shown in (4.18). It is used the traditional notation for decision analysis (Utility Theory), where p (from payoff) denotes an element of the set of outcomes D, whereas P (capital P) refers to a probability (which is a probabilistic payoff).

$$L(\theta, s_i) = -\int_p P(p \mid \theta, s_i) U(p) dp \tag{4.18}$$

Therefore, the expected risk can be calculated for each section of pipeline under each criterion, applying (4.18) in (4.11), then (4.19) is obtained.

$$r(s_i) = -\sum_{\theta} \pi_i(\theta) \cdot \int_p P(p \mid \theta, s_i) u(p) dp \tag{4.19}$$

As previously discussed, the ELECTRE TRI method is more appropriate than MAUT for undertaking the inter-criterion pipeline risk evaluation. Another issue related to the DM's structure of preferences is the observation that not all hypotheses required by MAUT are always accepted in the case of inter-criterion evaluation (among risk dimensions). This may happen even when these hypotheses are appropriate in the intra-criterion evaluation. To be precise, the DM accepts the Utility Theory hypotheses when he evaluates separately each risk dimension.

The use the utility functions is justified because the model can incorporate the DM's behavior regarding risk (averse, prone or neutral). The utility function is also appropriate because the results occur in an interval scale rather that an ordinal scale for comparison with the profiles categories in the sorting problematic. Furthermore, this interval scale is explored in the process of eliciting preferential parameters for ELECTRE TRI method, including the profile for each category defined and the thresholds. In other words, the DM knows the amount of risk differences to be considered in the ELECTRE TRI method for building the credibility index. In this manner, the integration of the utility theory and the ELECTRE is seen as a useful (Brito et al. 2010; de Almeida 2005; de Almeida 2007).

The decision model proposed by Brito et al. (2010) presents the procedure steps for problem resolution and to construct multicriteria models, as shown in Chap. 2. This application aims to build an MCDM/A model for the multicriteria risk assessment of pipeline sections and for their assignment into risk categories.

Initially, the pipeline system was segmented into 12 different sections. These sections were divided according to several technical factors such as age of the pipeline section, pressure, land occupation, soil characteristics, degree of third-party interference and demographic concentration on the surface area surrounding each section.

In addition, it was considered 10 hazard scenarios (θ): Detonation/Deflagration; Fireball/Jet Fire; Confined Vapor Cloud Explosion (CVCE); Flash Fire; Gas Dispersion to both failure modes: rupture and puncture.

The accidental scenario probabilities, $\pi_i(\theta)$, were based on EGIG report, because of its ability to distinguish between pipeline failures modes, and also because it gives more conservative estimates for the scenario of probabilities than other databases, such as those from the United States Department of Transportation (Brito et al. 2010).

The payoffs used in this application involve the human (H), environmental (M) and financial (N) consequences of an accident caused by the release of gas. The payoff of the human consequences considers injuries to human beings. Generally, it is dealt as the number of fatalities due to thermal radiation (Jo and Ahn 2005). The use of monetary values for estimating this type of consequence is not appropriate to represent the consequence in a decision-making problem (Brito et al. 2010). Therefore, this model adopts a more conservative criterion for analyzing the human consequences (*H*) than monetary estimates or the number of deaths. These consequences are estimated as the number of people exposed, at least, to second degree burns. According to Brito et al. (2010), although very conservative, this reasoning is appropriate when dealing with impacts on human beings, assuming that any type of physical harm to the population should be avoided.

The environmental impacts (*M*) are given by the area that is exposed to the atmospheric pollution and to the effects of scorched vegetation on animal and vegetable species. Similarly, as in the case of human consequences, it cannot be expressed by monetary values. Therefore, it is used the area of the vegetation destroyed (in square meters) as measurement (Alencar et al. 2014). According to Brito et al. (2010), although this is not a very complete way to interpret these types of consequences, this measurement is useful and is reasonably related to the extent of environmental impacts caused by natural gas pipeline accidents.

The financial consequences (*N*) are associated to operational losses that a pipeline accident may cause, such as: expenses on labor, equipment and raw material to substitute pipes, expected loss in revenues from supply interruptions, refunds to customers for interrupted production, and compensation for damage caused to others.

The one-dimensional utility functions $U(h)$, $U(m)$ and $U(n)$ may be obtained from the elicitation of some utility values in each dimension, using a lottery procedure (Keeney and Raiffa 1976). Thus, a regression curve over plotted values may be adjusted. Exponential functions are among functions that often present a best fit for utility functions (Berger 1985), as per (4.20).

$$U(p) = e^{-\mu_p \cdot p} \tag{4.20}$$

where $p = h$, m or n. The parameter μ_p is obtained by means of curve fitting. The following parameters were obtained for the utility functions, as given in (4.21): for $U(h)$: $\mu_h = 0.12$ ($R^2 = 0.91$); for $U(m)$: $\mu_m = 0.0017$ ($R^2 = 0.89$); and for $U(n)$: $\mu_n = 3.5x10^{-7}$ ($R^2 = 0.94$).

The calculation of consequence probabilities $P(p|\theta,s_i)$ is obtained for each pair (θ,s_i) of scenario and section of pipeline. In other words, this function is the probability of obtaining a consequence p given that θ happened. Depending on the mathematical models used, these consequence functions may assume different forms (Arnaldos et al. 1998; Jo and Ahn 2002). For Brito et al. (2010), this modeling can consider any type of probability distribution obtained for consequence functions, simply by adjusting the calculations of the consequences functions to another context or system. Thus, it is not limited to a single application.

Based on expected loss function (4.18), the combination of the probability density functions to the one-dimensional utility functions $U(h)$, $U(m)$ and $U(n)$ was undertaken in order to estimate the one-dimensional losses.

Next, it is necessary to estimate the risk values for each pipeline section. Whereas there is a state of nature (scenario) in which there is a probability associated with it of no failures occur (named by θ_N), then this section pipeline suffers no damage ($L(\theta,s_i) = 1$). Therefore, the human, environmental and financial risk values for each section of pipeline are given by (4.21). A linear scale transformation, $r'_p(s_i) = 100r_p(s_i) + 100$, was used to facilitate the handling of values by the DM. These risk values are shown in the Table 4.3 (Brito et al. 2010).

$$r_p(s_i) = \sum_\theta \pi_i(\theta) \cdot \left(-\int_p P(p \mid \theta, s_i) e^{-\mu_p \cdot p} dp \right) - (1)\pi_i(\theta_N) \tag{4.21}$$

Table 4.3 Human, environmental and financial risk values

Section pipeline	Human risk	Environmental risk	Financial risk
s_1	0.0093	0.0142	0.0080
s_2	0.0180	0.0199	0.0326
s_3	0.0249	0.0265	0.0101
s_4	0.0085	0.0270	0.0521
s_5	0.0104	0.0113	0.0282
s_6	0.0293	0.0181	0.0237
s_7	0.0379	0.0152	0.0242
s_8	0.0081	0.0128	0.0345
s_9	0.0104	0.0070	0.0233
s_{10}	0.0205	0.0245	0.0554
s_{11}	0.0565	0.0440	0.0467
s_{12}	0.0190	0.0201	0.0738

Subsequently, the DM wishes to sort those sections pipelines in risk categories, ordered by decreasing levels of risk, these are: High Risk (*C1*), Medium Risk (*C2*) and Low Risk (*C3*). For each defined category, the reference profiles (ELECTRE TRI parameters) are determined, as shown in Table 4.4.

Table 4.4 ELECTRE TRI parameters employed in the analysis

Parameter	r_h	r_m	r_n
b_1 (divides the High Risk from the Medium Risk category)	0.025	0.025	0.05
b_2 (divides the Medium Risk from the Low Risk group)	0.013	0.01	0.02
weight	0.60	0.10	0.30
q (indifference threshold)	0.001	0.001	0.005
p (strict preference threshold)	0.005	0.009	0.007

By analysis of the DM, the sections in the first category demand higher states of alert, and thus financial resources would be assigned preferentially to this category in order to increase measures of physical protection and to intensify the monitoring of the high risk sections. The Medium Risk category involves pipeline sections which, although they do not lay claim to such intensive care as those in the previous class, do demand more thorough planning for preventive measures in order to avoid neglect in relation to maintaining their safety levels. Finally, as to the sections assigned to the Low Risk category, the maintenance of routine inspection actions is planned in order to keep these sections with low risk levels within the human, environmental and financial dimensions of possible outcomes (Brito et al. 2010).

The analyst has to explain the meaning of ELECTRE TRI parameters in order to obtain the proper specification. It was decided not to use a veto threshold for any risk dimension. With regard to the cutting level, $k = 0.65$ has been applied. After applying the sorting model for each individual section of pipeline, the results in Table 4.5 were obtained.

Table 4.5 Final sorting

Section pipeline	Category
s_1	*C3*
s_2	*C2*
s_3	*C1*
s_4	*C3*
s_5	*C2*
s_6	*C2*
s_7	*C2*
s_8	*C3*
s_9	*C2*
s_{10}	*C2*
s_{11}	*C1*
s_{12}	*C2*

It was observed that, for this application under study, the results were intensely influenced, but not completely controlled, by the human risks, given their high weight value. Among the segments under study, 7 out of the total of 12 sections were assigned to the Medium Risk category (*C2*), for which more rigorous preventive measures should be established within 6 months. Sections s_3 and s_{11} were assigned to the High Risk category (*C1*), for they present risk levels worse than or very close to the profile b_1 in a more significant proportion of impact dimensions. Finally, sections s_1, s_4 and s_8 were assigned to the Low Risk category (*C3*) because they had more satisfactory performances than those presented by profile b_2.

A sensitivity analysis was conducted in order to analyze responses and opinion from the DM and to evaluate the robustness of the results with respect to imprecise data, and the way in which the model can be used by the DM.

The parameters were varied by 10% of the initial value specified by the DM. It was concluded to be robust for the majority of parameters, such as weights and profiles for environmental and financial risk criteria.

Nevertheless, it was observed changes for parameters related to weight and profiles for the human risk criterion (r_h). A particular change was found for the specification of the cutting level k, which is related to the weight for r_h. A reduction of less than 10% in k makes it less than the weight for r_h, which should be avoided. As a result, sections s_5 and s_6 change from category *C2* (Medium Risk) to *C1* (High Risk). According Brito et al. (2010), this happens precisely because the risk for human criterion is greater than the profile b_1 for this criterion. Since this analysis, the DM decided to maintain the previous results, classifying sections s_5 and s_6 as category *C2* (Medium Risk).

Another sensitivity was observed when k is increased by 10%. Only section s_3 changes to a lesser risk category. Into a more safety view, it was also decided to maintain the previous classification, so s_3 remained in *C1*.

4.4 Other MCDM/A Applications on Multidimensional Risk

In the next sub-sections, several other decision problems in the related to multidimensional risk analysis, using MCDM/A, are presented. These problems are grouped by its context, such as: power electricity systems and natural hazards.

4.4.1 Power Electricity Systems

The generation of electrical energy can be from various sources. Each energy source will generate different risks inherent in its own production and supply. Regős (2012) compared the general risk of the four most important energy chains (coal, nuclear, gas, hydro). For this, he applied an MCDM/A approach, and chose severe accidents, terrorism, environmental and health risks, risk of price changes as risk criteria.

Normally, generation and power supply systems are large and complex systems which society considers form a critical part of the infrastructure. Typically, several other systems or subsystems, such as water supply systems, telecommunication, traffic, health, food supply, etc. are dependent on power supply systems. Thus, failures in the electricity system can impact other systems and generate a chain of consequences, which is why it is critical for the infrastructure.

Moreover, the system of transmitting and distributing energy consists of networks in different settings, such as networks in rings, radial or redundant networks. These settings are intended: to distribute the loads, this creating redundancy in the system; to increase reliability; to minimize the loss in case faults occur; or to minimize the occurrence of failures in chains, which can cause multiple impacts. Therefore, analysis and risk management becomes very complex since several aspects have to be considered.

There are several reasons for failure in power systems. The most common technical failures are those which originate from: inadequate maintenance of the system; system overload; using design (dimensioning) and unsuitable equipment; conducting maneuvers in the wrong networks (human error); dimensioning loads poorly, etc.

Besides these factors, one of the causes of failures is due to the occurrence of extreme natural events such as storms, hurricanes, floods and earthquakes. Furthermore, there is an external pressure causing stress on the network because of the need to integrate new public services and the joint use of renewable energy, and hence, increasingly, power systems are operated closer to their stability limits (Haidar et al. 2010).

In order to evaluate risk and manage risk effectively, there must be a clear analysis. Consequently, in order to facilitate the process of decision-making, various aspects analyzed in this context need to be taken into consideration.

Faced with increasing pressure from society in general for a higher level of safety, risk management has become an arduous, complex and uncertain task. This because it can involve all of the following: a large number (hundreds or even thousands) of primary and secondary power systems with particular characteristics; the absence or incomplete historical data on failure modes and accidental events that have already occurred; the rarity of occurrence of accident scenarios; the magnitude of consequences; and the complexity of the area surrounding the hazard zone, etc. (Garcez and de Almeida 2014a; Garcez and de Almeida 2014b, Garcez and de Almeida 2014c).

Therefore, effective risk management plays a role of great importance to society, the public sector and the electricity distributors, since the impacts caused by accidents can adversely affect all three areas, directly or indirectly.

The importance of evaluating the risks comprehensively and realistically generates knowledge that can be applied to assist the distributor power company in choosing what preventive and mitigating actions to take, thus resulting in risk management that is effective and efficient (Garcez and de Almeida 2014b).

Furthermore, since the available resources (monetary, time available, work teams, technology, etc.) of power energy companies are limited and scarce, and regulators require power systems to demonstrate greater availability and system reliability, it is necessary to use decision-making that adds in the effects and uncertainties from multidimensional risks and to evaluate these together with the preference structure of the company. It is only by doing so that the problem will be dealt with more adequately (Garcez and de Almeida 2014a, Garcez and de Almeida 2014b).

In the area of asset management of energy companies in general, it is recognized that there is a need to use a more formal and structured analysis of the increasingly complex decisions. This is challenging. Current asset management practices focus primarily on risk quantification in monetary terms, and on the reliability of the system, combined with estimates of the condition of the components (estimated lifetime, etc.). The analysis of other aspects of risk, such as the risk to personal safety, the risk of environmental damage or the risk of a negative public response are usually "decoupled" from quantitative risk analysis. So for Catrinu and Nordgård (2011), it is necessary to improve the current practice of asset management, by making the best use of knowledge and data available from experts and adopting new methods of risk analysis and decision support, and moreover, the best ways to document decisions.

For this, Catrinu and Nordgård (2011) integrate the methods of risk analysis and decision support for advanced management under uncertainty in the assets of a power distribution system. The focus of this study was to incorporate different business objectives of risk analysis in a structured framework so as to decide how to deal with the physical assets of the electricity distribution network.

The growing importance of environmental issues at the global and regional levels including water and air pollution, the use of non-renewable energy sources, as well as outcomes such as global warming and climate change, have led to it being considered essential to take environmental factors into account when planning how and from where to generate and distribute power (Jozi and Pouriyeh 2011; Rezaian and Jozi 2012). Therefore, in the process for planning energy systems, uncertainties should be more carefully handled because of the increasing concern about the environmental impact of electricity generation and because this market sector is highly competitive.

Linares (2002) presents a multi-criteria model for planning electricity, which deals with uncertainty and risks associated with minimizing the environmental risk and performs a risk analysis (in a multicriteria view) to apply classical decision-making rules and therefore to select the best planning strategy under uncertainty. Linares emphasizes that incorporating additional criteria leads to more flexible and efficient strategies, which greatly reduces the environmental risk at a small incremental cost, while the process of risk analysis selects flexible and robust strategies for the scenarios analyzed.

In this context of risk management, the need to generate a risk hierarchy of the various subsystems of the electricity supply system is seen. Garcez and de Almeida (2014b) propose a form of risk assessment in an underground electricity distribution system under a multidimensional view (multicriteria), in which they generate risk measures, which can be ordered. The aim is to generate a priority list of issues to be considered when allocating additional resources to prevent and mitigate risks, such as conducting inspections and maintenance; modifying projects in order to increase safety; developing preventive and mitigating actions; modernizing and improving the subsystems (upgrade) (Garcez and de Almeida 2014c).

A mitigation measure widely used in case of faults in power systems to prevent failures in chain is Load Reduction (LR). It is considered a very effective emergency measure for stabilizing the power system (Dong et al. 2008). To implement load reduction it is necessary to disconnect certain areas of the power grid, so this technique generates direct impacts on the population, economy and local industry. Therefore, normally it is among the last measures to be applied and is usually only used to prevent the total collapse of the network.

However, to implement LR, it is first necessary to setup which areas should be disconnected. That choice alone is already a decision process, because not only operational aspects of systems are considered but also aspects of consequence covering a multidimensional view of the problem. LR has also been successfully implemented in Europe and USA. More recently, LR was applied successfully to manage: the impacts of Hurricane Sandy 2012; *The 2006 European Blackout* (Van der Vleuten and Lagendijk 2010); *The 2003 Northeast Blackout* (Andersson et al. 2005), and *The Italian 2003 Blackout* (Berizzi 2004).

For the LR method, from a decision analysis point of view, the areas of energy supply represent the alternatives of the model. To analyze the potential consequences resulting from the uncoupling of these areas, the vulnerability of each area must be analyzed.

4.4.2 Natural Hazards

According to natural hazard theory, risk appears wherever and whenever assets are subjected to hazards; it is usually defined as 'the expected potential loss due to a particular hazard for a given area and reference period' and can be mathematically defined as the combination of hazard and vulnerability (Merad et al. 2004).

For Nefeslioglu et al. (2013), it is fully acceptable for a natural event, such as a flood or an earthquake to become a natural hazard when people are affected by a natural hazard. Since the world population is increasing, the need to find habitable areas has increased considerably, which has led to people having to being caught up in these natural events more often.

Consequently, Viscusi (2009) states that the occurrence of natural disasters often generates a cluster of fatalities rather than just a single fatality. Hundreds or sometimes thousands of people could die from the occurrence of a single event.

Additionally, another important point is that deaths perceived to occur due to the probability of a natural disaster is a very heterogeneous concept and often, and its probability is much lower when compared to other risks associated other causes with fatality. However, the risk management of natural disasters should not just stick to the issue of people being killed or injured.

Depending on the type and scale of natural event occurring and its impact on society, other points may be incorporated in the analysis, such as the issue of safety, security and public health, population migration, cost estimation, information sharing, planning public and environmental aspects.

For example, when examining the risk of a flood, a check should be made on aspects that are part of this context, such as the question of the complexity of the event, broad spatial scales, intervals of time between events, vulnerability and social-psychological aspects such as depressions, anxieties and conflicts of interest, in addition to several conflicting aspects between them.

Additionally, taking into account the study described in Levy (2005) for the operation and management of reservoirs, there is a complex analysis of the tradeoffs between protection against flooding (i.e., minimizing the discharge of reservoirs during the peak periods of flooding) and energy production (meeting the goal of producing pre-defined levels of energy). On the one hand, flood protection means that the tank must be maintained at the lowest possible level so that the reservoir can accommodate the excess water coming from the period of flooding. On the contrary, the production of energy requires that there be the largest possible amount of water in the reservoir. In this case, the decision-making process will directly affect risk management, so what is needed is a more structured analysis that provides satisfactory results.

According to Nefeslioglu et al. (2013) the evaluation of the interaction between natural and human events in terms of hazards and risks has become a common topic for analysis in the last 20 years. Modeling consequences and probabilities is one of the main tools for assessing the impacts of natural hazards.

Given the uncertainty associated with the environmental context, Parlak et al. (2012) state that analysis based on multicriteria decision methods provides a systematic approach to managing the complexities and uncertainties associated with the occurrence of natural disasters, since multicriteria methods make use of stochastic approach which help to develop this modeling.

Levy (2005) points out the use of MCDM/A has increased in the last three decades due to a number of factors, including dissatisfaction with conventional methods that use only a single criterion, as well as ease of access to software and algorithms that enable the solution of complex environmental problems to be found. Thus, in the aforementioned study for the operation and management of reservoirs, MCDM/A is useful for eliciting and modeling stakeholders' preferences and to improve coordination between state agencies, organizations and the affected population in such a way as to minimize the risks associated with floods e.g., death of or injury to persons, damage to property and possible environmental impacts.

The need for multicriteria approaches can also be observed in planning the response to a disaster, where, according to Parlak et al. (2012) such planning requires the engagement of multiple disciplines such as engineering (infrastructure), management emergencies, health care, mass communication, water supply and food logistics. Planning the integration scenario by using multicriteria analysis, according to the authors, enables initiatives to be prioritized and that this contributes to plans (the response to a disaster) being better understood.

Several applications of multicriteria decision methods are to be found in the area of risk management for natural disasters, as will be explored in the next paragraphs.

For addressing territorial risk evaluation considering a group of DM's, Cailloux et al. (2013) proposed an MCDM/A model based on ELECTRE TRI method to evaluate the level of risk for territorial zones surrounded by a given industrial plant considering a natural hazard, such as flooding. Scawthorn (2008) indicates how to assess assets at risk in risk areas and in particular, the impact of an earthquake on social cohesion and peace; public confidence; political unity; education; and the mental health of the population affected. He includes physical assets and non-physical assets that can be given a monetary value. Subjective judgments may be necessary to compare the vulnerability of these different assets before finally obtaining an overall assessment of risk.

Nefeslioglu et al. (2013) propose a derivation of the AHP (Analytical Hierarchy Process) MCDM/A method called M-AHP to support decision-making problems in natural hazard areas, specifically snow avalanches in mountainous regions.

Karvetski et al. (2011) consider principles of MCDM/A to define a methodology that measure the impact of possible scenarios for engineering systems in the context of climate change.

Stefanidis and Stathis (2013) evaluate hazard areas associated with floods by using AHP and GIS (Geographic Information Systems) to assess the danger from both natural and anthropogenic aspects, thereby creating of two indices for floods.

Tamura et al. (2000) deal with a process of decision analysis to mitigate risks associated with natural disasters, which consider events of low probability and high consequence. The authors propose the use of a function value at risk (Value Function under Risk) instead of the expected utility theory.

In the context of landslides, the use of GIS and spatial multicriteria evaluation is widely used. Multiple indicators are processed, analyzed and weighted according to their contribution to the risk and vulnerability. To reduce losses from disasters, existing planning on being prepared for a disaster and the immediate response to it needs to be improved as does planning on how to reduce risks from disasters. This should be based on a multidimensional evaluation of risk at all levels of management. Abella and Westen (2007) in their study used four key indicators for a study on vulnerability from a landslide:

- living conditions and transportation indicators (physical vulnerability);
- population (indicator of social vulnerability);
- production (indicator of economic vulnerability), and;
- protected areas (indicator of environmental vulnerability).

Abella and Westen (2007) applied these indicators and the results obtained from the analysis led to the development of a plan for mitigating risks from landslides at the national level in Cuba, and this information being linked to the national system, which gives early warning of hurricanes, and warns and evacuates people from areas prone landslides.

Another context of natural hazard together with man's intervention arises from ending mining operations in populated regions. According to Merad et al. (2004), in the Lorraine region of France many landslides and subsidence have occurred, which led to the need to develop a specific methodology for risk zoning of the area. The authors propose a methodology based on a multicriteria decision support tool (ELECTRE-TRI), with the aim of assigning risk zones in predefined classes of inhabited regions. This approach enabled the knowledge of experts, multiple qualitative and quantitative criteria and uncertainties to be considered.

4.4.3 Risk Analysis on Counter-Terrorism

In recent decades, the fight against terrorism has been the focus of constant analysis worldwide. Security measures have been strengthened and new anti-terrorism policies are presented to the world by nations periodically to society. One goal of these policies is to establish the benefits of preventing a terrorist attack, such as reducing the number of deaths and injuries associated with the human dimension. More specifically, the risk management of terrorist attacks has intensified, especially after the terrorist attack on the World Trade Center on September 11, 2001 in the United States.

Risk management, according to Aven and Renn (2009) seeks to ensure that adequate measures are established to protect people, the environment and assets from harmful consequences arising from human activities or natural events. The extent to which risk reduction measures are justified depends on the balance between costs and benefits in terms of the security gain. Furthermore, several PRA models have been applied taking into account aspects such as infrastructure, food supply chains, population, etc., and considering risk as a product of three components: threat, vulnerability and consequence (Greenberg et al 2012).

Due to terrorism threats, several models and approaches have been proposed in order to mitigate the risks associated with such events (Merrick and Leclerc 2014; Shan and Zhuang 2014; Haphuriwat and Bier 2011; Ezell et al. 2010; Parnell et al. 2010; Ngange et al 2008; Leung et al 2004). Among these models, there are several studies considering MCDM/A approaches (Akgun et al. 2010; Sri Bhashyam and Montibeller 2012; Koonce et al. 2008; Patterson and Apostolakis 2007).

In their paper, Akgun et al. (2010) stress that assessing the vulnerability of critical assets (e.g.: airports, dams, chemical plants, nuclear power plants) to terrorist attacks is a highly complex strategic activity, requiring a methodology structured to support the decision-making process in defense planning. Their approach seeks to define the vulnerability of each critical defense asset against terrorist attacks taking into account multiple criteria. They use SMART in conjunction with Fuzzy Set Theory and Fuzzy Cognitive Maps in a group decision environment. Their model seeks to identify hidden vulnerabilities and to define the roles and most critical (or active) components of each system, and five criteria were established:

- Deterrence (implemented method of defense, perceived by terrorists as hard to penetrate);
- Detection (of a terrorist attack);
- Delay (the time during which an element of a physical protection system is designed to prevent terrorist invasions);
- Response (the time taken to respond to a threat), and;
- Recovery (the time taken to return the areas and people affected to their existing status prior to the event).

In the second study, Sri Bhashyam and Montibeller (2012) propose a framework that can be used to infer how the priorities of the terrorists may change over time and the impact that these changes may have on the choice of a harmful action. This is done based on a multicriteria model that uses MAUT. The objectives were visualized in three categories: revenge, reputation and reaction. The alternatives of the decision problem were established by: strikes by the terrorists, improvised explosive devices in a public place, explosions of portable nuclear devices in modes of mass transit, detonating bombs and biological weapons or dirty bombs that combine explosives and radioactive materials.

Therefore, the aim of modeling terrorists' priorities is to define the objectives that terrorists will use to evaluate the attack, providing the best tradeoff between the operational side of an attack (costs) and benefits (if goals are achieved).

4.4.4 Nuclear Power

Risk analysis in power systems is a crucial activity so as to ensure adequate security for society, especially with regard to the operation of power plants. More specifically, in recent years, different sectors of society have insisted on new discussions with regard to safety in nuclear power plants due to the Fukushima accident in Japan in 2011. In this context, according to Rogner (2013), accidents like Fukushima have created a greater climate of distrust with respect to society's view of nuclear energy. In contrast, industries have tried to increase their security level. Additionally, several issues have begun to dominate public debate on energy policy such as: energy security; the price of fossil fuels; climate change; the increase in the demand for electricity. As nuclear power has a mitigating role in several of these points, the societies of some countries once again have a higher level of tolerance for nuclear technology.

In this context, Papamichail and French (2012) points out that radioactive accidents have emphasized the requirement to provide support for all emergency management phases. Several decision support tools are currently being developed to prevent and mitigate the effects of radioactive accidents. Among these tools, multicriteria decision techniques stand out.

The literature describes some recent applications of multicriteria decision methods in the context of nuclear energy. Examples include the following:

- Atmaca and Basar (2012) use an Analytic Network Process (ANP) to evaluate 6 different alternatives of nuclear power plants taking into account criteria such as technological aspects and sustainability, economic viability, quality of life and socio-economic impacts.
- Hong et al. (2013) use a multi-criteria decision analysis to assess future scenarios for generating electricity in Japan electricity which take economic, environmental and social impacts into consideration. Their study is a response to the nuclear crisis caused by the Fukushima accident.
- Erol et al. (2014) define the location problem of a nuclear power plant in Turkey as a multicriteria decision problem using fuzzy logic, and consider qualitative and quantitative criteria. The primary criteria that they establish are: proximity to the existing electrical infrastructure; proximity to the transportation infrastructure; access to large amounts of cooling water. The authors also consider a number of secondary criteria: population density; geological issues; atmospheric conditions; cost factors; and risk factors.
- Beaudouin (2015) proposes an MCDM/A model that supports debate about nuclear power plants safety choices. Therefore, six safety criteria are considered in combination with cost-effectiveness analyses to point out the best portfolio of power plant design modifications, satisfying security requirements.

Thus, MCDM/A tools can be used at various stages in the context of nuclear power production in order to contribute to risk management, thereby making the decision-making process an important aspect when planning safety measures for nuclear power plants.

4.4.5 Risk Analysis on Other Contexts

A requirement for the building industry, both with regard to permission to build and to certification that the finished building meets regulatory requirements, is necessary an environmental management, where the identification and evaluation of risk to human and environmental health are the first stages. Topuz et al. (2011) propose an approach that integrates the assessment of risk to humans with environmental health in industries using hazardous materials, to support environmental DMs with quantitative and directive results. For this, the methodologies used multicriteria and fuzzy logic to deal with the problems arising from the complexity of the environment and uncertain data.

Specifically, in the context of bridges, the use of bridges is among the most important structural elements that reduce traffic problems. Risk management of bridges serves to determine the best allocation of resources. According to Adey et al. (2003), these systems are usually evaluated against the structural deterioration which bridges may suffer from as a result of traffic loads. However, these systems are affected by various other hazards, such as floods and earthquakes, not only the traffic load.

The destruction of big bridges are usually important and significant events, and may result in loss of lives, property and economic losses. According to Shetty et al. (1997), the consequences of the destruction of bridges can be summarized as:

- Human elements that impact the number of deaths and injuries, such as the high rate of vehicular traffic, the flow of pedestrians that pass over or under the bridge;
- Environmental consequences resulting from spills of hazardous substances, due to the intersection of transport between road, rail, etc.;
- Formation of traffic jams, increasing the volume of traffic at a particular site, causes overload on other transport routes;
- Economic factors, including the cost of taking construction material residuals away; reconstruction; indemnities payable on the destruction of vehicles; the environmental catharsis; and legal costs.

According to Wang et al. (2008), the risk assessment of bridges is essentially a multicriteria problem, which involves multiple assessment criteria such as safety (safety of the public), functionality (effects on the level of service/availability of the network for use), sustainability (expenditure and workload) and environment (effects on the environment, including the (aesthetic) appearance of the structures). Wang et al. (2008) propose an integrated AHP–DEA methodology to evaluate risks to or from bridges of hundreds or thousands of bridge structures, based on which maintenance priorities for the bridge structures can be drawn up.

Environmental risk assessment and decision-making strategies in recent decades have become increasingly sophisticated, and use intensive and complex information, including approaches such as expert opinions, cost-benefit analyzes and evaluation of the toxicological risk. According to Linkov et al. (2006), a tool that has been used to support environmental decision-making is comparative risk assessment (CRA), but CRA lacks a structured process to arrive at an alternative optimal design method. The approach of using multicriteria decision analysis fills this need by providing methods that give better support to comparing alternatives and also provides a structure which incorporates input from stakeholders of the project, the aim of which is to rank alternatives.

In the context of the hazard from forest fires, over past decades, in several regions, especially in tropical and Mediterranean regions, these fires are due to several underlying factors, which have received increasing attention because of the wide range of ecological, economic, social and political impacts. The more complex fire models require spatial information, which is done by remote sensing and GIS (Vadrevu et al. 2010; Arianoutsou et al. 2011).

According to Vadrevu et al. (2010), the integration of MCDM/A methods in the spatial domain provides a new framework for addressing many environmental problems, including quantifying "fire hazards". These authors conducted a study in a thickly-forested area (Indian region), where most of the stakeholders are the local people, and their dependence on forest resources is immense.

Moreover, the problem of forest fires in the study area is spatially diverse in nature and involves both biophysical and socioeconomic parameters, providing an ideal place to use an MCDM/A methodology. Combining these multiple parameters using decision-making methods in a collaborative framework may yield good results, so, the risk of fires in tropical deciduous forests, in India, was quantified as a function of topographic, vegetation, climatic, and socioeconomic attributes in order to evaluate the fire risk in the study area.

Still in the environmental context, the contamination of water resources on land has been a major environmental concern during the last decades, mainly due to public health concerns. According to Khadam and Kaluarachchi (2003), traditionally, environmental decision-making scenarios of subsurface contamination are guided by means of cost-benefit analysis.

This context, the risk assessment includes quantification of the risk to human health, as well as evaluating the importance of this risk. When the risk is determined unacceptable, potential remedial alternatives are identified and decision analysis is performed to choose the best corrective action. There is a tradeoff between individual risk and societal risk, the tradeoff between the residual risk and the cost of reducing this risk, and cost-effectiveness as a justification for remediation. The authors propose an integrated approach for the management of contaminated ground water using a multicriteria decision framework to assess the risk to health and to make an economic analysis.

Another current context to be analyzed is in the newish field of nanotechnology which is increasingly being embedded in innovations that can benefit humanity (Siegrist et al. 2007). However, there is a variety of factors involved in managing the development of nanomaterial, ranging from the technical specifications of the material to possible adverse effects in humans. Therefore, it is important to assess the benefits and risks inherent in issues of Environmental Health and Safety (EHS) related to nanotechnology. According to Linkov et al. (2007), there is currently no structured approach for making justifiable and transparent decisions with explicit trade-offs among the many factors.

Linkov et al. (2007) conceptualize the use of the MCDM/A as a powerful analytical framework and scientifically sound decision tool for assessing and managing risk when using nanomaterial. They seek a balance between social benefits and unintended side effects and risks. They also investigate how to gather multiple lines of evidence to estimate the likely toxicity and risks of nanomaterial, given limited information on its physical and chemical properties. An essential contribution of MCDM/A, highlighted by the authors, is to link this information on performance with decision criteria and weightings triggered from scientists and managers, thus enabling the trade-offs involved in the decision making process to be visualized and quantified.

Luria and Aspinall (2003) use expert opinions, complementary skills and expertise from different disciplines in conjunction with quantitative traditional analysis, in an approach to major industrial hazard assessment, based on a multi-criteria approach (Analytic Hierarchy Process - AHP). According to these authors, this approach is in line with the main concepts proposed by the European directive on major hazard accidents, which recommends increasing the participation of operators, taking the other players into account and, moreover, paying more attention to the concepts of urban control, subjective risk (risk perception) and intangible factors.

References

Abella EAC, Van Westen CJ (2007) Generation of a landslide risk index map for Cuba using spatial multi-criteria evaluation. Landslides 4:311–325

Adey B, Hajdin R, Brühwiler E (2003) Risk-based approach to the determination of optimal interventions for bridges affected by multiple hazards. Eng Struct 25:903–912

Akgun I, Kandakoglu A, Ozok AF (2010) Fuzzy integrated vulnerability assessment model for critical facilities in combating the terrorism. Expert Syst Appl 37:3561–3573

Alencar MH, Cavalcante CAV, de Almeida AT, Silva Neto CE (2010) Priorities assignment for actions in a transport system based on a multicriteria decision model. In: Bris R, Soares CG, Martorell S (eds) European safety and reliability conference, Prague, September 2009. Reliability, Risk, and Safety: Theory and Applications, Vol. 1-3. 2009. Taylor and Francis, London, UK, p 2480

Alencar MH, de Almeida AT (2010) Assigning priorities to actions in a pipeline transporting hydrogen based on a multicriteria decision model. Int J Hydrogen Energy 35(8):3610–3619

Alencar MH, Krym EM, Marsaro MF, de Almeida AT (2014) Multidimensional risk evaluation in natural gas pipelines: Environmental aspects observed through a multicriteria decision model. In: Steenbergen RDJM, VanGelder PHAJM, Miraglia S, Vrouwenvelder ACWMT (eds) 22nd Annual Conference on European Safety and Reliability (ESREL), Amsterdam, 2013. Safety, Reliability and Risk Analysis: Beyond the Horizon. Taylor & Francis Group, London, UK, p 758

Almeida-Filho AT de, de Almeida AT (2010a) Cost-effectiveness analysis and multicriteria approaches: two irreplaceable paradigms for different problems in risk and safety problems. In: Proceedings of the European Safety and Reliability Annual Conference, Rhodes, 2010. Reliability, Risk and Safety: Back to the Future, p 2293

Almeida-Filho AT de, de Almeida AT (2010b) Multiple dimension risk evaluation framework. In: Bris R, Soares CG, Martorell S (eds) European Safety and Reliability Conference (ESREL), Prague, Czech Republic, 2009. Reliability, Risk and Safety: Theory and Applications. CRC Press-Taylor & Francis Group, p 1049

Andersson G, Donalek P, Farmer R, et al. (2005) Causes of the 2003 major grid blackouts in North America and Europe, and recommended means to improve system dynamic performance. Power Syst IEEE Trans 20:1922–1928

Apostolakis GE, Lemon DM (2005) A Screening Methodology for the Identification and Ranking of Infrastructure Vulnerabilities Due to Terrorism. Risk Anal 25:361–376

Arianoutsou M, Koukoulas S, Kazanis D (2011) Evaluating Post-Fire Forest Resilience Using GIS and Multi-Criteria Analysis: An Example from Cape Sounion National Park, Greece. Environ Manage 47:384–397

Arnaldos J, Casal J, Montiel H, et al. (1998) Design of a computer tool for the evaluation of the consequences of accidental natural gas releases in distribution pipes. J Loss Prev Process Ind 11:135–148

Atmaca E, Basar HB (2012) Evaluation of power plants in Turkey using Analytic Network Process (ANP). Energy 44:555–563

Aven T, Renn O (2009) The Role of Quantitative Risk Assessments for Characterizing Risk and Uncertainty and Delineating Appropriate Risk Management Options, with Special Emphasis on Terrorism Risk. Risk Anal 29:587–600

Beaudouin F (2015) Implementing a Multiple Criteria Model to Debate About Nuclear Power Plants Safety Choices. Gr Decis Negot 1–29

Beaudouin F, Munier B (2009) A revision of industrial risk management: Decisions and experimental tools in risk business. Risk Decis Anal 1:3–20

Bedford T, Cooke R (2001) Probabilistic Risk Analysis: Foundations and Methods. Cambridge University Press, New York

Berger JO (1985) Statistical decision theory and Bayesian analysis. Springer Science & Business Media, New York

Berizzi A (2004) The Italian 2003 blackout. Power Eng Soc Gen Meet 2004 IEEE 1673–1679 Vol. 2

Brito AJ, de Almeida AT (2009) Multi-attribute risk assessment for risk ranking of natural gas pipelines. Reliab Eng Syst Saf 94(2):187–198

Brito AJ, de Almeida AT, Mota CMM (2010) A multicriteria model for risk sorting of natural gas pipelines based on ELECTRE TRI integrating Utility Theory. Eur J Oper Res 200:812–821

Cailloux O, Mayag B, Meyer P, Mousseau V (2013) Operational tools to build a multicriteria territorial risk scale with multiple stakeholders. Reliab Eng Syst Saf 120:88–97

Catrinu MD, Nordgård DE (2011) Integrating risk analysis and multi-criteria decision support under uncertainty in electricity distribution system asset management. Reliab Eng Syst Saf 96:663–670

Comes T, Wijngaards N, Hiete M, et al (2011) A Distributed Scenario-Based Decision Support System for Robust Decision-Making in Complex Situations. Int J Inf Syst Cris Response Manag 3:17–35

Cox LA Jr (2009) Risk analysis of complex and uncertain systems. Springer Science & Business Media, New York

Cox LA Jr (2012) Evaluating and Improving Risk Formulas for Allocating Limited Budgets to Expensive Risk-Reduction Opportunities. Risk Anal 32(7):1244–1252

Crowl DA, Jo Y-D (2007) The hazards and risks of hydrogen. J Loss Prev Process Ind 20:158–164

de Almeida AT (2005) Multicriteria Modelling of Repair Contract Based on Utility and ELECTRE I Method with Dependability and Service Quality Criteria. Ann Oper Res 138:113–126

de Almeida AT (2007) Multicriteria decision model for outsourcing contracts selection based on utility function and ELECTRE method. Comput Oper Res 34(12):3569–3574

de Almeida AT, Ferreira RJP, Cavalcante CAV (2015) A review of multicriteria and multi-objective models in maintenance and reliability problems. IMA Journal of Management Mathematics 26(3):249–271

Dong M, Lou C, Wong C (2008) Adaptive Under-Frequency Load Shedding. Tsinghua Sci Technol 13:823–828

Dziubiński M, Frątczak M, Markowski AS (2006) Aspects of risk analysis associated with major failures of fuel pipelines. J Loss Prev Process Ind 19:399–408

Erol İ, Sencer S, Özmen A, Searcy C (2014) Fuzzy MCDM framework for locating a nuclear power plant in Turkey. Energy Policy 67:186–197

Ezell BC, Bennett SP, Von Winterfeldt D, et al. (2010) Probabilistic Risk Analysis and Terrorism Risk. Risk Anal 30(4):575–589

Faber MH, Stewart MG (2003) Risk assessment for civil engineering facilities: critical overview and discussion. Reliab Eng Syst Saf 80:173–184

Figueira J, Greco S, Ehrgott M (eds) (2005) Multiple Criteria Decision Analysis: State of the Art Surveys. Springer Verlag, Boston, Dordrecht, London

Garcez TV, de Almeida AT (2012) Multiple Dimension Manhole Explosion in an Underground Electrical Distribution System. In: Proceedings of the 11th International Probabilistic Safety Assessment and Management Conference and the Annual European Safety and Reliability Conference 2012. Curran Associates, Inc. Helsinki, Finland, p 4893–4899

Garcez TV, de Almeida AT (2014a) A risk measurement tool for an underground electricity distribution system considering the consequences and uncertainties of manhole events. Reliab Eng Syst Saf 124:68–80

Garcez TV, de Almeida AT (2014b) Multidimensional Risk Assessment of Manhole Events as a Decision Tool for Ranking the Vaults of an Underground Electricity Distribution System. Power Deliv IEEE Trans 29:624–632

Garcez TV, de Almeida AT (2014c) Multidimensional risk assessment of underground electricity distribution systems based on MAUT. In: Steenbergen RDJM, VanGelder PHAJM, Miraglia S, Vrouwenvelder ACWMT (eds) 22nd Annual Conference on European Safety and Reliability (ESREL), Amsterdam, 2013. Safety, Reliability and Risk Analysis: Beyond the Horizon. CRC Press-Taylor & Francis Group, p 2009

Garcez TV, de Almeida-Filho AT, de Almeida AT, Alencar MH (2010) Multicriteria risk analysis application in a distribution gas pipeline system in Sergipe. In: Bris R, Soares CG, Martorell S (eds) Reliability, risk and safety: theory and applications vols 1-3. European safety and reliability conference (ESREL 2009), Prague, September 2009. Taylor and Francis, 1043-1047

Geldermann J, Bertsch V, Treitz M, et al. (2009) Multi-criteria decision support and evaluation of strategies for nuclear remediation management. Omega 37:238–251

Greenberg M, Haas C, Cox LA Jr, et al. (2012) Ten Most Important Accomplishments in Risk Analysis, 1980–2010. Risk Anal 32(5):771–781

Guedes Soares CG, Teixeira AP (2001) Risk assessment in maritime transportation. Reliab Eng Syst Saf 74(3):299–309

Haidar AMA, Mohamed A, Hussain A (2010) Vulnerability control of large scale interconnected power system using neuro-fuzzy load shedding approach. Expert Syst Appl 37:3171–3176

Haphuriwat N, Bier VM (2011) Trade-offs between target hardening and overarching protection. Eur J Oper Res 213(1):320–328

Henselwood F, Phillips G (2006) A matrix-based risk assessment approach for addressing linear hazards such as pipelines. J Loss Prev Process Ind 19:433–441

Hobbs BF, Meier P (2000) Energy Decisions and the Environment. A guide to the use of multicriteria methods (International Series in Operations Research & Management Science). Kluwer Academic Publisher, Norwell

Hong S, Bradshaw CJA, Brook BW (2013) Evaluating options for the future energy mix of Japan after the Fukushima nuclear crisis. Energy Policy 56:418–424

IEEE1584 (2002) IEEE Guide for Performing Arc-Flash Hazard Calculations. IEEE Std 1584-2002

Jo Y-D, Ahn BJ (2002) Analysis of hazard areas associated with high-pressure natural-gas pipelines. 2 J Loss Prev Process Ind 15:179–188

Jo Y-D, Ahn BJ (2005) A method of quantitative risk assessment for transmission pipeline carrying natural gas. J Hazard Mater 123:1–12

Jo Y-D, Crowl DA (2008) Individual risk analysis of high-pressure natural gas pipelines. J Loss Prev Process Ind 21:589–595

Jozi A, Pouriyeh A (2011) Health-safety and environmental risk assessment of power plants using multi criteria decision making method. Chem Ind Chem Eng Q 17:437–449

Karvetski CW, Lambert JH, Keisler JM, Linkov I (2011) Integration of Decision Analysis and Scenario Planning for Coastal Engineering and Climate Change. Syst Man Cybern Part A Syst Humans, IEEE Trans 41(1):63–73

Keeney RL, Raiffa H (1976) Decisions with multiple objectives: Preferences and Value Trade-Offs. Wiley Series in Probability and Mathematical Statistics. Wiley and Sons, New York

Khadam IM, Kaluarachchi JJ (2003) Multi-criteria decision analysis with probabilistic risk assessment for the management of contaminated ground water. Environ Impact Assess Rev 23:683–721

Koonce AM, Apostolakis GE, Cook BK (2008) Bulk power risk analysis: Ranking infrastructure elements according to their risk significance. Int J Electr Power Energy Syst 30:169–183

Leung M, Lambert JH, Mosenthal A (2004) A Risk-Based Approach to Setting Priorities in Protecting Bridges Against Terrorist Attacks. Risk Anal 24(4):963–984

Levy J (2005) Multiple criteria decision making and decision support systems for flood risk management. Stoch Environ Res Risk Assess 19:438–447

Linares P (2002) Multiple criteria decision making and risk analysis as risk management tools for power systems planning. Power Syst IEEE Trans 17:895–900

Linkov I, Satterstrom FK, Kiker G, et al. (2006) From comparative risk assessment to multi-criteria decision analysis and adaptive management: Recent developments and applications. Environ Int 32:1072–1093

Linkov I, Satterstrom FK, Steevens J, et al. (2007) Multi-criteria decision analysis and environmental risk assessment for nanomaterials. J Nanoparticle Res 9:543–554

Lins PHC, de Almeida AT (2012) Multidimensional risk analysis of hydrogen pipelines. Int J Hydrogen Energy 37:13545–13554

Lopes YG, de Almeida AT, Alencar MH, Wolmer Filho LAF, Siqueira GBA (2010) A Decision Support System to Evaluate Gas Pipeline Risk in Multiple Dimensions. In: Bris R, Soares CG, Martorell S (eds) European Safety and Reliability Conference (ESREL), Prague, Czech Republic, 2009. Reliability, Risk and Safety: Theory and Applications. Crc Press-Taylor & Francis Group, p 1043

Luria P, Aspinall PA (2003) Evaluating a multi-criteria model for hazard assessment in urban design. The Porto Marghera case study. Environ Impact Assess Rev 23:625–653

Merad MM, Verdel T, Roy B, Kouniali S (2004) Use of multi-criteria decision-aids for risk zoning and management of large area subjected to mining-induced hazards. Tunn Undergr Sp Technol 19:125–138

Merrick JRW, Leclerc P (2014) Modeling Adversaries in Counterterrorism Decisions Using Prospect Theory. Risk Anal. doi: 10.1111/risa.12254

Montiel LV, Bickel JE (2014) A Generalized Sampling Approach for Multilinear Utility Functions Given Partial Preference Information. Decis Anal 11(3):147–170

Morgan MG, Florig HK, DeKay ML, Fischbeck P (2000) Categorizing Risks for Risk Ranking. Risk Anal 20:49–58

Mousseau V, Slowinski R (1998) Inferring an ELECTRE TRI Model from Assignment Examples. J Glob Optim 12:157–174

Nefeslioglu HA, Sezer EA, Gokceoglu C, Ayas Z (2013) A modified analytical hierarchy process (M-AHP) approach for decision support systems in natural hazard assessments. Comput Geosci 59:1–8

Nganje W, Bier V, Han H, Zack L (2008) Terrorist Threats to Food: Guidance for Establishing and Strengthening Prevention and Response Systems. Am J Agric Econ 90(5):1265–1271

Papamichail KN, French S (2013) 25 Years of MCDA in nuclear emergency management. IMA J. Manag. Math. pp 481–503

Parlak AI, Lambert JH, Guterbock TM, Clements JL (2012) Population behavioral scenarios influencing radiological disaster preparedness and planning. Accid Anal Prev 48:353–62
Parnell GS, Smith CM, Moxley FI (2010) Intelligent Adversary Risk Analysis: A Bioterrorism Risk Management Model. Risk Anal 30:32–48
Patterson SA, Apostolakis GE (2007) Identification of critical locations across multiple infrastructures for terrorist actions. Reliab Eng Syst Saf 92:1183–1203
Radeva A, Rudin C, Passonneau R, Isaac D (2009) Report Cards for Manholes: Eliciting Expert Feedback for a Learning Task. Mach Learn Appl 2009 ICMLA '09 Int Conf 719–724
Raiffa H (1968) Decision analysis: introductory lectures on choices under uncertainty. Addison-Wesley, London
Regős G (2013) Comparison of power plants' risks with multi criteria decision models. Cent Eur J Oper Res 21:845–865
Rezaian S, Jozi SA (2012) Health- Safety and Environmental Risk Assessment of Refineries Using of Multi Criteria Decision Making Method. APCBEE Procedia 3:235–238
Rogner H-H (2013) World outlook for nuclear power. Energy Strateg Rev 1:291–295
Roy B (1996) Multicriteria Methodology for Decision Aiding. Springer US
Rudin C, Passonneau R, Radeva A, et al. (2010) A process for predicting manhole events in Manhattan. Mach Learn 80:1–31
Rudin C, Passonneau RJ, Radeva A, et al. (2011) 21st-century data miners meet 19th-century electrical cables. Computer (Long Beach Calif) 44:103–105
Rudin C, Waltz D, Anderson RN, et al. (2012) Machine Learning for the New York City Power Grid. Pattern Anal Mach Intell IEEE Trans 34:328–345
Salvi O, Merad M, Rodrigues N (2005) Toward an integrative approach of the industrial risk management process in France. J Loss Prev Process Ind 18:414–422
Scawthorn C (2008) A Brief History of Seismic Risk Assessment. In: Bostrom A, French S, Gottlieb S (eds) Risk Assessment, Model. Decis. Support. Springer Berlin Heidelberg, Berlin, Heidelberg, pp 5–81
Shan X, Zhuang J (2014) Subsidizing to disrupt a terrorism supply chain-a four-player game. J Oper Res Soc 65(7):1108–1119
Shetty NK, Chubb MS, Halden D (1997) An overall risk-based assessment procedure for substandard bridges. In: Das PC (ed) Saf. Bridg. Telford, London, Uk, pp 225–235
Siegrist M, Keller C, Kastenholz H, et al. (2007) Laypeople's and Experts' Perception of Nanotechnology Hazards. Risk Anal 27:59–69
Sklavounos S, Rigas F (2006) Estimation of safety distances in the vicinity of fuel gas pipelines. J Loss Prev Process Ind 19:24–31
Sri Bhashyam S, Montibeller G (2012) Modeling State-Dependent Priorities of Malicious Agents. Decis Anal 9:172–185
Stefanidis S, Stathis D (2013) Assessment of flood hazard based on natural and anthropogenic factors using analytic hierarchy process (AHP). Nat Hazards 68:569–585
Tamura H, Yamamoto K, Tomiyama S, Hatono I (2000) Modeling and analysis of decision making problem for mitigating natural disaster risks. Eur J Oper Res 122:461–468
Topuz E, Talinli I, Aydin E (2011) Integration of environmental and human health risk assessment for industries using hazardous materials: A quantitative multi criteria approach for environmental decision makers. Environ Int 37:393–403
Tweeddale M (2003) Managing Risk and Reliability of Process Plants. Gulf Professional Publishing. Burlington
Vadrevu K, Eaturu A, Badarinath KVS (2010) Fire risk evaluation using multicriteria analysis—a case study. Environ Monit Assess 166:223–239
Van der Vleuten E, Lagendijk V (2010) Transnational infrastructure vulnerability: The historical shaping of the 2006 European "Blackout." Energy Policy 38:2042–2052
Vincke P (1992) Multicriteria Decision-Aid. John Wiley & Sons, New York
Viscusi WK (2009) Valuing risks of death from terrorism and natural disasters. J Risk Uncertain 38:191–213

Walsh BP, Black WZ (2005) Thermodynamic and mechanical analysis of short circuit events in an underground vault. Power Deliv IEEE Trans 20:2235–2240

Wang J (2006) Maritime Risk Assessment and its Current Status. Qual Reliab Eng Int 22(1):3–19

Wang Y-M, Liu J, Elhag TMS (2008) An integrated AHP–DEA methodology for bridge risk assessment. Comput Ind Eng 54:513–525

Willis HH, DeKay ML, Fischhoff B, Morgan MG (2005) Aggregate, Disaggregate, and Hybrid Analyses of Ecological Risk Perceptions. Risk Anal 25:405–428

Yoe C (2012) Principles of Risk Analysis – Decision Making under uncertainty. CRC Press, Boca Raton

Chapter 5
Preventive Maintenance Decisions

Abstract: Technological advances in equipment and the increase in process automation has led to the maintenance function having a role in business competitiveness. The contribution of preventive maintenance is discussed, as an important part of this function, with some emphasis on methods for planning replacement, in the sense of time interval of preventive maintenance. The classical optimization approach is used to illustrate the original preventive maintenance problem, thereby enabling insights and discussion of the main features that require the use of MCDM/A approaches for these decisions, and thus considering the multidimensional consequence space. A structured framework to build a multi-criteria decision model for supporting the selection of time interval is presented. Two different MCDM/A methods are applied depending on the decision maker's (DM) preferences. The first illustrates the application of Multi-attribute Utility Theory (MAUT) as an example of compensatory method and; the second details the application of a non-compensatory PROMETHEE method, which considers outranking relations.

5.1 Introduction

In the face of growing competition, leading to an ever increasing need for higher productivity, there is a need for methods, tools and technologies that enable the producing systems to acquire competitive advantages. Preventive maintenance decisions are quite relevant to the strategic results of any business organization, in which a producing systems has to make products, may them be goods or services.

The type of product makes a great difference in the way that maintenance in general (and preventive maintenance in particular) is linked to business results. For instance, a service producing system has a feature of simultaneousness (Slack et al. 2010), which means that at the time the system is producing the product, the customer is being served. In such a context, when a failure in the system occurs, the maintenance has an immediate impact on the business competitiveness (de Almeida and Souza 2001). Therefore, preventive maintenance planning becomes a more strategic decision that is linked to a higher level of the hierarchical organizational structure. For given decision context, the consequences are characterized by

A.T. de Almeida et al., *Multicriteria and Multiobjective Models for Risk, Reliability and Maintenance Decision Analysis*, International Series in Operations Research & Management Science 231, DOI 10.1007/978-3-319-17969-8_5

multiple and less tangible objectives, which may require an MCDM/A support. These issues are discussed in Chap. 1, presenting the peculiarities of two different types of systems, service and goods producing systems.

This chapter addresses preventive maintenance planning by the selection of the preventive maintenance time interval. This kind of decision is applied to a component or item (or device) and it is not applicable to a system, unless this system is replaced as whole, using the failure behavior of the system.

The next section presents a classical optimization approach, followed by a general MCDM/A preventive maintenance model. The last two sub-sections deal with two different MCDM/A approaches to support the preventive maintenance time interval.

It is important to note that there are many different mathematical models related to preventive maintenance (Shafiee and Finkelstein 2015), although similar process to build MCDM/A models can be applied considering the required adaptations.

5.2 A General MCDM/A Model for Preventive Maintenance

One of the most important problems in the maintenance area is the definition of the frequency at which preventive maintenance actions should be performed. In both producing systems, this decision has a great impact.

In a literature review on MCDM/A models in maintenance, around 22 % of research found is related to preventive maintenance (de Almeida et al. 2015). Multiobjective optimization in preventive maintenance has been considered since the late 1970. Inagaki et al. (1978) considered three objectives: mission reliability, total cost, system weight in a multiobjective nonlinear mixed-integer problem and proposed a procedure based on interactive optimization and a nonlinear programming algorithm, called ICOM (Interactive Coordinatewise Optimization Method). Hwang et al. (1979) formulated a scheduled-maintenance policy problem and set three objectives: minimum replacement cost-rate, maximum availability, and lower-bound on mission reliability. Four multicriteria methods were analyzed: strictest-selection; lexicographic; Waltz lexicographic, and the sequential multiple-objective problem-solving technique (SEMOPS). Jiang and Ji (2002) consider four attributes: cost, availability, reliability, and lifetime, via a multiple attribute value theory (MAVT).

Before the presentation of a general MCDM/A preventive maintenance model, a classical optimization approach is presented in the next subsection.

5.2.1 Classical Optimization Problem of Preventive Maintenance

Glasser (1969) presents age replacement and block replacement as two methods of planning replacement in a program of preventive maintenance. Although these methods have been previously described by other authors (Barlow and Hunter 1960; Cox 1962), the main contribution of Glasser (1969) is the focus on the managerial impacts of these methods. These issues have been presented in many other texts in the literature (Scarf et al. 2005).

According to Glasser (1969), the main problem of preventive maintenance is associated with the uncertainty about the exact time at which an item will fail. This uncertainty establishes a difficulty in guarantee the effectiveness of the replacement, which in some times could happen earlier than failure, in others only after a failure takes place.

Glasser (1969) structures a two-phase process to model the problem of replacement planning. The phases consist of: 1) the description of the pattern of failures of the item over time, in terms of a probability density function *f(t)*; and 2) the development of an equation that describes the expected cost per time of following a particular policy of planned replacement.

A general description of this model is given by (5.1).

$$cr(t) = \frac{E(c(t))}{E(v(t))} \tag{5.1}$$

where:
cr(t) is the cost rate;
E(c(t)) is the expected cost;
E(v(t)) is the expected cycle length.

The final expression of *cr(t)* depends on the assumptions of the model, which may be related to the influence of the action on the system.

A great number of papers deal with these different aspects, almost all following the general structure of Glasser (1969). Although these models have a great potential to support the maintenance manager, they may not be sufficient to describe the consequence space of failures. So, the rate cost as a criterion should be considered together with other criteria. This is described in a general framework that could be used to address the problem via the MCDM/A approach, given in the next sub-section.

The assumptions related to the simplest case of the replacement model (Cox 1962) are valid for the MCDM/A models in the subsequent sub-sections. The assumptions of the simplest replacement age-based model are (Cox 1962):

1. The state of the item is known;
2. The alternatives set is defined as opportunistic intervals, which can be days, weeks, months, or other period;

3. The failure probability density function *f(t)* of the item is IFR (an increasing failure rate);
4. The system can only be in one of two states, failed or operational;
5. Replacement prior a failure is worthwhile, there are saving in avoiding a failure by doing preventive replacements;
6. The item replacement restores the system to the as good as new state;
7. The equipment failure times can be modeled by a known probability density function *f(t)*;
8. The time necessary to perform a replacement is negligible compared to the time between failures, so it is not considered into a cycle.

From these assumptions (5.1) becomes (5.2).

$$cr(t) = \frac{c_a(1 - R(t)) + c_b R(t)}{\int_0^t x f(x)dx + t(R(t))} \tag{5.2}$$

where
$cr(t)$ is the cost rate;
$R(t)$ is the reliability function;
$f(t)$ is the density probability function;
c_a is the cost replacement after a failure;
c_b is the cost replacement before a failure.

As already stated, most of the models restrict the analysis to only the cost rate criterion $cr(t)$. Therefore, it is important to understand the behavior of this aspect.

To illustrate the different behaviors of the cost rate, a Weibull function is assumed to *f(t)*. Also, different values for its parameters may be applied to give an idea of how the cost rate may change with different failure data patterns.

Assuming a Weibull distribution in (5.2), (5.3) can be obtained.

$$cr(t) = \frac{c_a\left(1 - e^{\left[-\left(\frac{t}{\eta}\right)\right]^{\beta}}\right) + c_b e^{\left[-\left(\frac{t}{\eta}\right)\right]^{\beta}}}{\int_0^t x\frac{\beta}{\eta}\left[\frac{x}{\eta}\right]^{\beta-1} e^{\left[-\left(\frac{x}{\eta}\right)\right]^{\beta}} dx + t e^{\left[-\left(\frac{t}{\eta}\right)\right]^{\beta}}} \tag{5.3}$$

It is clear that preventive maintenance is only effective if the *f(t)* function is IFR. In practice, this means that a time-based preventive maintenance action, is effective only if the failure mechanism is associated with time.

Some behaviors of the cost rate function (5.3) for different values of parameter β, of the Weibull density function $f(t)$ are shown in Fig. 5.1. This parameter is associated with the intensity at which the failure rate function increases.

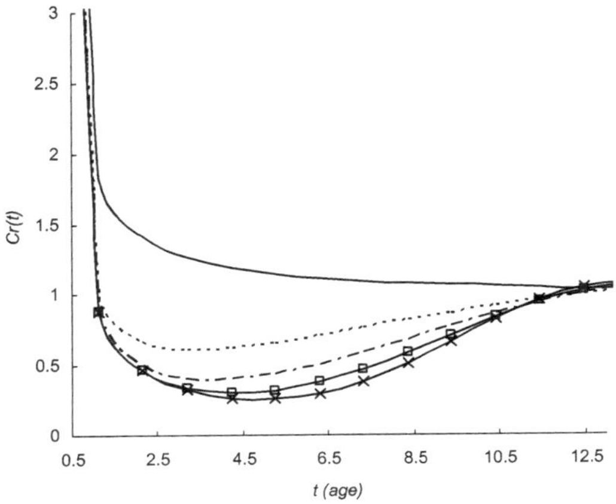

Fig. 5.1 The cost rate function for c_a=10, c_b=1, η =10 and different values of β: β =1 (—); β=2 (·······); β=3 (- · -); β=4 (- □ -); β=5 (- × -)

In a particular case, when $\beta = 1$, the Weibull distribution corresponds to an exponential distribution. In this case, Fig. 5.1 shows that there is not optimum point for age replacement. In other words, a preventive maintenance plan is not indicated, and the replacement only should happen when the item fails.

Another possibility is that the cost rate function presents a flat curve, as for small values of β. Although there are advantages in doing preventive maintenance at the optimum point, there is not a great difference in terms of cost, versus when this action is taken at points other than the optimum.

Variations in costs (c_b and c_a) affect the time (t^*) of the minimum cost, as shown in Fig. 5.2. The greater the ratio c_a/c_b, the smaller is the time t^*. This is exactly what is necessary to guide the activities of maintenance manager. To avoid failures, the management guideline should mandate conducting preventive maintenance actions more often as the cost ratio increases.

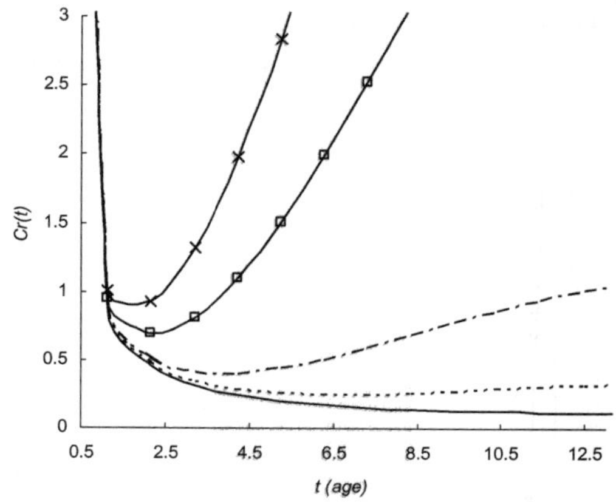

Fig. 5.2 The cost rate function for c_b=1, η =10 and different values of c_a: c_a=1(—); c_a=3 (⋯⋯);c_a=10 (-˙-); c_a=50 (-□-); c_a=100 (-×-)

It is important to note that in some situations the cost rate does not provide information about the best time to carry out preventive maintenance because the different alternatives (t ages) have almost the same evaluation in terms of cost rate. Therefore, for the purpose of making a decision, this aspect does not help the DM. It should not be considered, for example, in the curve for c_a=3 (⋯⋯) when considering $t > 4.5$.

Alternatives with almost the same evaluation according to one specific criterion could have very different evaluations in terms of others. That is why it is essential to make sure that the DM has as broad a view as possible, to make consistent decisions.

In the next sub-section, other criteria are introduced into the decision problem. This includes one step from the MCMD/A framework to build the multicriteria decision problem to support the selection of the preventive maintenance interval.

5.2.2 MCDM/A Framework for the General Model for Preventive Maintenance

This sub-section is organized in accordance with the decision structure presented in Chap. 2. For the sake of clarity some of the steps discussed in detail in that chapter are omitted or superficially considered in this section.

Identifying Objectives and Criteria

As stated in Chap. 2, this is one of the most important steps, as the objectives influence every step in the decision process. In the preventive maintenance context, the cost is only part of the maintenance objective. As discussed in Chap. 3, the main objectives of the maintenance function are: to extend the useful life of assets, to ensure satisfactory levels of availability, to ensure operational readiness of systems, and to safeguard the people who use the facilities. These objectives are pursued by the maintenance function as a whole.

It is not necessary to emphasize that for service producing systems, the system availability is even more important. When failures lead to interruptions of these systems, they are easily perceived by the customer. Thus, an increase in the availability may increase the level of user satisfaction. The downtime provides an indirect measure of this objective. The availability is also related to the reliability, the capability of the system to work without interruption. The behavior of the reliability function is shown in Fig. 5.3.

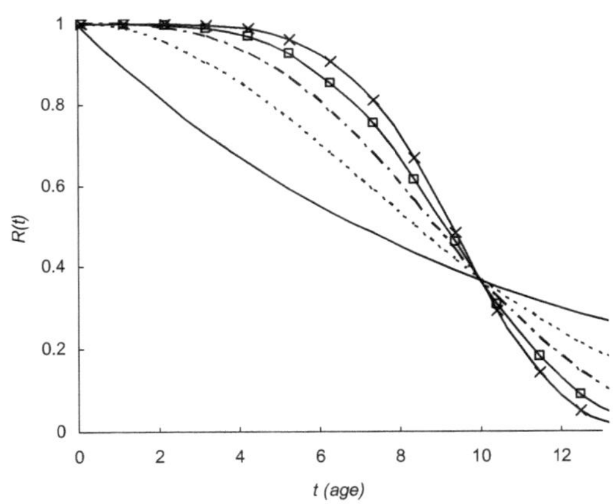

Fig. 5.3 Reliability function for, η =10 and different values of β: β=*1* (—)*;* β=*2* (…….)*;* β=*3* (- —)*;* β=*4* (_ □ _)*;* β=*5* (_ × _)

Sometimes, the reliability is used as a constraint. However, it may be useful to distinguish the alternatives even beyond the constraint level. The DM's preference structure with respect to this aspect should also be considered to be reflected in the MCDM/A results.

Availability, cost rate, downtime, mean-time between operational failures are possible criteria related to the decision context in preventive maintenance. In a recent literature review on MCDM/A models in maintenance, several criteria are described as having been considered in previous works, including those discussed above (de Almeida et al. 2015).

Establishing a Set of Actions and a Problematic

As presented in Chap. 2, this step addresses four topics: a) establishing the structure of the set of alternatives, b) establishing the problematic to be applied to this set, c) the generation of alternatives; and d) establishing the matrix of consequences. In preventive maintenance, some of these topics are either not necessary or straightforwardly defined. In this decision problem, the solution is related to the time interval for preventive maintenance, and therefore the problematic is a straightforward choice. The generation of alternatives also need not be considered. Therefore, only two topics need to be discussed, the structure of the set of alternatives and the matrix of consequences.

The kind of the set of alternatives may completely change the MCDM/A methods to be applied. For the selection of the preventive maintenance interval, the two kinds of sets of alternatives (discrete or continuous) require different methods. As already mentioned, a set of alternatives consists of the different possible intervals of time at which the maintenance activities within a maintenance policy could be performed.

This problem is associated with the classic optimization problem, in which the set of alternatives is already well defined and consists of a continuous set of time interval for preventive maintenance t. This time interval t may be seen as days in a calendar, such that the set of alternatives becomes discrete: $A = \{d_1, d_2, d_3, \ldots, d_n\}$. This model is more realistic because there is no need to use a continuous time t that includes any time by day or night. Making a choice of day d_i is a reasonable approximation for the context of preventive maintenance because a variation of 24 hours does not have a relevant difference in the consequences related to the decision problem, as shown in Fig. 5.1.

At this stage, with the criteria and the set of alternatives established, the matrix of consequences can be built, collecting the necessary cost data and other relevant data associated with the criteria. The construction of this matrix, for this particular problem, is somewhat straightforward.

Identifying State of Nature

As stated in Chap. 2 the state of nature (θ) corresponds to aspects that could not be controlled by the DM and influence the outcome. In fact, they may change randomly, and consequently, may deeply influence the consequences of the decision process. The modeling process for this ingredient uses decision theory, which includes MAUT.

A typical θ is the reliability (de Almeida and Souza, 2001), which influences the outcomes, such as the availability, a usual criterion in the preventive maintenance decision problem. That is, the reliability is not a consequence, although it may be considered as such, as a simplification of the model (Cavalcante and de Almeida 2007; de Almeida 2012).

Similar to the set of alternatives, the set of states of nature may be discrete or continuous and incorporate prior probabilities $\pi(\theta)$ on θ.

In the maintenance preventive problem, as the data on failure are scarce, the probability density function that models the time to failure may be completely unknown or otherwise have undetermined parameters. Thus, if prior probabilities $\pi(\theta)$ are incorporated, a probabilistic modeling task complements the preference modeling.

Preference Modeling

This step provides information for choosing the MCDM/A method, aligned with the DM's preference structure, which may consider, among other factors, compensatory or non-compensatory rationality. The main question of this factor considers which of these classes of methods would be more appropriate for a particular problem. This process could use the model building procedure in Chap. 2. The analysis of the DM's rationality is essential to ensure that the results from the MCDM/A model truly reflect the DM's preferences.

In the next two sections, applications illustrate the use of compensatory and non-compensatory methods to support the problem of selecting intervals of preventive maintenance.

Intra-Criterion Evaluation

For a specific problem, this step consists of the elicitation of the value function $v_j(x)$, or utility function $u_j(x)$, related to the values of different performances of outcomes of criterion j, for any $j = 1, 2, \ldots, n$.

For a non-compensatory method, an ordinal scale is enough, so the intra-criterion evaluation is easily quantified. Furthermore, for an outranking method, the parameters related to indifference, preference and the discordance threshold may be addressed.

For the compensatory methods, the usual results consist of an overall value for each alternative that reflects a synthesis of all the criteria for that alternative. This overall value arises from the aggregation of the utility functions related to each criterion $u_j(x)$. The assessment of the $u_j(x)$ relies on the elicitation procedure. The utility function reflects the preferential structure from the DM for uncertainty contexts, considering his behavior with respect to risk. The DM could be risk neutral, averse and prone. Each of these standard behaviors is reflected by a specific form of the utility function $u_j(x)$.

Inter-Criteria Evaluation

The inter-criteria evaluation is a fundamental step in the MCDM/A problem. The inter-relationship among criteria is what distinguishes the results from any other approaches even when multiple aspects are considered, such as availability and cost. The essence of the MCDM/A approach is how the conflicts between availability and cost criteria, for instance, are reflected in the preference domain.

The inter-criteria evaluation includes the process of defining the criteria weights by means of an elicitation procedure. This process and the meaning of the criteria weights depend on the type of method.

Evaluating Alternatives and Sensitivity Analysis

For the preventive maintenance interval selection, the alternative evaluation results in the time interval to be applied.

To evaluate how the results provided by the model vary with the parameters and whether the assumed simplifications affect the results, a sensitivity analysis is essential.

This step provides further insight to the DM. Some non-obvious behaviors may be identified during this process providing the DM with the broad view that is needed for a consistent decision.

Elaborating Recommendation

Given the insights and the view achieved by the application of an MCDM/A method, a complete report should include the essentials of the whole decision process, as well as, the main aspects that came up during this process. It should provide any detail that would be requested during the explanation of the results and the recommended decisions.

Assumptions, simplifications and changes to the original problem should be explicit and clear, to aid in transmitting an understanding of the results of the model and their limitations.

5.3 Compensatory MCDM/A Model for Preventive Maintenance

A compensatory method deals with the DM's preference structure by means of a tradeoff amongst criteria, with features that where discussed in Chap. 2. In this section, a compensatory method is applied to illustrate an MCDM/A model for selecting preventive maintenance intervals. This model is based on MAUT (de Almeida and Souza 2001; de Almeida 2012), and illustrates a real study in an electric power company.

One insight of this model is the analysis of the consequence space for the preventive maintenance decision problem. When this consequence space cannot be reduced to only one dimension, the classical optimization approach is not useful. Additionally, when the consequences are multidimensional, the DM's preference for each criterion has to be treated very carefully because any misconception or mistake in this process may waste the effort to bring the DM to the center of the problem.

5.3.1 The Context, the Set of Alternatives and the Criteria

The context of this problem is an electric power company and it considers the cost rate and reliability criteria (de Almeida 2012). The underlying model that this application takes as its base is the model of age based replacement, so all assumptions and expressions that were presented before are valid in this application.

The set of alternatives corresponds to a discrete set of time intervals. For instance, months or days may be applied as usual intervals. For a month interval any alternative is a multiple of 30 days, so that an element of the set of alternatives could be represented by *30i*, where *i* is any positive integer from 1 to *N* and *N* is the number of alternatives. A quantile of the probability distribution of the time interval could be used to choose *N*, although it is not explicit in the model.

It is worth noting that for the age replacement based, whenever a first failure happens, the timing counting should be restarted. This means that the planning of the preventive maintenance actions should be performed very carefully, because the calendar time is not useful to help the manager schedule a particular preventive maintenance action, as once the schedule is put into action, a failure can force the schedule to be rearranged. In this way, the use of the base time does not mean that the action will necessarily happen each *30i* days, but rather it means that *30i* is the maximum number of days that a specific item will run until it is replaced by another. The calendar logic is valid for the block replacement based policy, in which it is not necessary to keep a register of times to failure.

As already stated the criteria are the cost rate *cr(t)* and reliability *R(t)*, and their parameters are presented in Table 5.1.

Table 5.1 Data of the cost and reliability functions

Weibull	β	3
	η	1200
Costs	Replacement Cost C_b	600
	Failure Cost C_a	1200

It is possible to build a consequence matrix applying the models for *cr(t)* and *R(t)*, as shown in Fig. 5.4.

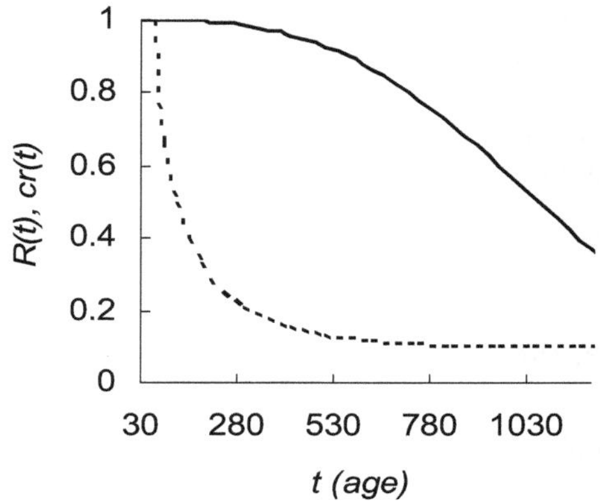

Fig. 5.4 The criteria in function of *t*: *R(t)* (—) and *cr(t)* (······)

5.3.2 Preference Modeling and Intra-Criteria and Inter-Criteria Evaluations

For the intra-criteria evaluation, a logistic function was found for the reliability attribute *U(R)* and an exponential function for the cost *U(cr)* attribute. The logistic utility function for reliability shows that the DM considers the variation at $R > 0.9$ to be small, and views only changes at $R < 0.8$ to be important. However, the higher the cost is, the less the utility is, which reveals the risk averse behavior of the DM.

With regard to the inter-criteria evaluation, the elicitation process (Keeney and Raiffa 1976) includes the validation of some axioms about the DM's preferential structure. The mutual utility independence between the two attributes was confirmed, and a multilinear utility function is therefore applied, as given by (5.4).

$$U(cr, R) = K_1 U(cr) + K_2 U(R) + K_3 U(cr) U(R) \tag{5.4}$$

where:
U(cr) is the utility function for the cost rate criterion;
U(R) is the utility function for the reliability criterion;
U(cr,R) is the multiattribute utility function;
K_1, K_2 and K_3 are the scale constants with $K_1+K_2+K_3 = 1$.

Following the elicitation procedure, the values obtained for these scale constants were $K_1 = 0.35$, $K_2 = 0.45$, and $K_3 = 0.20$.

5.3.3 Results and Discussion

The results show the highest value for the overall utility function for $t = 600$ days, corresponding to 20 months. Applying the classical optimization model, which considers only the cost rate criterion, the time of the minimum cost rate (t^*) is $t=780$, which corresponds to 26 months.

The reliability for the time of the minimum cost rate (t^*), $R(780)$ is around 0.75. Therefore, it may be too risky to follow the policy that minimizes the cost rate, because the probability of a failure of this item may be considered too high to risk interrupting the supply of electricity. Consequently, the reliability for this item should not be neglected.

For a large time interval [540, 660] the overall utility function varies over a range of less than 0.009. In practical terms, this means that the DM has flexibility regarding the time interval for preventive maintenance without a considerable decrease in the overall utility value.

Another interesting insight can be recognized when analyzing the change on the utility form of the cost criterion. This analysis shows that when a linear function is used to model the cost rate instead of an exponential function, the overall utility is affected and has its highest value at $t^*=360$ days. In this case, the DM is not averse to risk to slightly increase the cost, so the smaller time intervals that were judged unfavorable when using the exponential functions have improved results using a linear utility function.

The DM may view one of the characteristics of the compensatory method not suitable. In this approach, alternatives with very poor performance in some criteria can compensate by good performance in other criteria, and this could happen in unlimited way. This feature does not apply to the non-compensatory methods. However, this would not be a reason to change the approach, which should be based only on the DM's compensatory rationality. Another way to address this issue is using the compensatory method with veto (de Almeida 2013).

5.4 A Non-Compensatory MCDM/A Model for Preventive Maintenance

Among non-compensatory approaches, the outranking methods are the main group of methods following this rationality. In these methods, an outranking relation is built by a pairwise comparison between alternatives, and incomparability may be considered. The methods of the PROMETHEE family have been applied in this case (Chareonsuk et al. 1997; Cavalcante and de Almeida 2007; Cavalcante et al. 2010).

Two applications are presented at this section. There are some similarities to previous models thus some steps of the modeling process are omitted.

5.4.1 First Application

The criteria considered were cost rate and reliability, as in the previous study. Let the set of alternatives be $A=\{t_i\}$, where $t_i= 720\ i$, for $i = 1 \ldots 12$.

The parameters for the criteria are given in Table 5.2.

Table 5.2 Cost and reliability data

Weibull	β	1.4
	η	1800
Costs	Replacement Cost C_b	300
	Failure Cost C_a	1800

The consequence matrix is given in Table 5.3.

Table 5.3 Consequence matrix for the decision problem

Alternatives	T	$R(t)$	$C_m(t)$
T1	720	0.9638	0.2526
T2	1440	0.9072	0.1633
T3	2160	0.8421	0.1401
T4	2880	0.7733	0.1327
T5	3600	0.7038	0.1315
T6	4320	0.6354	0.1333
T7	5760	0.5075	0.1413
T8	7200	0.3957	0.1522
T9	7920	0.3466	0.1582
T10	8640	0.3022	0.1644

The intra-criterion evaluation, for the PROMETHEE method, as described in Chap. 2, produces the preference functions $P_j(a,b)$, which leads to $\Pi(a,b)$. Because in this section $\pi(\theta)$ represents the prior probability function, the notation $\Pi(a,b)$ is used for the preference index, although in Chap. 2, as in the general literature it is represented by $\pi(a,b)$.

$\Pi(a,b)$ is based on $P_j(a,b)$, as shown in (5.5).

$$\begin{Bmatrix} \Pi(a,b) = \sum_{j=1}^{N} P_j(a,b) w_j \\ \Pi(b,a) = \sum_{j=1}^{N} P_j(b,a) w_j \end{Bmatrix} \tag{5.5}$$

The weight for a criterion j (w_j) has to be established for each criterion based on the DM's preferences.

The scores of the alternatives are based on the outcome and income flows, as shown in Chap. 2, and recalled in (5.6) and (5.7).

$$\phi^{+}(a)=\frac{1}{n-1}\sum_{x\in A}\Pi(a,x)=\frac{1}{n-1}\sum_{x\in A}\sum_{j=1}^{k}P_j(a,x)w_j=\sum_{j=1}^{k}\phi_j^{+}(a)w_j \tag{5.6}$$

$$\phi^{-}(a)=\frac{1}{n-1}\sum_{x\in A}\Pi(x,a)=\frac{1}{n-1}\sum_{x\in A}\sum_{j=1}^{k}P_j(x,a)w_j=\sum_{j=1}^{k}\phi_j^{-}(a)w_j \tag{5.7}$$

The weights and intra-criterion parameters are given in Table 5.4.

Table 5.4 Preference function and criteria characteristics

Characteristics	R	C_m
Max/Min	Max	Min
Weight	0.34	0.66
Preference function	Type V	Type V
Indifference threshold	0.001	0.00062
Preference threshold	0.07	0.032

The next step consists of building the outranking relations, as shown in Table 5.5.

Table 5.5 Alternatives flows

Alternatives	T	ϕ^{+}	ϕ^{-}
T1	720	0.242142	0.660000
T2	1440	0.307792	0.318536
T3	2160	0.444011	0.022754
T4	2880	0.514990	0.037727
T5	3600	0.505728	0.092916
T6	4320	0.459229	0.164855
T7	5760	0.323753	0.248657
T8	7200	0.180039	0.409705
T9	7920	0.111208	0.533040
T10	8640	0.073333	0.674034

The PROMETHEE I method is applied as given by (5.8), in which P^{I}, I^{I}, and R^{I} correspond to preference, indifference and incomparability, respectively.

$$aP^I b \Leftrightarrow \begin{cases} aS^+b \text{ e } aS^-b \\ aS^+b \text{ e } aI^-b \\ aI^+b \text{ e } aS^-b \end{cases}$$

$$aI^I b \Leftrightarrow aI^+b \text{ e } aI^-b \tag{5.8}$$

$$aR^I b \quad \text{on the other cases}$$

The best alternatives have been found to be T3 and T4, which correspond to replacing the components every 2160 or 2880 hours, respectively. These alternatives are not comparable, indicating that the DM must reflect further when choosing between them, given that there is not sufficient information or reason, through the comparison, to have a particular preference for one or the other.

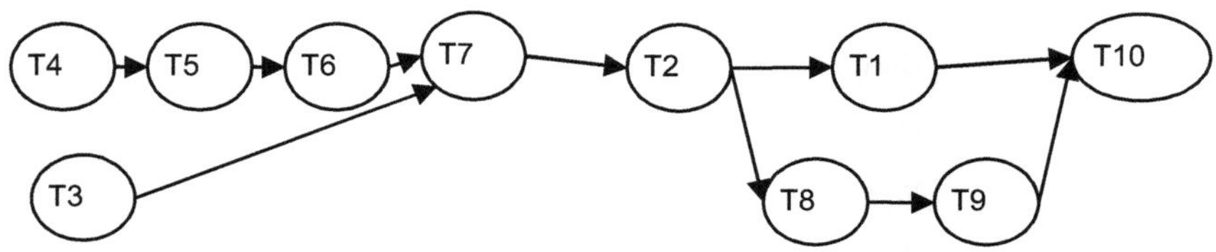

Fig. 5.5 Partial pre-ranking among the alternatives for actions

As explained in Chap. 2, the PROMETHEE II method provides a complete pre-order, forcing a comparison between T3 and T4. It should be noticed that the result of using PROMETHEE I is more informative than that of using PROMETHEE II because the incomparability is known. Fig. 5.6 shows the result of PROMETHEE II.

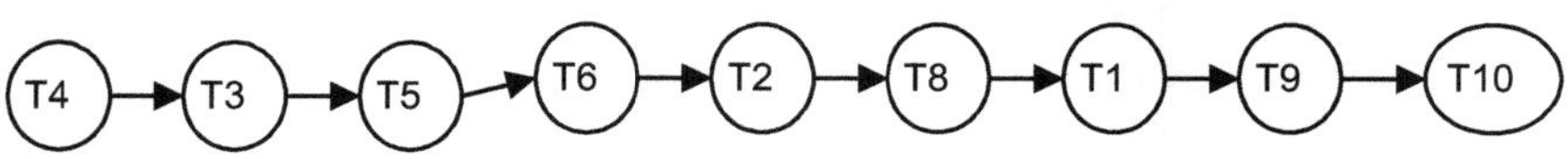

Fig. 5.6 Complete pre-ranking among the alternatives for action

5.4.2 Second Application

This model takes different assumptions than the previous application, as following (Cavalcante et al. 2010):

- The time spent on maintenance actions, whether preventive replacement, or corrective replacement, is non-negligible and known;
- The distribution of the time to failure is known, but its parameters are not.

With these two basic changes the expressions for $R(t)$ and $cr(t)$ are different. Despite this increasing in complexity, this model is realistic because it is common the absence of time to failure data, which makes the parameters of the distribution of time unknown.

The alternatives that are considered are times multiples of 100; inside the interval [200, 3000]. These time intervals in units of days.

The data for this problem are given in Table 5.6.

Table 5.6 Preference function and criteria data

Weibull	$\pi(\beta)$	$\beta_1 = 3.4$
		$\eta_1 = 4.5$
	$\pi(\eta)$	$\beta_2 = 2.8$
		$\eta_2 = 2200$
Costs	Replacement Cost C_b	*$250*
	Failure Cost C_a	*$1000*
Times	Preventive replacement	*0.5* day
	Corrective replacement	*3* days

The preference function for the criteria are both type V, as presented in Chap. 2, which corresponds to the case in which the difference in performance increases as the difference of evaluation in a criterion increase in a linear relationship. In addition there are two thresholds: the indifference and preference threshold.

Different values of weight are used to give to the DM more information about the sensitivity of these values. From the variations applied the first solution indicated by the PROMETHEE II rank changed from 600 days to 800 days. The sensitivity analysis provides the DM with more information, indicating the level of variation that is expected (Cavalcante et al. 2010).

References

Barlow R, Hunter L (1960) Optimum Preventive Maintenance Policies. Oper Res 8:90–100

Cavalcante CAV, Almeida AT de (2007) A multi-criteria decision-aiding model using PROMETHEE III for preventive maintenance planning under uncertain conditions. J Qual Maint Eng 13:385–397

Cavalcante CAV, Ferreira RJP, de Almeida AT (2010) A preventive maintenance decision model based on multicriteria method PROMETHEE II integrated with Bayesian approach. IMA J Manag Math 21:333–348

Chareonsuk C, Nagarur N, Tabucanon MT (1997) A multicriteria approach to the selection of preventive maintenance intervals. Int J Prod Econ 49:55–64

Cox DR (1962) Renewal theory, vol 4. Methuen & Co, London

de Almeida AT (2012) Multicriteria Model for Selection of Preventive Maintenance Intervals. Qual Reliab Eng Int 28:585–593 de Almeida AT (2013) Additive-veto models for choice and ranking multicriteria decision problems. Asia-Pacific J Oper Res 30(6):1-20

de Almeida AT, Ferreira RJP, Cavalcante CAV (2015) A review of multicriteria and multi-objective models in maintenance and reliability problems. IMA Journal of Management Mathematics 26(3):249–271

de Almeida AT, Souza FMC (2001) Gestão da Manutenção: na Direção da Competitividade (Maintenance Management: Toward Competitiveness) Editora Universitária da UFPE, Recife

Glasser GJ (1969) Planned replacement- Some theory and its application (Probability theory applied to age and block replacement models in preventive maintenance of parts, noting inspection cost distribution). J Qual Technol 1:110–119.

Hwang CL, Tillman FA, Wei WK, Lie CH (1979) Optimal Scheduled-Maintenance Policy Based on Multiple-Criteria Decision-Making. Reliab IEEE Trans R-28:394–399

Inagaki T, Inoue K, Akashi H (1978) Interactive Optimization of System Reliability Under Multiple Objectives. Reliab IEEE Trans R-27:264–267

Jiang R, Ji P (2002) Age replacement policy: a multi-attribute value model. Reliab Eng Syst Saf 76(3): 311-318

Keeney RL, Raiffa H (1976) Decisions with multiple objectives: Preferences and Value Trade-Offs. Wiley Series in Probability and Mathematical Statistics. Wiley and Sons, New York

Scarf PA, Dwight R, Al-Musrati A (2005) On reliability criteria and the implied cost of failure for a maintained component. Reliab Eng Syst Saf 89:199–207

Shafiee M, Finkelstein M (2015) An optimal age-based group maintenance policy for multi-unit degrading systems. Reliab Eng Syst Saf 134:230-238

Slack N, Chambers S, Johnston R (2010) Operations management, 6th ed. Pearson Education, Harlow

Chapter 6
Decision Making in Condition-Based Maintenance

Abstract: Predictive maintenance modeling is a tool that can provide many benefits to the maintenance management area. This chapter suggests a multi-criteria approach for modeling condition based maintenance decisions. Whereas the Predictive maintenance uses condition information of assets, the preventive maintenance is time-based only. This chapter discusses decision making in condition-based maintenance (CBM) and presents some useful approaches for building multiple objective models in this context. Initially, a summary is given of the fundamentals of CBM and several concepts of monitoring and inspection activities. Basic concepts of delay time are presented and discussed within an MCDM/A approach. Then, a structure of a multicriteria model to determine inspection intervals of condition monitoring based on Multi-attribute utility theory (MAUT) is introduced. The aspects of preference modeling, scale constant elicitation, utility theory and a DM's behavior to risk (prone, neutral, averse) are included in the decision model. An illustrative example of an MCDM/A model is given in the context of an electric power distribution system. Thus, the multicriteria model presented, which is based on MAUT and which has an axiomatic structure, aims to answer the needs identified and enables the tradeoff among the costs, downtime and frequency of breakdowns to be dealt with.

6.1 Introduction

Equipment manufacturers often state that periodic inspection and preventive maintenance activities must be done in accordance with certain recommendations so that the warranty remains valid and the equipment operates properly. Maintenance teams in various industries have to follow predetermined schedules for most of their maintenance activities. The advantage of these programs is that they are simple. However, information on the condition of the equipment in operation is not considered with regard to modifying the maintenance schedule. Thus, some facilities are not maintained optimally. Consequently, maintenance policies that are based on condition rather than the age of the equipment have been developed with the aim of improving the efficiency of maintenance actions and of extending the useful life of the assets.

A.T. de Almeida et al., *Multicriteria and Multiobjective Models for Risk, Reliability and Maintenance Decision Analysis*, International Series in Operations Research & Management Science 231, DOI 10.1007/978-3-319-17969-8_6

Nowlan and Heap (1978) stated that an item is said to be maintained by condition monitoring if it is permitted to remain in service without preventive maintenance until a functional failure occurs. Condition-Based Maintenance (CBM) is considered a broader subject and includes condition monitoring. CBM is stated by Jardine et al. (2006) as being a maintenance program that recommends maintenance actions based on the information collected using condition monitoring. Besides, CBM attempts to avoid unnecessary maintenance tasks by taking maintenance actions, only when there is evidence of abnormal behaviors of a physical asset. This topic was found in 6.5% of publications considering MCDM/A approaches in a literature review (de Almeida et al. 2015).

Wang and Gao (2006) indicate that there is an increased need for more effective and efficient techniques that monitor machine conditions in real time, detect the inception and progression of defects, and enable flexible maintenance scheduling, before a defect results in unexpected machine downtime. Jardine et al. (2006) state that diagnostics and prognostics are two important aspects in a CBM program. While diagnostics deal with the detection, isolation and identification of defects when they occur, prognostics deal with the prediction of defect before it occurs.

Diagnostic techniques are tools that have an increasing applicability in companies due to the potential financial returns, especially when compared with the policies of corrective and preventive maintenance. Included among the main techniques are procedures for measuring or analyzing: Vibration, Acoustic emission, Oil analysis, Ultrasonic, Thermography, Temperature, Speed, Performance, Corrosion, Output power, Pressure, Electric Current. Guidelines on collecting and analyzing condition monitoring data can be found in ISO 17359 (2011).

The diagnostics problem can be described as a hypothesis test problem, since the state of the system is unknown. An efficient diagnosis is one that has a minimal error rate. As in hypothesis testing, there may be two types of errors in diagnosis. It can mistakenly claimed that a product is defective when in fact it is not, and this type of error is also called a false negative. On the other hand, it can be said that a product is not defective, when in fact there is a bug also called false positive. Berrade et al. (2012) presented two models in order to deal with error in the inspection process such as false positives (false alarms) and false negatives of protection systems.

Martin (1994) distinguishes between hard and soft faults. These faults are shown in Fig. 6.1. The soft fault leads to a predictable situation; it lends itself to condition monitoring while the hard fault takes place instantaneously. He states that predictive maintenance involves periodic monitoring on the health of the machine and scheduling maintenance only when a functional failure is detected. This allows for trends of the machine component to be constructed and time to failure to be estimated.

Prognostic techniques aim to estimate the residual life of to a piece of equipment, taking condition monitoring into consideration. The concept of prognostics goes beyond diagnostics. First the problem is detected; then a diagnosis is made about the failure mode and its severity. It is also important to predict the evolution of the failure in order to estimate the remaining useful life of the machine (Ben-Daya et al. 2000; Do Van and Bérenguer 2012).

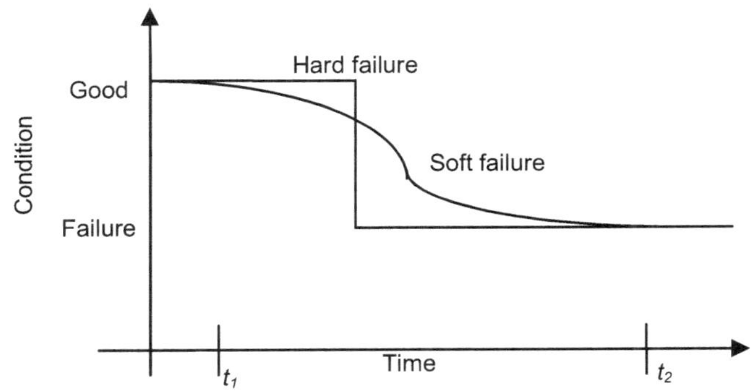

Fig. 6.1 Condition evolution from hard and soft failures

The development of monitoring and diagnostic systems for aircraft and other complex systems has led to the recognition that predictive prognosis is desired and technically possible. The most used prognostics technique is concerned with predicting how much time is left before a failure occurs given the current machine condition and past operation profile, usually called its remaining useful life (RUL). Jardine et al. (2006) classified research on prognostics in three areas: remaining useful life; prognostics incorporating maintenance policies and condition monitoring interval.

According to Wang (2008), current prognostic approaches can be classified into three basic groups: a model-based approach, a data-driven approach, and a hybrid approach. Residual life modelling via stochastic filtering is a relevant modelling technique used in CBM (Wang and Christer 2000). Jardine et al. (1989) proposed proportional hazards modelling (PHM) in order to incorporating explanatory variables into a model for estimating the failure rate. Vlok et al. (2002) used a Weibull proportional-hazards model to determine the optimal replacement for an item which is subject to vibration monitoring. Wang (2011) outlines a semi-stochastic filtering-based residual life prediction approach for the items monitored in CBM. In general, prognostics information is a key element in modeling the decision-making aspect of CBM.

6.2 Monitoring and Inspection Activities

Barlow and Prochan (1965) modeled inspection policies which assume that failure is discovered only by actual inspection and, in general, only after some time has elapsed since the occurrence of the failure and evaluated schedules of inspection times which minimize the total expected cost resulting from both inspection and failure. They assume that failure evolve from a deterioration process. This process is assumed stochastic, and the condition of the system is known only through inspection.

In inspection policies, no replacement or repair is recommended before detection of failure. Each inspection has a cost that implies being unfeasible inspect very often. However, a long lapse of time between failure and detection implies a penalty cost. The main challenge is to find the best inspection policy in order to minimize expected total cost. For example, there are systems subject to precedence constraints and a sequence of inspections should be determined (Chiu et al. 1999).

Barlow and Prochan (1965) proposed a model that minimizes expected cost until detection of failure, a model that minimizes expected cost assuming renewal at detection of failure and an opportunistic replacement strategy of a single part in the presence of several monitored parts.

Ben-Daya et al. (2000) pointed out that for some systems, a continuous monitoring of their operating states is not economically justifiable, and inspections are useful in monitoring the condition of the system at predetermined times in order to reduce the probability of its malfunctioning. Once an out-of-control state is detected, a repair is carried out to restore the system to its in-control state.

There are systems that symptoms of failure are not apparent and the level of degradation (or deterioration) can be known only through inspection (Fouladirad and Grall 2014; Chelbi and Ait-Kadi 2009; Grall et al. 2002; Huynh et al. 2012). Some examples are alarm and stand-by systems. An inspection strategy establishes the time at which one or more operating parameters have to be controlled, in order to determine if the system is in an operating or a failure state (Chelbi and Ait-Kadi 2009).

Two general situations were identified by Chelbi and Ait-Kadi (2009). Firstly, inspections consist simply in assessing if the equipment is working or in a failed state. Secondly, the equipment condition can be assessed through direct or indirect control and preventive actions can be done before failure occurrence, it is known as CBM.

A basic inspection model suggested by Chelbi and Ait-Kadi (2009) assumed that equipment inspected at instants x_i. If an inspection reveals that the equipment is in a failed state, a new identical one immediately replaces it. The sequence of inspection instants is shown in Fig. 6.2.

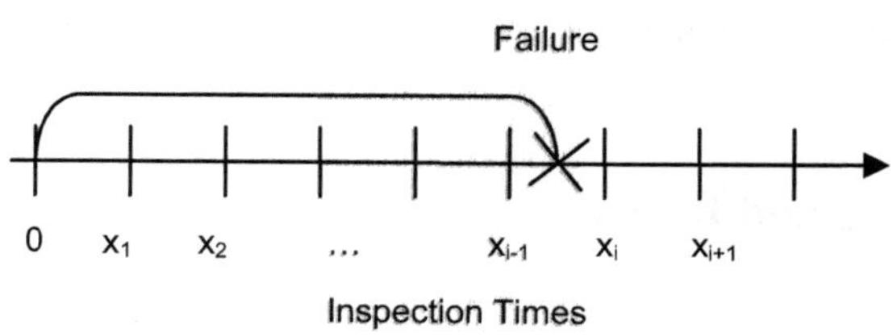

Fig. 6.2 The sequence of inspection times

6.3 Delay Time Models to Support CBM

Christer and Waller (1984) introduced an inspection model based on two stage failure process called Delay Time (DT). They called the initial point u of the defect the first opportunity where the presence of a defect might reasonably be expected to be recognized by an inspection, and the time h to failure from u called the delay time of the defect as shown in Fig. 6.3. Delay time represents a time window for preventing a failure after a defect occurred (Wang 2008).

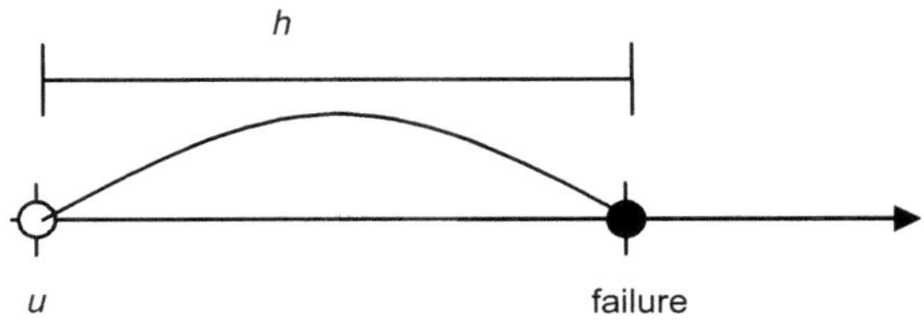

Fig. 6.3 Delay Time h of a failure

In DT models, the failure process is assumed to be a non-homogeneous Poisson process. The operational cost of applying an inspection policy can be measured. The inspection cost denoted by C_i represents the value of the resources needed to perform an inspection task. The inspection repair cost, denoted by C_r consists of the costs necessarily incurred to repair a fault identified in the inspection. Basically, the cost of a breakdown is a penalty cost and is associated with the cost of the consequences caused by a failure, which is, at bottom, the cost related to loss of production. The breakdown repair cost is denoted by C_b. Based on the delay time concept and admitting that the most important assumption that is brought from this fundamental approach is cost, according to Christer and Waller (1984) the expected cost of an inspection policy, for a basic inspection model, is given in (6.1).

$$C(T) = \frac{\lambda T\{C_b b(T) + C_r[1 - b(T)]\} + C_i}{(T + d)} \tag{6.1}$$

where:
T – time between inspections;
$f(h)$ – the probability density function of delay time;
λ is the arrival rate of defects per unit time.
C_b – the average breakdown repair cost.
C_r – the average repair cost.
C_i – the average inspection cost.
d_b – the average downtime to repair a breakdown.

d – the expected duration of the inspection time (d<< T).
$b(T)$ – the probability of a fault causing a breakdown:

The initial instant at which a defect may be assumed to first arise within the plant is uniformly distributed over time since the last inspection and independent of h, as given by (6.2).

$$b(T) = \int_0^T \left(\frac{T-h}{T}\right) f(h)dh \tag{6.2}$$

In terms of availability, this criterion is related to the non-monetary aspects associated with the failures. This reflects the ability of the system to perform under the influence of an inspection policy. Availability is about the percentage of time that the system is available. So, in terms of a service system, this availability corresponds to the percentage of time that the service is provided to the client. Thus, the lower the availability is, the greater the client's dissatisfaction.

Additionally, a critical factor for assessing the performance of an inspection policy is system availability. The downtime can be evaluated in order to represent this factor. According to Christer and Waller (1984), the expected downtime for a basic inspection model is given by (6.3):

$$D(T) = \frac{\lambda \cdot T \cdot d_b \cdot b(T) + d}{(T+d)} \tag{6.3}$$

6.4 Multicriteria and Multiobjective Models in CBM

Diagnostic techniques have been developed which are able to identify the most incipient defects and prognostic techniques are able to estimate residual life more accurately, these two areas of CBM have been based on mathematical and statistical estimation techniques. Jardine et al. (2006) presented data acquisition, data processing and maintenance decision-making as three key steps of a CBM program. Suppose a set of equipment of the same type have the same residual useful life. Depending on the impact of the failure, a piece of equipment can be inspected more frequently or preventive maintenance activity can be anticipated due to the probability of failure. The DM's preferences for the performance of a maintenance policy can be modeled during the decision process. When more than one objective is considered, a MCDM/A model can be developed to support a CBM program. Some examples of MCDM/A models in CBM programs are discussed.

Carnero (2006) proposed an evaluation system of setting up a predictive maintenance program using the Analytic Hierarchy Process (AHP), Bayesian techniques and decision rules. Sasmal and Ramanjaneyulu (2008) analyze a methodology for assessing the condition of bridges using the AHP process in a fuzzy environment. Visual, general and detailed assessments were the criteria considered. Tanaka et al. (2010) present a procedure for assessing the health of equipment for substation maintenance and for planning upgrades based on the AHP and they supply reliability, hardware integrity and regulation as criteria for the model. In terms of inspection planning in the electric power industry, a fuzzy model was defined by Sergaki and Kalaitzakis (2002) for ranking the criticality of components and they incorporated criteria concerning aspects of safety and reliability, economy, variable operational conditions and environmental impacts. Ferreira et al. (2009) proposed a model for multiobjective optimization based on the delay time concept, in which cost and downtime are objective functions of the model.

Kim and Frangopol (2010) developed a multiobjective model with two objectives (monitoring cost and availability), from which Pareto solutions associated with the duration of monitoring and of predictions are obtained. Liu and Frangopol (2005) used a multiobjective genetic algorithm in order to balance the objectives of the maintenance costs of the life-cycle with the condition and safety levels of deteriorating bridges. Marseguerra et al. (2004) present a multi-objective optimization approach, based on genetic algorithms. They considered the probability of system failure and its variance as objectives. Martorell et al. (2006) demonstrate a double-loop Multiple objective genetic algorithm to perform the simultaneous optimization of periodic Test Intervals (TI) and Test Planning (TP) in optimizing surveillance requirements which have the mean unavailability, the maximum time-dependent unavailability and the cost of the system cost as objective functions.

Podofillini et al. (2006) provide a multiobjective genetic algorithm to optimize inspection and maintenance procedures with respect to both the economic and safety-related aspects of railway tracks. Dependent probability of failure on demand, spurious trip rate and lifecycle cost were the objective functions used by Torres-Echeverria et al. (2009) in an optimization model for proof testing policies for safety instrumented systems. Their model was integrated with the NSGA-II genetic algorithm. Zio and Viadana (2011) developed a Multiobjective Differential Evolution so as to optimize the inspection intervals of a High Pressure Injection System. Unavailability, cost, exposure time were the objectives of the model presented.

6.5 A MCDM/A Model on Condition Monitoring

The most common assumption is that any failure is detected at the time of the next check and a replacement is immediately made (Barlow and Prochan 1965; Nakagawa 2005). A derivative approach analyzes the delay time. Delay time, a two-step failure process, is the time lapse from when a system defect could first have been noticed until the time when its repair can no longer be delayed because of unacceptable consequences such as a serious catastrophe which might arise due to failure (Christer 1999). The importance of delay time in maintenance management applications was investigated by Wang (2012). Delay Time models have been applied in several contexts. In a manufacturing industry, Jones et al. (2009) evaluate a subjective measure of the failure consequences based on delay time analysis in terms of cost to the environment, in monetary value to the company and the damaging effect to the company image. A Delay Time model is also proposed to determine inspection intervals on fishing vessels (Pillay et al. 2001).

Inspection tasks are able to identify intermediate states before failure. A MCDM/A decision model in order to aid maintenance planning in inspection models was developed based on MAUT by Ferreira et al. (2009). This model takes the DM's preferences into account, as well as, the most important aspects, when considering setting the inspection intervals for periodic condition monitoring, and these are the cost and downtime associated with the inspection policy.

Inspection can be defined as a task of examining and observing in order to classify an inspected item in terms of its features and properties. When inspections are well defined, a company can minimize maintenance costs and improve the availability of systems. In general, the interest is in discovering the state of an asset i.e., whether it is defective or not.

Managers are interested in balancing the costs of the inspection policy with savings arising from improving the performance of the system. In fact, depending on the production process that is supported by a complex system, a very short interruption due to a failure can cause a substantial financial loss.

Since the model is based on the concept of delay time, the construction of the model should consider the common assumptions defined in the delay time approach.

A MAUT model is proposed which considers that attributes of cost and availability are additive independent if and only if the two-attribute utility function is additive. For these criteria, the additive form may be written as (6.4):

$$\max u(C(T), D(T)) = k_c u_c(C(T)) + k_d u_d(D(T)) \tag{6.4}$$

where:

k_c – scale constant for the cost criterion;

$u_c(C(T))$ – conditional utility function for the cost criterion;

k_d – scale constant for the downtime criterion;

$u_d(D(T))$ – conditional utility function for the downtime criterion.

According to Keeney and Raiffa (1976), the assessment process for a MAU Function consists basically of five steps: (1) introducing the terminology and idea; (2) identifying relevant independence assumptions (3) assessing conditional utility functions; (4) assessing scaling constants; (5) checking for consistency and reiterating.

The first step consists of making the DM understand the main purpose of the utility function, and especially so that he/she understands the consequence space. For our particular case, it is important to notice that t (time to inspection) is considered a feasible alternative and T is the set of all time-of-inspection alternatives. In this case, the set of all times-of-inspection alternatives is defined by $T = [0, \infty)$. For each inspection time, t, there is a consequence in terms of cost, $C(t)$, and downtime, $D(t)$. For example, the point $(C(t_1), D(t_1))$ belongs to the consequence space. An example of the consequence space of this problem is shown in Fig. 6.4.

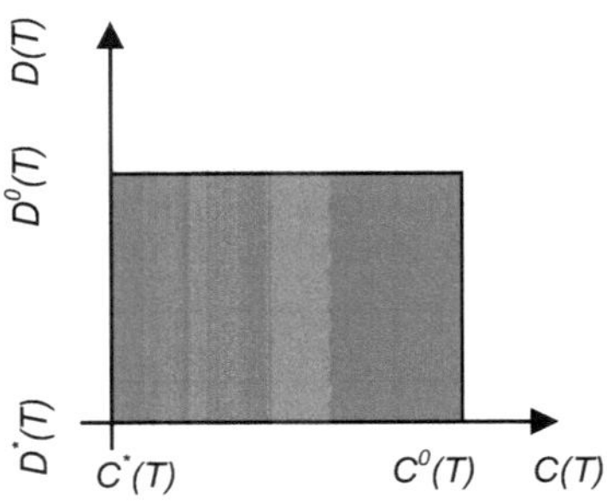

Fig. 6.4 Consequence space for cost and downtime

The second step consists of identifying some independence assumptions that are true for the DM with regard to the criteria that are being considered. Utility independence needs to be checked in order to evaluate the hypothesis of the theory. If this independence occurs, then the MAU function can be simple, otherwise more complex functions are necessary to represent this function.

Once additive independence is observed, the strategy of divide and conquer could be thoroughly explored when assessing the MAU function. Therefore, each one-dimensional utility function for each respective attribute should be elicited. Alternatively, in some cases, a specific analytic function could be used where its shape gives an interesting description of a specific instance of the DM's behavior for a given attribute.

The third step consists basically of eliciting conditional utility functions for each criterion. The conditional utility functions can be performed basically by direct assessment, or estimation of the utility function. In some cases, there is an analytical expression which is known to be a good model for representing utility functions of some specific criteria.

As to the fourth step, assessing the scaling constants depends mainly on the second step. If additive independence is confirmed, we can use the lottery shown in Fig. 6.5 to identify the value of the constant. Finally, the MAU function $u(C(T),D(T))$ should be maximized in order to find the best option for inspection time in terms of cost and downtime.

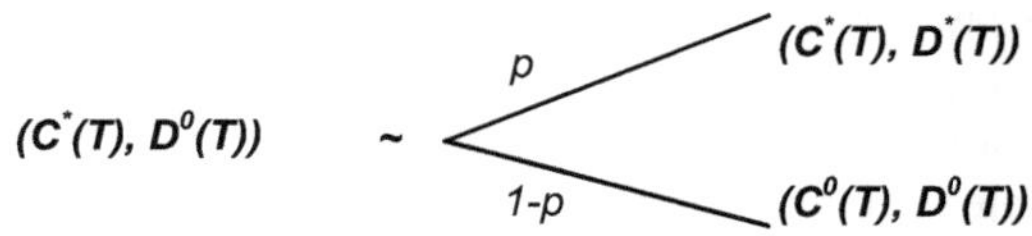

Fig. 6.5 Lottery to find the scaling constant k_c

In Fig. 6.6, the MAUT application phase is summarized, showing part of the modeling procedure, which is based on the general procedure proposed for building the MCDM/A model, as given in Chap. 2. A peculiarity in this specific procedure is the use of the Pareto front identification for the generation of the set of alternatives.

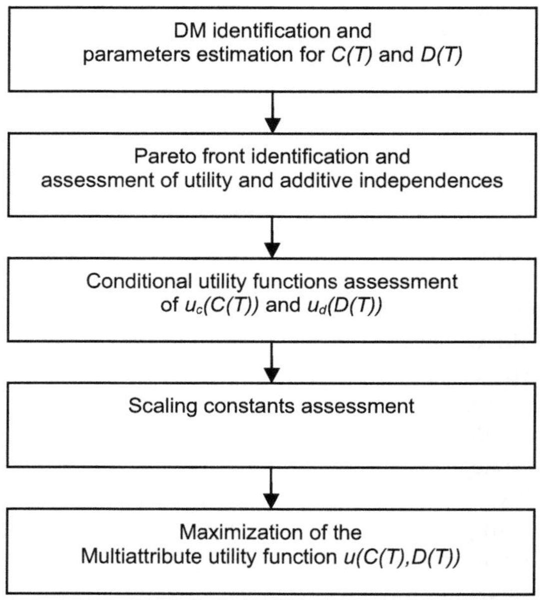

Fig. 6.6 Structure of the decision model for inspection intervals of condition monitoring

6.6 Building an MCDM/A Model on Condition Monitoring for a Power Distribution Company

Maintenance management is a business function that aims to ensure the availability of production resources to enable the operation of an organization. In the context of electric power distribution companies (EPDCs), the availability of electric power is essential for society. For this reason, government regulatory agencies perform an essential role in controlling the quality of service of these companies. Regulatory agencies provide favorable conditions for the electricity market to develop in a balanced environment amongst agents, for the benefit of society. In this section, a case study is carried out to evaluate and apply the performance of a decision model based on data from a Brazilian power distribution company (Ferreira and de Almeida 2014).

In several countries, electric power distribution companies are in an environment controlled by government regulatory agencies. Various performance indicators were developed with the aim of ensuring the quality of service which is monitored by regulatory agencies. The profitability of companies is directly related to such goals. Two relevant distribution reliability indices that measure the duration and frequency of the average interruption of a system are known as System Average Interruption Frequency Index (SAIFI) and System Average Interruption Duration Index (SAIDI) (Čepin 2011). Therefore, a measure that can be used to assess the number of customers affected by an outage is derived from the expected number of failures $N_f(T)$, a SAIFI equivalent estimate, for an inspection policy as defined by Wang (2008) in (6.5):

$$E[N_f(T)] = \int_0^T \lambda F(t)dt \tag{6.5}$$

where:
$F(t)$ – cumulative distribution function of the delay time

Based on the delay time concept, the equivalent estimate that represents the *SAIDI* is the downtime, $D(T)$, given in (6.6).

$$D(T) = \frac{d_f \cdot E[N_f(T)] + d_s}{T + ds} \tag{6.6}$$

where:
Failure will be repaired immediately at an average cost c_f and downtime d_f.

An inspection takes place every T time units, costs c_s units and requires d_s time units, where $d_s << T$.

The cost of an inspection policy can be determined by the delay time concept using (6.7) (Christer 1999):

$$C(T) = \frac{c_f \cdot E[N_f(T)] + c_s}{T + ds} \tag{6.7}$$

When making decisions about the frequency of inspections in an electric energy system, several factors should be taken into account, such as the availability of the system and the number of interruptions. In the electrical energy sector, consequences associated with failure should be avoided due to the high impact on the concession of the service.

A multicriteria model can be used in order to define strategies of inspection intervals that will meet three objectives, namely to minimize the number of experiences a customer has of sustained interruption over a predefined period of time, the length of an interruption, and the cost to the system.

MAUT (Keeney and Raiffa 1976) was chosen to model the problem. The main reason for this choice is based on the assumption that the DM's reasoning for this problem can be represented by the axiomatic structure of this theory. In this theory, the compensation between the criteria implies the use of a synthesis function, the goal of which is to aggregate all criteria in one analytic function. Therefore, the DM's preference structure should be based on the notion of compensation.

It is assumed that there is a maintenance manager whose responsibility it is to determine the inspection intervals of condition monitoring. Thus, the model proposed through MAUT is developed in order to meet his/her requirements so as to assess the MAU Function. In this case, the set of all times-of-inspection alternatives is defined by $T = [0, \infty)$. For each inspection time, t, there is a consequence in terms of cost, $C(t)$, downtime, $D(t)$ and Expected Number of failures, $N_f(t)$. For example, the point $[C(t_1), D(t_1), N_f(t_1)]$ belongs to the consequence space.

An example of the consequence space of this problem is shown in Fig. 6.7. The DM wishes to minimize the three dimensions of the consequence space simultaneously. The problem consists of choosing the best inspection time, but there is a conflict between the three dimensions of the consequence.

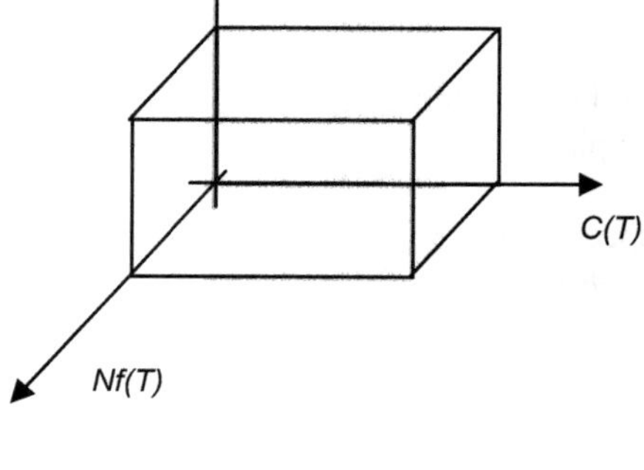

Fig. 6.7 Consequence space for the three criteria: downtime, cost and number of failures

The consequence space defined in Fig. 6.7 shows all possible combinations of the three consequences, but generally this space presents a lot of unfeasible points such as the optimal point in each dimension, the point $(C^*(T), D^*(T), Nf^*(t))$. If this point is feasible, then it is not necessary to model the problem using a multicriteria approach, because this point dominates all other alternatives and so it should be chosen.

Preferential, utility and additive independences are three important concepts to be explored in the use of MAUT. Utility independence is a concept in MAUT equivalent to that of probabilistic independence in multivariate probability theory. Utility independence conditions imply that the MAU function must be of a specified form. In general, independence assumptions greatly simplify the assessment of the original utility function. The three attributes are additive independent, if the paired preference comparison of any two lotteries, defined by two joint probability distributions, depends only on their marginal probability distributions (Keeney and Raiffa 1976).

Based on the additive utility concept, it can be concluded that the attributes are additive independent, if and only if the three-attribute utility function is additive. For these criteria, the additive form may be defined as (6.8):

$$u(C(T), D(T), Nf(T)) = k_c u_c(C(T)) + k_d u_d(D(T)) + k_n u_n(Nf(T)) \qquad (6.8)$$

where:

k_c – scale constant for the cost criterion;
$u_c(C(T))$ – conditional utility function for the cost criterion;
k_d – scale constant for the downtime criterion;
$u_d(D(T))$ – conditional utility function for the downtime criterion;
k_n – scale constant for the number of failures criterion;
$u_n(N_f(T))$ – conditional utility function for the number of failures criterion.

This application is based on a confidential case study in an electric power distribution company. Although the figures and other aspects of this application are not the real data, they have been appropriately altered in order to represent a realistic and consistent context.

The objective of the company is to minimize the cost, downtime and the expected number of interruptions of the system brought about by an inspection policy. Based on the model proposed in the previous section, the first step towards applying the model is to identify the DM. A maintenance manager responsible for making such decisions in the company was identified. The necessary parameters of the model were estimated and are illustrated in Table 6.1.

Table 6.1 Parameters of the model

Parameter	Value
λ	0.005 faults per Day
h	Weibull(1.9;200)
ds	0.01 days
d_b	2 days
C_b	US$ 500.00
C_i	US$ 25.00

As a result of the MAU function, the maximum utility of the inspection time is 0.8981 for 44 days. In Fig. 6.8, the utility of the inspection intervals is shown.

This section presented a MAUT to support the planning of an inspection policy in an electric power distribution company. The number of times a customer experienced a sustained interruption over a predefined period of time, the length of interruption, and cost of system are three objectives considered. The concept of delay time was used to model the failure process.

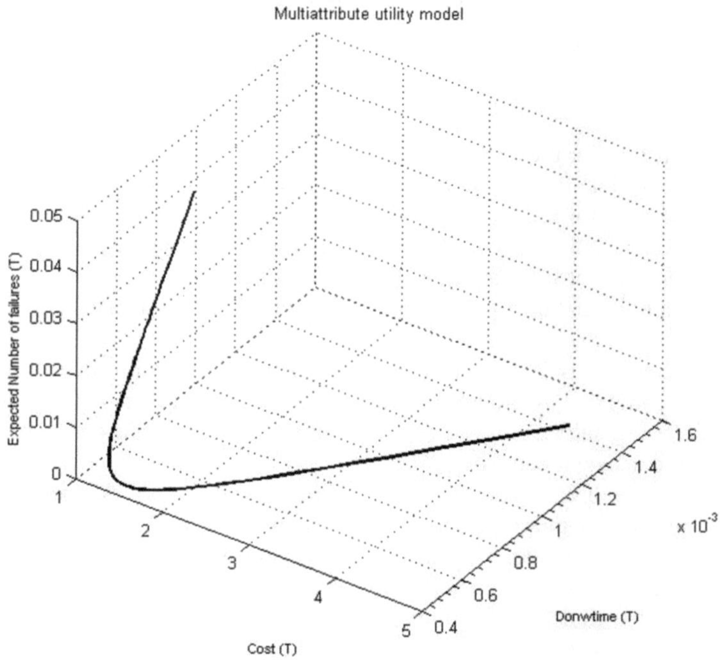

Fig. 6.8 MAU function

MAUT was chosen to model a DM's preferences for the cost, SAIDI and SAIFI criteria in accordance with regulatory laws of this sector and in a suitable way to deal with the tradeoff of probabilistic consequences. This model was evaluated and validated by managers from a Brazilian company.

The modeling of predictive maintenance and monitoring is a tool that can provide many benefits to the area of maintenance management. This chapter suggests a multicriteria approach for modeling CBM decisions. Thus, the proposed multicriteria model aimed to answer this need based on the MAUT which has an axiomatic structure and allows to deal with the conflict between the expected and the cost of an inspection policy downtime.

References

Barlow RE, Proschan F (1965) Mathematical theory of reliability. John Wiley & Sons, New York

Ben-Daya M, Duffuaa S, Raouf A (eds) (2000) Maintenance, Modeling, and Optimization. Kluwer Academic Publishers, Norwell

Berrade MD, Cavalcante CAV, Scarf PA (2012) Maintenance scheduling of a protection system subject to imperfect inspection and replacement. Eur J Oper Res 218:716–725

Carnero MC (2006) An evaluation system of the setting up of predictive maintenance programmes. Reliab Eng Syst Saf 91:945–963

Čepin M (2011) Assessment of Power System Reliability: Methods and Applications. Springer London

Chelbi A, Ait-Kadi D (2009) Inspection Strategies for Randomly Failing Systems. In: Ben-Daya M, Duffuaa SO, Raouf A, et al. (eds) Handb. Maint. Manag. Eng. SE - 13. Springer London, pp 303–335

Chiu SY, Cox LA Jr, Sun X (1999) Optimal sequential inspections of reliability systems subject to parallel-chain precedence constraints. Discret Appl Math 96 - 97:327–336

Christer AH (1999) Developments in delay time analysis for modelling plant maintenance. J Oper Res Soc 50:1120–1137

Christer AH, Waller WM (1984) Delay time models of industrial inspection maintenance problems. J Oper Res Soc 35:401–406

de Almeida AT, Ferreira RJP, Cavalcante CAV (2015) A review of multicriteria and multi-objective models in maintenance and reliability problems. IMA Journal of Management Mathematics 26(3):249–271

Do Van P, Bérenguer C (2012) Condition-Based Maintenance with Imperfect Preventive Repairs for a Deteriorating Production System. Qual Reliab Eng Int 28(6):624–633

Ferreira RJP, de Almeida AT (2014) Multicriteria model of inspection in a power distribution company. Reliab Maintainab Symp (RAMS), 2014 Annu 1–5

Ferreira RJP, de Almeida AT, Cavalcante CAV (2009) A multi-criteria decision model to determine inspection intervals of condition monitoring based on delay time analysis. Reliab Eng Syst Saf 94:905–912

Fouladirad M, Grall A (2014) On-line change detection and condition-based maintenance for systems with unknown deterioration parameters. IMA J Manag Math 25(2):139–158

Grall A, Bérenguer C, Dieulle L (2002) A condition-based maintenance policy for stochastically deteriorating systems. Reliab Eng Syst Saf 76(2):167–180

Huynh KT, Castro IT, Barros A, Bérenguer C (2012) Modeling age-based maintenance strategies with minimal repairs for systems subject to competing failure modes due to degradation and shocks Eur J Oper Res 218(1):140-151

ISO 17359:2011 (2011) Condition monitoring and diagnostics of machines - General guidelines. International Organization for Standardization

Jardine AKS, Lin D, Banjevic D (2006) A review on machinery diagnostics and prognostics implementing condition-based maintenance. Mech Syst Signal Process 20:1483–1510

Jardine AKS, Ralston P, Reid N, Stafford J (1989) Proportional hazards analysis of diesel engine failure data. Qual Reliab Eng Int 5:207–216

Jones B, Jenkinson I, Wang J (2009) Methodology of using delay-time analysis for a manufacturing industry. Reliab Eng Syst Saf 94(1):111–124

Keeney RL, Raiffa H (1976) Decisions with multiple objectives: Preferences and Value Trade-Offs. Wiley Series in Probability and Mathematical Statistics. Wiley and Sons, New York

Kim S, Frangopol D (2010) Cost-Effective Lifetime Structural Health Monitoring Based on Availability. J Struct Eng 137:22–33

Liu M, Frangopol D (2005) Multiobjective Maintenance Planning Optimization for Deteriorating Bridges Considering Condition, Safety, and Life-Cycle Cost. J Struct Eng 131:833–842

Marseguerra M, Zio E, Podofillini L (2004) Optimal reliability/availability of uncertain systems via multi-objective genetic algorithms. Reliab IEEE Trans 53:424–434

Martin KF (1994) A review by discussion of condition monitoring and fault diagnosis in machine tools. Int J Mach Tools Manuf 34:527–551

Martorell S, Carlos S, Villanueva JF, et al. (2006) Use of multiple objective evolutionary algorithms in optimizing surveillance requirements. Reliab Eng Syst Saf 91:1027–1038

Nakagawa T (2005) Maintenance Theory of Reliability. Springer, London

Nowlan FS, Heap HF (1978) Reliability-centered Maintenance. Dolby Access Press

Pillay A, Wang J, Wall AD, Ruxton T (2001) A maintenance study of fishing vessel equipment using delay□time analysis. J Qual Maint Eng 7:118–128

Podofillini L, Zio E, Vatn J (2006) Risk-informed optimisation of railway tracks inspection and maintenance procedures. Reliab Eng Syst Saf 91:20–35

Sasmal S, Ramanjaneyulu K (2008) Condition evaluation of existing reinforced concrete bridges using fuzzy based analytic hierarchy approach. Expert Syst Appl 35:1430–1443

Sergaki A, Kalaitzakis K (2002) A fuzzy knowledge based method for maintenance planning in a power system. Reliab Eng Syst Saf 77:19–30

Tanaka H, Tsukao S, Yamashita D, et al. (2010) Multiple Criteria Assessment of Substation Conditions by Pair-Wise Comparison of Analytic Hierarchy Process. Power Deliv IEEE Trans 25:3017–3023

Torres-Echeverría AC, Martorell S, Thompson HA (2009) Modelling and optimization of proof testing policies for safety instrumented systems. Reliab Eng Syst Saf 94:838–854

Vlok PJ, Coetzee JL, Banjevic D, et al. (2002) Optimal component replacement decisions using vibration monitoring and the proportional-hazards model. J Oper Res Soc 53:193–202

Wang L, Gao RX (2006) Condition Monitoring and Control for Intelligent Manufacturing. Springer Series in Advanced Manufacturing, Springer, New York

Wang W (2008) Delay time modelling. In: Kobbacy KAH, Murthy DNP (eds) Complex Syst. Maint. Handb. SE - 14. Springer London, pp 345–370

Wang W (2011) Overview of a semi-stochastic filtering approach for residual life estimation with applications in condition based maintenance. Proc Inst Mech Eng Part O J Risk Reliab 225:185–197

Wang W (2012) An overview of the recent advances in delay-time-based maintenance modelling. Reliab Eng Syst Saf 106:165–178

Wang W, Christer AH (2000) Towards a general condition based maintenance model for a stochastic dynamic system. J Oper Res Soc 51(2):145–155

Zio E, Viadana G (2011) Optimization of the inspection intervals of a safety system in a nuclear power plant by Multi-Objective Differential Evolution (MODE). Reliab Eng Syst Saf 96:1552–1563

Chapter 7
Decision on Maintenance Outsourcing

Abstract: This chapter presents key aspects of multicriteria (MCDM/A) approaches for decisions on maintenance outsourcing regarding maintenance contract, which includes contract selection (e.g. repair contract) and supplier selection. Contract design is a multi-objective task that leads the maintenance manager (or decision maker - DM) to decide amongst a combination of contracts and suppliers' bids for the service. Given the multiple objective nature of this kind of problem, this chapter presents models that include maintainability, dependability, quality of repair and other aspects besides cost. The decision models presented consider methods such as Multi-attribute utility theory (MAUT) to address compensatory preferences and ELECTRE for preferences that require an outranking method. The DM's behavior to risk (prone, neutral and averse) is considered by using Utility Theory and Decision theory foundations in order to include the state of nature in decision models. Thus, most of the problems are related to supplier and contract selection, which may be modeled into a single problem when considering all combinations of contracts and suppliers as alternatives, including the possibility of in-house maintenance being undertaken by a maintenance service supplier. Depending on the organization and in how strategic its maintenance function may be, decisions in maintenance outsourcing may be approached in different stages. Thus, a key performance indicator (KPI) for such problems are defined depending on the type of organization, its capabilities and the number of maintenance activities, while the tradeoff amongst strategic objectives is balanced in order to assure the system's availability.

7.1 Introduction

Ever since management theory took shape, there has been extensive discussion with regard to downsizing, core competences, business process re-engineering and other managerial trends, that are deployed into general outsourcing. Such discussion is also applicable to maintenance. According to Buck-Lew (1992), a company outsource when it requests the services of an outside party to fulfill a function or functions in the organization. Decisions on outsourcing are very close related to contract selection (de Almeida 2001b), contract design and supplier selection decisions.

A.T. de Almeida et al., *Multicriteria and Multiobjective Models for Risk, Reliability and Maintenance Decision Analysis*, International Series in Operations Research & Management Science 231, DOI 10.1007/978-3-319-17969-8_7

According to de Almeida (2007) the outsourcing decisions are requiring more and more attention since the contract price is not the only aspect to be considered by a DM. Therefore, MCDM/A techniques include the most appropriate tools for evaluating the costs of the contract and the associate service performance. This topic was found in 2.7% of publications reviewed by de Almeida et al. (2015), considering MCDM/A approaches for reliability and maintenance.

As noticed in other organizational areas such as Information Systems (IS), the trend towards outsourcing in maintenance has been greatly affected by rapid changes in technology. Murthy and Jack (2008) pointed out the role of technological advances, which have resulted in more complex and expensive equipment, have increased the level of specialties and techniques needed to repair such equipment and have led to a variety of work force specialties and diagnostic tools that require constant upgrading.

Murthy and Jack (2008) also remind that the maintenance of governmental infrastructures was traditionally maintained in-house. This changed in line with these managerial trends in order to have a second party performing activities such as road or rail maintenance services, for example.

Thus, outsourcing is a trend followed by many organizations that wish to focus on their core competences, and due to technological advances, it has increased and inspired a wide range of articles in the literature on maintenance outsourcing decisions, especially because most of the organizations do not view maintenance as a core business activity.

Based on the literature, this chapter presents some of the main MCDM/A maintenance outsourcing decision problems, and criteria that might be considered to guide these decisions. Also, repair outsourcing decision problems are approached, including contract selection.

Outsourcing decisions are strategic and most of them include defining which functions and activities are candidates for outsourcing and which should be kept in-house. Secondly, there is a need to establish the criteria and key performance indices to be followed by the maintenance service supplier, which should be structured in the outsourcing contract.

Moreover, there is a choice problem, when selecting the service supplier. These decisions are based in multiple factors that emphasize an MCDM/A approach inherent in such decisions.

These decisions can be modeled into a single decision problem, by evaluating all combinations of available contracts and suppliers as alternatives, including the in-house maintenance service if the organization is capable to carry out its maintenance activities. Thus, in-house maintenance service is also one of the service supplier alternatives, therefore the final recommendation will reflect if the activity should be outsourced or not, and also which contract should be selected.

In these decisions it is clear that there is a need to consider MCDM/A, especially when facing problems with characteristics discussed in Chap. 2 that turns maintenance problems more strategic and relevant. First of all, when selecting if an activity should be outsourced, strategic aspects are observed as key performance indices regarding the impacts of outsourcing such activity considered. With regard to the outsourcing requirements established in outsourcing contracts, there are several attributes that are considered in terms of the objectives that shall be used to evaluate these contracts. Same applies for deciding upon a list of service suppliers. Traditionally, costs are considered to be among these criteria.

The need to consider an MCDM/A approach arises from several factors related to these decisions, which require methodological support to ensure that the manager (DM) will be supported in order to evaluate these factors properly according to his/her preferences. During the following sections, specific problems are tackled, and the criteria found in the literature to address outsourcing decisions.

This chapter focuses on the maintenance and repair contracts problems with an MCDM/A perspective, which can be adapted to supplier selection. There are other problems in maintenance outsourcing decisions related literature considering warranties (Wu 2013), extended warranties and maintenance contract design (Wang 2010). These topics are not addressed in this chapter.

7.2 Selection of Outsourcing Requirements and Contract Parameters

The strategy for outsourcing maintenance goes together with the management of contracts. The relationship between the contractor (the company outsourcing one or more of its services – client company) and contracted firms (suppliers of such services) is regulated by a contract in which the parties involved define the rules of the service to be performed for an agreed length of time.

Service contracts in the area of maintenance typically emphasize the legal aspects in clauses (terms) that deal with price, forms of readjusting price, payment terms, quality and warrant provisions of the service to be provided, technical aspects, transfers of responsibilities to third parties, retention/fines/damages, termination, period (deadline), exchanges of information (communication channel) and other important aspects of this relationship.

According to Brito et al. (2010), selecting contracts is a very important stage in the process of outsourcing maintenance given the current trend towards reducing costs and increasing competitiveness by focusing on core competences. Many studies have been carried out on outsourcing and maintenance contracts, most of which deal with qualitative aspects (Kennedy 1993; de Almeida 2005). Thus, MCDM/A, plays an important role supporting DMs to deal with multiple and conflicting criteria, and associated uncertainties in the process for selecting outsourcing contracts (Brito et al. 2010).

Wideman (1992) suggests that when companies consider outsourcing, they need to make prior enquiries about bidding companies at the start of the hiring process. According to Martin (1997), maintenance contractors are very interested in developing new types of contracts that promise to offer them higher profitability and increased flexibility and lower maintenance costs to them.

From a historical standpoint, the initial practice of the industrial sector was to hire maintenance services in the form of manpower, i.e., paid for in terms of man-hours worked. In this type of contract, it is the sole responsibility of the maintenance service providers to ensure the presence of their staff in the industrial plants of their customers, and therefore the suppliers is paid for the total number of hours their staff worked. Main weakness of manpower contract are: less-skilled personnel; low productivity of services; low quality of services; higher accident rates; noncompliance with labor legislation.

Although still widely applied, this type of contract does not require the commitment of outsourced staff to produce good results and, invariably, the consequences for the industry may be negative in the medium and long term. Therefore, this type of contract results in a relatively high business risk and should not be entered into if the reason is the company's vision is one of global optimization. This is because although there may be an apparent reduction in the cost of maintenance, undesired effects on the overall results can be generated.

This type of contract is practically a unilateral relationship. From the perspective of game theory, one can assume that the policy contract is "win-lose" in the short term, but in reality, in the medium or long term it can become a policy of "lose-lose". Therefore, sometimes this model proves to be bad for both the contractor and the supplier.

Due to the problems discussed above, industry developed a different type of contract: hiring for specific maintenance jobs or for special maintenance servicing of specified equipment and machinery. This type of contract occurs in an isolated form or as part of a hybrid contract (more than one contract type), the latter being widely used in the industrial sector. Some advantages of this type of contract are: better-qualified manpower; increased productivity; better quality of work.

The process of outsourcing maintenance activities evolved into hiring a single supplier or a few suppliers, who are highly specialized and qualified and are made responsible for the overall maintenance process. At this stage, the relationships that have been established by the partnership between companies and their subcontractors in the maintenance area mature.

In this type of contract, contractors must give support to the activities outsourced, and make the staff of the contracted company feel a bond with the contracting company and as if they were an integral part of a single organism, which for its best performance needs to keep its basic functions operating in a healthy way. Achieving this maximum mutual commitment remains the greatest challenge for obtaining the best results in the process of maintenance outsourcing.

Another type of contract, with emphasis on both the client and the supplier, is the type of contract that includes tracking results for performance. Typically, this type of contract involves greater commitment from both sides of the contract, and formalizes partnerships in the medium and long term (Wideman 1992).

Alternatively, Tsang (2002) discusses an alternative form of contracts by lease, in which the contractor is a user of the final product produced by the supplier (besides the active maintenance company being an investor).

Wideman (1992) discusses various types of contracts in the area of project management, which can be widely replicated for the reality of maintenance management. He considers that there are four main areas of risk in different types of contracts from the customer's perspective:

- Lump Sum (Global price) - the final price is based on the sum of all costs involved, considering the contingencies, risks, overheads, profit margin, or any parameter that can be expected to help form the contract price;
- Unit Price - should consider all direct and indirect costs involved, as well as the overall price, and divided by number of events occurred;
- Target Cost (Based on a goal of total cost) - costs are defined transparently between the parties involved and the final price contract is established together with the target contract value;
- Reimbursable Costs (Variable remuneration) - pays the actual costs involved and is based on full transparency and trust (partnership) between the parties to the contract. A strategic alliance and a high level of maturity between the client and the supplier need to be established.

Alternatively, Martin (1997) develops an analysis of the types of maintenance contract in terms of operational criteria and knowledge retention. He divides them into three classes: work package contract, performance contract and facilitator contract, described in Table 7.1:

Table 7.1 Features of maintenance contracts

Type of contract	Description
Work package contract	Most basic type of contract. The contract is simple, in which the payment of the contracted services is based on the unit rate or lump sum. The service request is made by the client. The contractor can focus on the supplier selection of the cheapest. The level of relationship with the contractor is minimal. The knowledge about the operation system remains almost entirely with the contractor.
Performance contract	Based on performance targets, the contractor and the contracted company assume shared responsibility. The complexity of the contract is high, due clause contracts that are defined to assess the outcome of the contract (the conflicts of interests of performance indicators). The relationship between the parties should be close and usually long-term. The knowledge is shared between both parties.
Facilitator contract	It is a type of contract where the service supplier is fully responsible for the result to be achieved, consequently the complexity of the contract is less. It is also known as a lease contract.

The relationships of these contract types to each other and issues of contract complexity, client-contractor relationship and client maintenance knowledge base are shown in Fig. 7.1. The three types of contract discussed in Martin (1997) are extreme cases. However, contractors may develop different (hybrid) contracts for different sets of production systems, the skills involved and to split the financial risk between the client and supplier.

The potential impact of maintenance on equipment and systems in terms of quality, flexibility, cost, availability, and safety is increasingly evident within the maintenance management system. Therefore, the need for measuring the performance of maintenance is evident, and this means that maintenance as a function generates profitability for the firm.

Therefore, what is critical is the process of setting performance indicators, which will serve as regulatory elements of quality, variable remuneration (depending on the chosen type of contract) or other indicators.

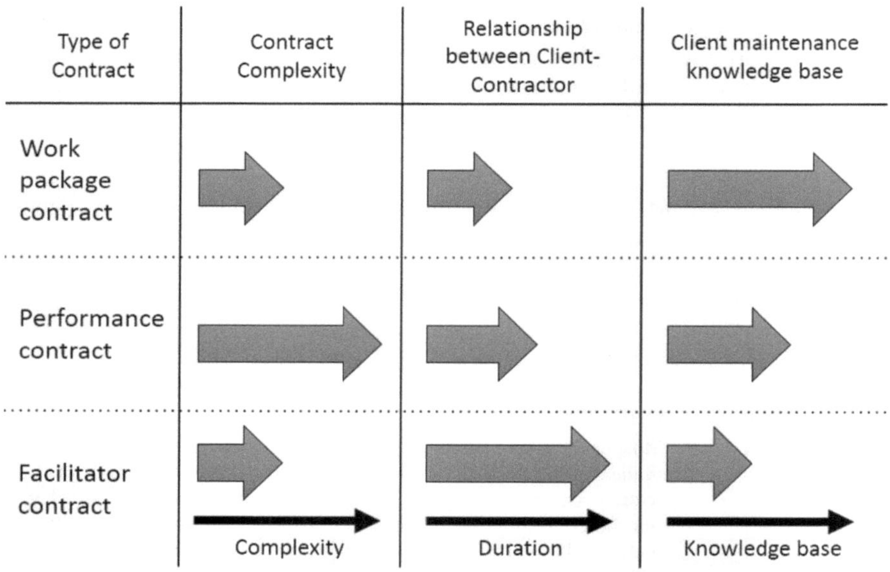

Fig. 7.1 Relationships of the various contract types and contract complexity, client-contractor relationship and client's knowledge of maintenance

It is noteworthy that there is no single standardization for the development of indicators for the different segments of production systems. Their ways of evaluating 'productivity' are different from each other, e.g.: one can consider 'productivity' can be considered as only about improving profits or as improvements in such matters as availability or production rate or products or inventory management or safety, or a combination of several of these.

In addition to selecting which performance indices will be used, the ranges of performance using these indices must be considered by the parties involved. For example, both parties to the contract must negotiate on the implications of having ranges for performance indicators. For example, they must agree on from what point in the range of reduced costs (cost of contract) that a high handling time (availability of resources for the outsourced activities) and from what point in the range that the high cost of the contract will result in it taking a short time to complete the outsourced service.

Within several contract templates, one can identify the type of partnership agreement based on indicators of availability and of the reliability of the production system (using the Mean Time Between Failure – MTBF - and Mean Time to Repair – MTTR - indicators), where the company to which services have been outsourced increases its profitability as it improves the availability and reliability of the client enterprise system (de Almeida and Souza 2001). Therefore, this type of contract no longer remunerates services (grants bonuses to), but rather solutions that will improve the levels of availability and reliability of systems.

However, some factors can disturb this type of contract. One that stands out is the alignment amongst the strategic objectives of the interested parties. In fact, there is a conflict of interest because the company to which services have been outsourced also has difficulties of surviving in the competitive market, and it needs to be competitive. For outsourced companies this type of service is a core activity, while for contracting companies, it is a means of supporting an end activity, so the maintenance services performed by subcontractors is the only source of funds. This conflict is quite evident when contracts are mainly short and medium term in length.

Therefore, assuming that the aspects of reliability were properly dealt with in the phase of the designing the production system (i.e., both parties are aware that the maintenance service will not have the ability to improve system reliability beyond that already specified in the design), it would remain for a maintainability study to be included in maintenance agreements (de Almeida 2002). In this regard, one has the administrative time (*TD*), the effective time to repair (*TTR*), the availability of spare parts and the level of training of the outsourced teams. However, all these previous aspects have a cost (*C*) associated with obtaining the levels desired.

TD is the time that it takes to notify a maintenance company of a failure and the time it takes this company to go to the client to deal with it. *TD* basically consists of: the time spent in selecting and making the technical staff ready to perform the service; the time taken to provide tools and the budget necessary to perform the service; and the commuting time between the service provider company and the location of the system to be repaired.

Therefore, *TD* can be a negotiated in a contract because it directly affects the interruption time (*TI*) of the client system. As a counterpoint to this, it has to be remembered that since outsourced firms have many clients, they try to keep the idle time of their work teams within certain levels in order to meet the demands of their diverse clients and to satisfy the times of visits agreed to by contract.

The time taken for the maintenance team from the start of the repair process to putting the production system back into a normal state of operation is the *TTR* . Normally, this time is directly related to the technical skills of the team, team training, the team's learning curve, the modularity of the system/equipment, the availability of repair spare parts and other variables.

Therefore, when the contract is modeled as a function of the *TD* and *TTR*, the maintenance contract must adequately compensate for the cost of the maintenance structure of the company providing services that ought to be in a state of readiness to meet sudden demands from the client company. However, keeping a large contingent of maintenance staff available to the client and a high level of inventory of spare parts, so as to guarantee an adequate level of system availability, becomes very and does not fit the competitive market model that both the client and the supplier find themselves in.

In reality, what is required is that the company providing maintenance services has a firm commitment to ensuring the availability of the system (hence the need for the outsourced team to be available at short notice) and not increasing the cost of service. Thus, the best choice would to make a contract using a decision model that incorporates the DM's preference structure represented by the utility function of the attributes cost (C), of the interruption time (*TI*=*TD*+*TTR*) and of the maintainability of the system as modeled in probabilistic terms.

Thus, the problem faced by the manager or DM becomes how to proceed with a decision process that allows the various performance indices considered in the contract to be optimized, since these various indices can conflict with each other. Brito et al. (2010) state that contracts that present a lower cost might present a less satisfactory performance concerning criteria related to quality and availability, which creates a complex frame of trade-offs.

The DM faces several options for maintenance contracts, each implying different system performances and related costs. de Almeida (2001a) identifies that the selection of repair contracts is a non-trivial process since the consequences of a wrong choice may be critical, for instance, in services where availability is fundamental, as in telecommunications and electric power distribution services.

With regard to selecting contracts, little work has been conducted on exploring a multi-criteria decision-making approach. de Almeida (2001b) has presented MCDM/A models based on MAUT for selecting repair contracts, which aggregate interruption time and related cost through an additive utility function. A different approach can be found in de Almeida (2002), where the ELECTRE I method has been combined with utility functions regarding a repair contract problem.

Brito and de Almeida (2007) and Brito et al. (2010) propose a MCDM/A methodology to support the selection of maintenance contracts in a context where in-formation is imprecise, when DMs are not able to assign precise values to the importance parameters of criteria used for contract selection. Utility theory is combined with the Variable Interdependent Parameters method (VIP) to evaluate alternatives using an additive value function regarding interruption time, contract cost and maintenance service supplier's dependability.

In general, in the context of drawing up outsourced maintenance contracts, the DM should choose the option most preferred, the one with the best combination of contract conditions (de Almeida 2002). For de Almeida (2005) what variables are used may vary depending on the market that the company is in and its strategy, and may involve: delivery speed or response time, quality, flexibility, dependability and obviously, cost.

From an overview of the types of outsourcing maintenance contracts, it is important to emphasize that there is no model for an optimal contract.

In reality, there are certain types of contract that are best suited to certain types of relationships and partnerships between client and supplier, the type of service contract, financial relations, economic issues, etc.

Therefore, in order to draw up complex contracts, it is necessary to have a large amount of consistent information and knowledge. Wideman (1992) recommends starting with the model for a simple and traditional maintenance contract. Later, when a closer and systematic relationship between the parties has been established, more advanced analysis of contracts involving performance evaluation criteria and evaluator can be used.

Another factor that must be taken into account in entering into a contract is the exchange of cultures between those involved. There is always resistance to a change in culture when the culture between the parties is initially quite divergent. The adaptation process can be time consuming and have a direct adverse effect on the expected results from the contract.

Furthermore, companies tend to think that maintenance contracts can be compiled quickly and easily which often leads to their being entered into precipitately as a result of which invalid assumptions are made that will disturb relationships in the partnership between the parties in the long term. The parties should strive to reach a common point of view in order to generate a win-win game, which will lead to their enjoying a transparent long-lasting relationship with a high level of satisfaction for both parties. Therefore, permanent maintenance contracts (with shared responsibilities) should be regarded as exemplifying the strategic alliance between the two parties.

The next section presents some of the literature regarding maintenance service supplier selection based on multiple criteria.

7.3 MCDM/A Maintenance Service Supplier Selection

This is an important decision problem for the outsourcing process; therefore, it should be a compromise between costs and the performance required from service suppliers. Specifically there is a need to address such problems with tools that enable conflicting criteria to be dealt with that are usually followed by uncertainties when referring to the consequences of maintenance decisions.

There are some decision models and applications in the literature that consider MCDM/A techniques which will be discussed in this section.

7.3.1 Maintenance Service Supplier Selection with Compensatory Preferences

To address a maintenance service supplier selection problem when the DM has a compensatory preference structure, the literature presents two decision models based in MAUT (de Almeida 2001a; de Almeida 2001b).

These decision models are based on the MCDM/A approach described in Chap. 2, dealing with the following objectives: Interruption time and Cost.

Although both models deal with the same objectives, different assumptions characterize each model, reflecting different situations that may be faced by a DM.

As described in Sect. 7.2, the interruption time is represented by the time spent with administrative activities and the time spent executing the repair.

The first model assumes that during the interruption time the administrative time is deterministic (de Almeida 2001a), while the second model assumes that the administrative time follows an exponential distribution (de Almeida 2001b).

Deterministic Administrative Time Model

Despite considering the administrative time deterministic, the model presented by de Almeida (2001a) follows the MCDM/A approach described in Chap. 2 and considers the uncertainties related to the states of nature inherent to this problem.

de Almeida (2001a) considered the following assumptions:

- TI is explained by TD and TTR; where $TI=TD+TTR$;
- TTR follows an exponential distribution for all service suppliers in all contracts, given by (7.1), where u is $MTTR^{-1}$, therefore this parameter u represents the state of nature.

$$f(TTR) = ue^{-uTTR} \tag{7.1}$$

- There is prior knowledge $\pi(u)$ about u that can be assessed from experts.
- TD is deterministic and assumes different values according to the service supplier and contract.
- DM's preference structure fits MAUT axiomatic requirements to be represented as an additive utility function $U(TI,C)$, given by (7.2), where k_{TI} and k_C are the respective scale constants:

$$U(TI,C)=k_{TI}U_{TI}(TI)+k_{C}U_{C}(C) \tag{7.2}$$

- DM's preference structure fits in both attributes an exponential utility function, $U_{TI}(TI)$ and $U_C(C)$, to represent DM's one-dimensional preferences, given by (7.3) and (7.4). It means that for such a DM higher values of time or cost are undesirable, which is a reasonable assumption, one of the reasons for assuming this kind of utility function in many practical applications.

$$U_{TI}(TI)=e^{-A_1 TI} \tag{7.3}$$

$$U_{C}(C)=e^{-A_2 C} \tag{7.4}$$

Considering the uncertainties referring to the states of nature (MTTR), the DM shall maximize his/her expected utility value $E_uU(u,a_i)$, where $U(u,a_i)$ is the utility of the state of the nature u and the action a_i, which refers to a specific maintenance service supplier and contract representing the consequence (TI,C), consequently $U(u,a_i)$ is obtained. The value of $E_uU(u,a_i)$ is given by (7.5) (de Almeida 2001a).

$$E_uU(u,a_i)=\int_u U(u,a_i)\pi(u)du \tag{7.5}$$

In order to maximize the expected utility from (7.5), it is required to obtain $U(u_i,a_i)$ from $U(TI,C)$. By considering the assumption that TD deterministic, is possible to include TD as a constant into TTR, then, TI is reduced to TTR, thus, $U(TI,C)$ is equivalent to $U(TTR,C)$, and $U(TI)$ becomes $U(TTR)$.

Thus, as pointed by de Almeida (2001a), $U(u,a)$ is the expected value of $U(TTR,C)$, given by (7.6):

$$U(u\ ,a_i)=\int_{TTR} U(TTR,C)\Pr(TTR\mid u\ ,a_i)dTTR \tag{7.6}$$

Since $\Pr(TTR\mid u\ ,a_i)$ corresponds to $f(TTR)$, then (10.6) can be rewritten as (7.7):

$$U(u\ ,a_i)=\int_0^{\infty}\left[k_{TI}U_{TI}(TI)+k_{C}U_{C}(C)\right]u\ e^{-u\ (TTR)}dTTR \tag{7.7}$$

By replacing (7.3) into (7.7), (7.8) is obtained:

$$U(u,a_i)=\frac{k_{TI}u}{A_1+u}+k_C U_C(C) \tag{7.8}$$

Finally, replacing (7.4) in (7.8) and (7.5), there is (7.9):

$$E_u U(u,a_i)=\int_u\left[\frac{k_{TI}u}{A_1+u}+k_C e^{-A_2C}\right]\pi_i(u)du \tag{7.9}$$

Thus, for each distribution of *TTR* there will be an implied cost for the respective service supplier contract, which means that the DM is deciding upon the *TTR* pdf and its respective cost (*C*) in order to maximize his/her multi attribute utility function. The alternatives for this problem are the existing combination of maintenance service suppliers and its contract. Solving this problem consists in solving (7.9) for all alternatives, which are all the existing combination of maintenance service suppliers and its contract. Therefore, k_{TI} and k_C represents the tradeoff between cost and time to repair according to DM's preferences.

Stochastic Administrative Time Model

The model proposed by de Almeida (2001b) enables consideration to be given to different types of contract, thereby seeking to select the best alternative in terms of cost and system performance given the decision maker's preferences represented also by an additive function.

This model differs from the model presented in the last section for considering the TD as a stochastic variable. This feature allows incorporating specific conditions that appears in many real problems, that includes significant variation and uncertainty on TI.

de Almeida (2001b) exemplifies situations that require to consider TD as a stochastic variable, such as those associated with spares provisioning.

The assumptions considered by de Almeida (2001b) for this model are:

- *TI* is explained by *TD* and *TTR;* where *TI*=*TD*+*TTR*.
- *TTR* follows an exponential distribution for all service suppliers in all contracts, given by (7.1).
- There is prior knowledge $\pi(u)$ about u that can be assessed from experts.
- *TD* follows an exponential distribution for all service suppliers in all contracts, given by (7.10), where ω is a parameter defined according to the service supplier contract service level and spare provisioning.

$$f(TD)=\varpi e^{-\varpi TD} \tag{7.10}$$

- *TD* and *TTR* are independent random variables.
- DM's preference structure fits MAUT axiomatic requirements to be represented as an additive utility function as given by (7.2).
- DM's preference structure fits in both attributes an exponential utility function to represent DM's one-dimensional preferences, as given by (7.3) and (7.4).

Since this model considers a stochastic *TD*, *TI* is now the sum of two independent random variables. The pdf of *TI* is obtained by (7.11):

$$f(TI) = \int_{-\infty}^{\infty} f(TD) f(TI - TD) dTD = \int_{-\infty}^{\infty} f(TTR) f(TI - TTR) dTTR \tag{7.11}$$

Thus, from (7.1) and (7.10), follows that (7.12) results in (7.13), considering that this result would be positive if, and only if $TD \geq 0$ and $TI \geq TD$, thus $TI \geq TD \geq 0$, from this result the integer from (7.12) turns into (7.13):

$$f(TI) = \int_{-\infty}^{\infty} \varpi e^{-\varpi TD} u e^{-u(TI-TD)} dTD \tag{7.12}$$

$$f(TI) = \varpi u e^{-uTI} \int_{0}^{TI} e^{-TD(\varpi - u)} dTD \tag{7.13}$$

Hence, developing (7.13), it is possible to find (7.14) for all $TI \geq 0$ as:

$$f(TI) = \frac{\varpi u}{u - \varpi}(e^{-\varpi TI} - e^{-uTI}) \tag{7.14}$$

Thus, each maintenance service supplier contract a_i is associated with a cost c_i and a specific probability function for *TD* represented by the parameter ω_i . Therefore, as in the previous model, the expected utility is given by (7.5) and shall be maximized considering the prior knowledge $\pi(u)$ over (7.14) instead of (7.1) (de Almeida 2001b). Thus, similar to the previous model, $U(u,a)$ is the expected value of $U(TI,C)$, given by (7.15) by applying the utility functions linearity property (de Almeida 2001b):

$$U(u, a_i) = \int_{TI} U(TI, C) \Pr(TI \mid u, a_i) dTI \tag{7.15}$$

Given that $\Pr(TI \mid u, a_i)$ corresponds to (7.14), then (7.15) can be rewritten as (7.16):

$$U(u, a_i) = \int_{TI} \left[k_{TI} e^{-A_1 TI} + k_C U_C(C)\right] \frac{\varpi u}{u - \varpi}(e^{-\varpi TI} - e^{-uTI}) dTI \tag{7.16}$$

By developing (7.16) into (7.17), and then replacing (7.14) in (7.17), (7.18) is obtained, and developed into (7.19), (7.20) and (7.21):

$$U(u\ ,a_i) = k_{TI}\frac{\varpi u}{u-\varpi}\int_{TI} e^{-A_1 TI}(e^{-\varpi TI} - e^{-uTI})dTI + k_C U_C(C)\int_{TI}\frac{\varpi u}{u-\varpi}(e^{-\varpi TI} - e^{-uTI})dTI \tag{7.17}$$

$$U(u\ ,a_i) = k_{TI}\frac{\varpi u}{u-\varpi}\left\{\int_0^{\infty} e^{-(A_1+\varpi)TI}dTI - \int_0^{\infty} e^{-(A_1+u)TI}dTI\right\} + k_C U_C(C)\int_0^{\infty} F(TI)dTI \tag{7.18}$$

$$U(u\ ,a_i) = k_{TI}\frac{\varpi u}{u-\varpi}\left\{\int_0^{\infty} e^{-(A_1+\varpi)TI}dTI - \int_0^{\infty} e^{-(A_1+u)TI}dTI\right\} + k_C U_C(C) \tag{7.19}$$

$$U(u\ ,a_i) = k_{TI}\frac{\varpi u}{u-\varpi}\left\{\left.\frac{e^{-(A_1+\varpi)TI}}{-(A_1+\varpi)}\right|_0^{\infty} - \left.\frac{e^{-(A_1+u)TI}}{-(A_1+u)}\right|_0^{\infty}\right\} + k_C U_C(C) \tag{7.20}$$

$$U(u\ ,a_i) = k_{TI}\frac{\varpi u}{u-\varpi}\left\{\frac{1}{(A_1+\varpi)} - \frac{1}{(A_1+u)}\right\} + k_C U_C(C) \tag{7.21}$$

Hence, by developing (7.21) and applying (7.4), *U(u,a_i)* is given by (7.22):

$$U(u\ ,a_i) = k_{TI}\frac{\varpi u}{(A_1+\varpi)(A_1+u)} + k_C e^{-A_2 C_i} \tag{7.22}$$

Thus, each service supplier contract (alternative or action) will be characterized by the distribution of *TI,* considering the random variables *TD* and *TTR* , and the respective implied cost.

Therefore the DM is deciding upon the a *TI* distribution and its respective cost (*C*) in order to maximize the expected value of (7.22), given $\pi(u)$ according to (7.5).

Applying (7.22) in (7.5) gives the expression of $E_u U(u,a_i)$ for the stochastic *TD* model as given in (7.23):

$$E_u U(u,a_i) = \int_u \left[k_{TI} \frac{\varpi u}{(A_1+\varpi)(A_1+u)} + k_C e^{-A_2 C_i} \right] \pi(u) du \tag{7.23}$$

Hence, for the stochastic *TD* model (7.23) should be maximized, similarly to (7.9) in the deterministic *TD* model.

The main assumptions of these models previously presented are that:

- Time has an exponential distribution.
- DM's preferences fits MAUT requirements for an additive utility function regarding system's performance and cost.

These are very realistic assumptions, since there are many practical situations in which both assumptions are confirmed during applications (de Almeida 2001a, de Almeida 2001b).

The application of such decision models allows to reasonably measure the response time of a maintenance service supplier contract, allowing also to consider in-house maintenance service to be compared with other maintenance service suppliers, and evaluate which activities would be better performed if outsourced by considering an additive utility function for modeling the DM's preferences with regard to cost and the performance of the system.

Depending on the context of the problem the MCDM/A framework given in Chap. 2 should be applied in order to build a more accurate decision model by considering different MCDM/A methods and/or different probabilistic assumptions, hence the choice among these will depend on the context of the problem as discussed in Chap. 2.

7.3.2 Maintenance Service Supplier Selection with Non Compensatory Preferences

In order to provide a more suitable model for a DM that has non-compensatory rationality, a decision model considering non compensatory preferences is presented. It adapts the decision model based in MAUT for using a compatible method with non compensatory preferences for the maintenance service supplier selection problem (de Almeida 2002). This illustrates a situation related to the step 6 of the building model procedure presented in Chap. 2.

This decision model associates Utility Theory with the ELECTRE I method. The use of the ELECTRE I adds to the MCDM/A decision model a pairwise dominance approach based on concordance and discordance indices that builds outrank preference relations for selecting the best maintenance service supplier contract. Thus, this decision model uses one-dimension utility functions values as the performance of alternatives for each criterion.

This decision model was built for a repairable system considering the implications of each alternative in terms of two aspects: Interruption time (or response time) and Costs.

Similarly to the decision model previously presented, this decision model evaluates the benefit of the maintenance service supplier contract in terms of maintainability and the associated cost of the service.

The maintenance service supplier contract performance in the response time reflects its specific condition for spare provisioning and repair capability.

The assumptions of the decision model are (de Almeida 2002):

- *TI* is explained by *TD* and *TTR;* where $TI=TD+TTR$.
- *TTR* follows an exponential distribution for all service suppliers in all contracts, given by (7.1).
- There is prior knowledge $\pi(u)$ about u that can be assessed from experts.
- *TD* follows an exponential distribution for all service suppliers in all contracts, given by (7.10).
- *TD* and *TTR* are independent random variables.
- DM's preference structure fits a non compensatory rationality and requires an MCDM/A approach compatible with outranking relation preferences according to Chap. 2.
- DM's preference intra criterion preference structure fits for both attributes an exponential utility function for representing DM's one-dimensional preferences as in the deterministic administrative time model (de Almeida 2001a) and in the stochastic administrative time model (de Almeida 2001b), these functions are given by (7.3) and (7.4).

Since this model considers *TTR* and *TD* as two independent random variables, given by (7.1) and (7.10) respectively, *TI* is given by (7.14).

Despite the fact of this decision model is not considering tradeoffs, the DM behavior facing subjected to uncertainties is being modeled in the intra criterion valuation by the utility functions given by (7.3) and (7.4).

Due to the assumptions of this particular decision model, the costs are not affected by the state of nature, therefore each maintenance service supplier contract has its particular cost definition not affected by uncertainties, thus the cost criterion is evaluated directly by (7.4) according to each alternative's cost (de Almeida 2001b).

In the other hand, the response time represented by *TI* cannot be evaluated directly for each alternative as the cost, due to the interference of state of the nature uncertainties over its consequences.

For dealing with this situation, de Almeida (2002) considered the parameter ω, related to *TD*. Applying the utility function linearity property as in the previous models, $U_{TI}(\omega)$ is given by (7.24).

$$U_{TI}(\varpi) = \int_{TI} U_{TI}(TI)\Pr(TI \mid \varpi)dTI \qquad (7.24)$$

$\Pr(TI \mid \varpi)$ corresponds to (7.14), then (7.24) can be rewritten as (7.25):

$$U_{TI}(\varpi) = \frac{u\varpi}{(A_1 + \varpi)(A_1 + u)} \qquad (7.25)$$

Based in the intra criterion alternatives evaluation given by (7.4) and (7.25), the ELECTRE I method builds outranking relations based in concordance index *C(a,b)* and in a discordance index *D(a,b)*.

The concordance index is given by (7.26), and measures the relative advantage of each alternative *a* compared with an alternative *b* (Vincke 1992).

$$C(a,b) = \frac{\sum(W^{+} + 0.5W^{=})}{\sum(W^{+} + 0.5W^{=} + W^{-})}, \qquad (7.26)$$

where W^{+} corresponds to the sum of weights in which *a* is preferable to *b*, $W^{=}$ is the sum of the weights in which *a* is equal to *b*, and W^{-} is the sum of the weights in which *b* is preferable than *a*.

The discordance index is given by (7.27) for measuring the relative disadvantage of each alternative *a* compared to an alternative *b* (Vincke 1992).

$$D(a,b) = \max\left[\frac{(Z_{bk} - Z_{ak})}{(Z_k^{*} - Z_k^{-})}\right], \qquad (7.27)$$

where Z_{ak} is the evaluation of alternative *a* related to the criteria *k*, Z_{bk} is the evaluation of alternative *b* related to the criteria *k*, Z_k^{*} is the best degree of evaluation obtained for criteria *k*, and Z_k^{-} is the worst degree of evaluation obtained for criteria *k*.

Due to DM's preference structure assumed for this decision model, it was necessary to change the approach for evaluating the maintenance service suppliers contracts on the response time, differently than the previous approaches using MAUT.

Besides the adaptations required since different assumptions are made, is important to highlight that the set of parameters representing DM's preferences for each decision model has different meanings, thus the measurements to represent DM's preferences such as the weights used for ELECTRE method are incompatible with the required "weights" for MAUT, namely scale constants.

Therefore, building a decision model considering a different preferential paradigm is important to improve the accuracy of the available decision models for this class of problems, in order to give more flexibility for different types of DM as discussed in Chap. 2.

7.3.3 Maintenance Service Supplier Selection with Non Compensatory Preferences Including Dependability and Service Quality

A maintenance service supplier selection problem using a non compensatory MCDM/A approach is addressed (de Almeida 2005). This model approaches a situation that includes three criteria besides cost, namely the repair time, dependability and service quality using the ELECTRE I method.

The definition of dependability given by Slack and Lewis (2002) is that dependability is related to measuring the performance of the promised deliveries accomplishments. Therefore it represents a measurement about the chances of a service supplier succeed in keep its service level beneath pre established limits.

Thus, for a maintenance service supplier it is associated with the probability d_i of succeeding to perform the service under a response time faithful to the contract proposal *i*.

Service quality may have several definitions. The definition adopted (de Almeida 2005) for the decision model is that the service quality reflects the degree of mistakes introduced once a repair has been performed. Thus, it is represented by the probability q_i that no fault has being introduced during the repair service according to the expected conditions defined in the contract *i*.

With these extensions, this decision model was built to address the maintenance service supplier selection problem including these four criteria:

- Interruption time or response time (*TI*);
- Cost (*C*);
- Dependability (d_i);
- Service quality (q_i).

Similarly to the decision model presented in the previous sections, this decision model (de Almeida 2005) evaluates the benefits of a maintenance service supplier contract in terms of these three criteria and the cost related to the service contract.

Therefore, the maintenance service supplier contract performance now includes not only the response time as a reflect of its specific condition for spare provisioning and repair capability, but also the reliability of the maintenance service team in order to avoid introducing failures in the system, and also that its sizing would be enough to provide the service under the response time settled in the contract.

Thus, the model considers (de Almeida 2005) the following assumptions:

- *TI* is explained only by *TTR,* following the maintainability approach given by Goldman and Slattery (1977);
- *TTR* follows an exponential distribution for all service suppliers in all contracts, given by (7.1);
- Although there is prior knowledge $\pi(u)$ about u that can be assessed from experts, it is assumed that there is an uncertainty about the real value of u_i, withregard to the respective contract i;
- Based on the last assumption, d_i is defined as the probability that $u_i \geq u_{ie}$ for action a_i, therefore u_{ie} is the value committed by contract for u_i. Therefore, d_i is given by (7.28):

$$d_i = \int_{u_{ie}}^{\infty} \pi(u_i) du_i \tag{7.28}$$

- DM's preference structure fits a non compensatory rationality and requires an MCDM/A approach compatible with outranking relation preferences according to Chap. 2.
- DM's preference intra criterion preference structure fits for an exponential utility function for the attributes repair time and cost, given by (7.3) and (7.4), respectively. Once the higher is d_i , higher is $U_d(d_i)$, the DM utility function for dependability. Thus, a logarithm utility function is assumed for the dependability given by (7.29) (de Almeida 2005). Same applies to service quality, therefore assuming also a logarithm utility function, given by (7.30).

$$U_d(d_i) = B_3 + C_3 \ln(A_3 d_i) \tag{7.29}$$

$$U_q(q_i) = B_4 + C_4 \ln(A_4 q_i) \tag{7.30}$$

From the assumptions of this decision model costs are also not affected by the state of nature as in the previous section. Therefore the cost criterion is evaluated directly by (7.4) according to each alternative's cost (de Almeida 2005). Same applies to dependability and service quality, evaluated respectively by (7.29) and (7.30). From the assumption of the prior knowledge over *TTR* , the state of nature must be considered for evaluating the consequences on repair time by considering the parameter u_i instead of *TTR*. Similarly to the previous, the decision model proposed by de Almeida (2005) uses the linearity property to obtain $U_{TI}(u_i)$ from (7.31).

$$U_{TI}(u_i) = \int U_{TI}(TTR) \Pr(TTR \mid u_i) dTTR \tag{7.31}$$

Since $\Pr(TTR \mid u_i)$ is given by (7.1), then, applying (7.1) and (7.3) to (7.31), (7.32) is achieved.

$$U_{TI}(u_i) = \frac{u_i}{(A_1 + u_i)} \tag{7.32}$$

By assuming that *TI* is explained only on *TTR* is a simplification that may be adopted if necessary for a particular organizational condition. It depends on the specificities of each application. Such simplification in the decision model allows to deal with more accuracy with the parameters included in the evaluation for addressing the preferences over the attributes dependability and service quality of the maintenance service supplier contracts.

Another point to emphasize is that for the response time one may be interested in assessing directly over u_i given (7.32), however it is easier for a DM to have its preferences elicited directly over *TTR* than in u_i.

7.3.4 Maintenance Service Supplier Selection with Preference's Partial Information

In some situations, the DM may not feel comfortable about setting precise values for the decision model parameters, and thus an approach suitable for dealing with this situation should be used.

The decision models presented by Brito and de Almeida (2007) and Brito et al. (2010) for selecting maintenance service supplier contracts addressed such a particular situation, using an approach that enabled a recommendation to be made based on imprecise statements with regard to the decision maker's preferences and this was supported by an elicitation procedure.

Many DMs have difficulties to fix constant values for criteria 'weights' that must represent not only the importance of the criteria but also the compensation rates between criteria in additive value functions.

There may be several reasons for a decision maker to avoid precise statements. One of these may be that he is unsure if a parameter should be 0.75 or 0.7. Thus if a range of values can be used for such parameters, the decision maker can give more confident statements regarding the decision problem.

Brito and de Almeida (2007) considered three basic criteria: interruption time, applicant's dependability and contract cost in this model.

The dependability criterion is used to assess alternatives of contract in relation to "deadlines" being met. It is a measure related to keeping delivery promises, which it is represented by the probability of the company selected achieving the time to repair under a specified probability distribution, as set out in the contract proposal of the maintenance service supplier, similarly as in the model presented in last section.

To Brito and de Almeida (2007), these three criteria may be conflictive among alternatives. Usually, lower interruption times (times to repair) are related to better resource conditions, better spares provisioning and higher professional skills, and they often imply higher costs.

Besides, the dependability of the alternative is not directly related to the proposal conditions associated with interruption time, but it is assessed by the contracting company taking into consideration other aspects such as the applicant's reputation, previous services, the structure of repair facilities, etc.

The approach used in the decision model by Brito et al. (2010) considers utility functions aggregated by variable interdependent parameters for an additive function and uses the following criteria for evaluation:

- Mean time to repair (MTTR).
- Service supplier cost.
- Geographical spread of the service supplier network.
- Service supplier reputation.
- Compatibility of company cultures.

The specific problem considered by Brito et al. (2010) was related to power distribution services, which may also be extended to the telecommunications context.

The service supplier's performance on MTTR indicates its structure and capabilities, thereby reflecting its maintenance staff's skills, transportation resources, facilities and spares inventory.

The geographical spread of the service supplier reveals its logistical network structure, and relates to the number and spread of local branch offices, which gives flexibility and speed with regard to performing repairs. This is an important point for companies with widespread local branch offices, and it is directly related to the speed of service response and flexibility offered to the contracting organization or its several units

The service supplier's reputation is another important factor to be considered since this may avoid bad experiences from past services or even service level inconsistencies being repeated during the time span of the contract. Evaluating the service supplier in this respect may be from external sources, such as other companies that had previous experiences with the service supplier, verifying if payment of taxes to the government is up-to-date and possession of the due certifications in quality and/or safety norms.

Cultural compatibility is an issue that has become more and more relevant, since many organizations are seeking to establish long-term relations by building strategic partnerships. Allied to such strategic factors, many companies have added undertaking social and environmental responsibility activities to their organizational objectives. This includes their seeking sustainability and requiring this commitment also from their partners and suppliers.

By using variable interdependent parameters, Brito et al. (2010) considered the range for each parameter, assessing a lower and an upper bound. Another kind of imprecise information was the order (ranking) of the parameters. The assessment of these imprecise statements given by decision maker enabled dominance relations among the service suppliers to be established, based on the decision maker's assessed preferences.

According to Brito et al. (2010), in order to assess the performances of alternatives of contracts for the first two criteria, since MTTR and contract cost can be directly represented by values, utility values should be elicited using the due procedures. However, the last three criteria present a less objective feature; in this case, each candidate may be evaluated after completion of a questionnaire, which is constructed so as to obtain all the information required by the contracting organization in order to assess the candidates on each of the three criteria.

7.4 Other Approaches for Supplier Selection

The problem of supplier selection has been studied in many contexts, rather than the RRM context. Also, studies have been found in a broader way, therefore MCDM/A and other approaches to supplier evaluation and selection problem have been widely studied.

There are various decision making approaches proposed in the literature. Ho et al. (2010) presented a literature review on this topic emphasizing which approaches were frequently applied, which criteria were most considered, and investigated inadequacy with regard to the studies of the approaches found in the literature within their review of articles published in international journals from 2000 to 2008.

Among other approaches widely used for supplier evaluation and selection, Ho et al. (2010) highlight the use of: Analytic hierarchy process (AHP), Analytic network process (ANP), Case-based reasoning (CBR), Data envelopment analysis (DEA), Fuzzy set theory, Genetic algorithm (GA), Mathematical programming, Simple multi-attribute rating technique (SMART), and hybrid approaches.

Ho et al. (2010) evaluated MCDM/A approaches versus traditional cost based approaches. The advantages of applying MCDM/A approaches enable consideration to be given to important and relevant factors for the decision process other than cost.

Another recent literature review on supplier selection was presented by Chai et al. (2013), considering articles published in journals from 2008 to 2012 that presented applications of decision making techniques for supplier selection.

From the literature review conducted by Chai et al. (2013) many decision making approaches have been applied to these problems recently. Chai et al. (2013) identified twenty six decision making techniques applied for supplier evaluation and selection, and grouped these techniques into three categories: MCDM/A, mathematical programming and artificial intelligence techniques.

Supplier selection is an important topic, studied and tackled with many approaches, although most of them are not related to RRM context. The specific techniques applied to supplier selection problems are listed below by each of the categories considered by Chai et al. (2013):

- MCDM/A: AHP, ANP, ELECTRE, PROMETHEE, TOPSIS, VIKOR, DEMATEL, SMART, Multiobjective programming, Goal programming.
- Single Objective Mathematical programming: DEA, Linear programming, Nonlinear programming, Stochastic programming.
- Artificial intelligence: Genetic algorithm, Grey system theory, Neural networks, Rough set theory, Bayesian networks, Decision tree, Case-based reasoning, Particle swarm optimization, Support vector machine, Association rule, Ant colony algorithm, Dempster-Shafer theory of evidence.

The choice of a maintenance service supplier may be addressed by using different criteria and different techniques depending on the decision context, although on the particular context of maintenance there still a scarce literature, this is an important complex decision problem which includes strategic organizational objectives and consequences subjected to different kinds of states of nature.

References

Brito AJ de M, Almeida-Filho AT de, de Almeida AT (2010) Multi-criteria decision model for selecting repair contracts by applying utility theory and variable interdependent parameters. IMA J Manag Math 21:349–361

Brito AJ de M, de Almeida AT (2007) Multicriteria decision model for selecting maintenance contracts by applying utility theory and variable interdependent parameters. In: Carr M, Scarf P, Wang W (eds) Model. Ind. Maint. Reliab. Proc. Mimar. 6th IMA Int. Conf. Manchester, United Kingdom, pp 74–79

Buck-Lew M (1992) To outsource or not? Int J Inf Manage 12:3–20

Chai J, Liu JNK, Ngai EWT (2013) Application of decision-making techniques in supplier selection: A systematic review of literature. Expert Syst Appl 40:3872–3885

de Almeida AT (2001a) Repair contract decision model through additive utility function. J Qual Maint Eng 7:42–48

de Almeida AT (2001b) Multicriteria decision making on maintenance: Spares and contracts planning. Eur J Oper Res 129:235–241

de Almeida AT (2002) Multicriteria modelling for a repair contract problem based on utility and the ELECTRE I method. IMA J Manag Math 13:29–37

de Almeida AT (2005) Multicriteria Modelling of Repair Contract Based on Utility and ELECTRE I Method with Dependability and Service Quality Criteria. Ann Oper Res 138:113–126

de Almeida AT (2007) Multicriteria decision model for outsourcing contracts selection based on utility function and ELECTRE method. Comput Oper Res 34(12):3569–3574

de Almeida AT, Ferreira RJP, Cavalcante CAV (2015) A review of multicriteria and multi-objective models in maintenance and reliability problems. IMA Journal of Management Mathematics 26(3):249–271 de Almeida AT, Souza FMC (2001) Gestão da Manutenção: na Direção da Competitividade (Maintenance Management: Toward Competitiveness) Editora Universitária da UFPE. Recife

Goldman AS, Slattery TB (1977) Maintainability: a major element of system effectiveness. Robert E. Krieger Publishing Company, New York

Ho W, Xu X, Dey PK (2010) Multi-criteria decision making approaches for supplier evaluation and selection: A literature review. Eur J Oper Res 202:16–24

Kennedy WJ (1993) Modeling in-house vs. contract maintenance, with fixed costs and learning effects. Int J Prod Econ 32:277–283

Martin HH (1997) Contracting out maintenance and a plan for future research. J Qual Maint Eng, 3:81–90

Murthy DNP, Jack N (2008) Maintenance Outsourcing. In: Kobbacy KAH, Murthy DNP (eds) Complex Syst. Maint. Handb. SE - 15. Springer London, pp 373–393

Slack N, Lewis M (2002) Operations Strategy. Prentice Hall, London

Tsang AHC (2002) Strategic dimensions of maintenance management. J Qual Maint Eng 8:7–39

Vincke P (1992) Multicriteria Decision-Aid. John Wiley & Sons, New York

Wang W (2010) A model for maintenance service contract design, negotiation and optimization. Eur J Oper Res 201:239–246

Wideman RM (1992) Project and program risk management: a guide to managing project risks and opportunities. Project Management Institute

Wu S (2013) A review on coarse warranty data and analysis. Reliab Eng Syst Saf 114:1–11

Chapter 8
Spare Parts Planning Decisions

Abstract An important issue related to maintenance management is the problem of sizing the amount of spare parts. An excess number of spare parts results in financial losses. However, a lack of spare parts is also negative, because this may result in a loss of production due to the increased downtime of equipment. Therefore, spare parts should be available in quantity and at the right time. Spare part planning decisions need to evaluate multidimensional objectives, such as costs, profitability, reliability, availability and probability of stockout. Typically, these objectives are conflicting. Unlike a single objective approach, which often implies the poor performance of other objectives desired by the decision maker (DM), a multicriteria (MCDM/A) approach provides a spectrum of compromise solutions, which reflect the tradeoffs represented by DM's preference structure, by using a multi-attribute utility function. Another relevant aspect is the management of uncertainties about the reliability or maintainability of the system, using the concepts of Decision Theory and a Bayesian approach, which incorporate experts' prior knowledge. This chapter presents a model, based on Multi-attribute Utility Theory (MAUT), for spare parts sizing that considers aspects of the risk of inventory shortages and cost. Furthermore, an NSGA-II multi-objective model for multiple spare parts sizing is discussed. Finally, a model considering condition-based maintenance (CBM) is presented.

8.1 Introduction

Management of spare parts certainly has a positive influence on maintenance management, since this leads to the higher reliability and availability of equipment and therefore has a direct impact on business profitability. Therefore, one of the most important issues related to maintenance management is the problem of sizing the number of spare parts to be held in stock, bearing in mind that this affects the performance of maintenance, because the number of spare parts available directly affects the downtime or interruption to the full operation of a given piece of equipment (system). Spare parts should be available in quantity and at the right time. Just as stocking an excess number of spare parts results in losses or foregoing funds that a company could have applied elsewhere, a lack of spare

A.T. de Almeida et al., *Multicriteria and Multiobjective Models for Risk, Reliability and Maintenance Decision Analysis*, International Series in Operations Research & Management Science 231, DOI 10.1007/978-3-319-17969-8_8

parts is also negative, as this may well result in a loss of production due to increased downtime of equipment while awaiting delivery of the spare parts needed. Therefore, sizing the number of spare parts that optimally need to be held strongly influences a company′s costs and profitability. Consequently, the management of this resource is one of the most critical tasks in maintenance management (de Almeida and Souza 2001). This topic is relevant for many contexts, may they be related to individual plants, such as refinery (Porras and Dekker 2008), or to a logistic network (Syntetos et al. 2009).

When comparing with other types of inventory models, such as raw material for manufacturing processes, sizing and managing spare parts inventory is a far more complex task, considering that manufacturing inputs are usually is easier to forecast its demand, especially if comparing its turnover. Production inventories usually follows market rules, but spare parts are required based on failure rates and the system reliability design.

Thus, spare parts are sized according to its relative importance according to the system reliability. A bad decision on spare part sizing may lead to high losses, compromising company's profitability as well as its system availability.

According to British Standards 3843-1:1992, Terotechnology is the study that allows the maintenance of assets in optimal manner by the combination of management, financial, engineering and other practices applied to physical assets such as equipment considering its life cycle costs.

The number of spare parts sized must consider time to repair or time of service disruption. This decision must assure that parts required will be available when requested. Thus, the spare parts sizing problem has conflicting goals that are the increase of spare parts available as a contribution to increase equipment availability by the reduction of service disruption time. As in the other hand, the goal to reduce the inventory and purchase costs of spare parts.

Considering that manufacturing inventory model are not suitable to manage spare parts inventory due to the differences in the items demand for both cases, spare parts inventory considers that the demand for each item follows a stochastic process, represented by the random variable of equipment failure.

According to Marseguerra et al. (2005), to avoid risks to plant and costly plant being unavailable due to a shortage of spares parts, the latter are often over-stocked, thus leading to huge losses due to having invested unnecessarily in an excessive number of them or to too many of them becoming obsolete.

For Roda et al. (2014), spare parts management plays a relevant role for equipment-intensive companies. They review the type of criteria applied to spare parts classification. An important step of such a process is that of classifying spare parts (criticality) with a view to enabling different items to be properly managed by taking into account their peculiarities. Many advantages can be achieved by proper classification, e.g. an organization may align its policy for the stock management system with the criticality of the need for holding spare parts (Macchi et al. 2011); demand forecasting may be driven by data collected on the parts for different classes while improving the performance of equipment and the system overall may concern critical classes, thus making the work of the analyst easier by allowing him/her to concentrate on tests of inventory control policies currently in force. Forecasting spare parts demand is an important issue (Boylan and Syntetos 2010) for building related decision models.

Some studies have addressed the problem of determining the optimal spare parts inventory, such as by using gradient methods, dynamic programming, integer programming, mixed integer and nonlinear programming. Unfortunately, as mentioned by Marseguerra et al. (2005), such optimization techniques typically entail the use of simplified plants 'or systems' models about which predictions may be of questionable realism and reliability.

In general, spare parts management has at least two main objectives that are conflicting: to contribute to increased system availability by acquiring and stocking spare parts; i.e., ensuring the supply of spare parts in the proper amount to reduce interruption times and; to reduce the cost of buying and stocking spare parts (de Almeida 2001).

A decision model on provisions for spares assumes that at least one spare item is held in stock (de Almeida 2001; de Almeida 1996). Normally, when a failure occurs, in due time, the failed item is replaced by a spare part, which should be available in the depot. The faulty item is sent out for repair and after being repaired (good-as-new), it is shipped back to the plant depot where it serves as a spare. This decision problem uses an MCDM/A model in order to define how many additional spares should be provided in accordance with what criteria. de Almeida (2001) applies a Bayesian approach, based on prior probability distributions. Aronis et al. (2004) also uses prior distributions of the failure rates to forecast demand.

Some techniques for planning and inventory control were developed for the context of manufacturing systems (goods producing systems), and were later extended to the service producing systems. An example of this is Just-In-Time, which aims to meet the instantaneous demand i.e. only the amount necessary for the customer at that moment of need. These techniques are suitable for systems that have predictable demand and are determined by the client. However, in studies on reliability (in the maintenance context), demand is a probabilistic event, represented by the number of failures (a random variable). For this reason, the literature on sizing stocks of spare parts in the maintenance area addresses the question in very specific ways.

Other noteworthy conditions that make them different from production inventories are (Kennedy et al. 2002; Macchi et al. 2011): the number of spare parts in stock is often too large; the sourcing of spare parts is often limited to one or a few suppliers, causing constraints regarding procurement lead time and the costs; or in the opposite case of multiple sourcing, the related risk of variations in the quality of materials supplied can occur; obsolescence may be a problem; indeed, it is difficult to determine how many units of a spare part to stock for an obsolescent machine; the high variety in the characteristics of spare parts can normally be observed (the rates of consumption for some parts are very much higher than for others; some parts are characterized by being cheap to buy, while others are very expensive; often, procurement lead times vary greatly and may be lengthy, especially in the case of specific parts or those that have to be placed on order); and, the management process often lacks information visibility, due to poor inventory data record-keeping, inefficient or ineffective ordering processes and inventory management information being hidden in separated “silos”, these being only some typical reasons for such low visibility.

Duchessi et al. (1988) propose a top-down methodology, which classifies spares into distinct categories and associates appropriate controls with each category. This methodology identifies spare parts that do not have to be stocked. By eliminating these spare parts from the inventory, the manager can reduce costs and thus improve profits. Thereafter, it identifies critical spare parts that, if not in stock when needed, result in excessive downtime costs. Moreover, avoidance of downtime reduces production lead time and improves performance regarding on time delivery to customers. Finally, it displays a logical framework so that the need for and stock of spare parts can be matched with formal control policies, procedures and techniques.

Molenaers et al. (2012) propose a spare part classification method based on the criticality of an item, using an MCDM/A model. Starting from a multicriteria analysis, the proposed model converts relevant criteria on such criticality into a single score which then is considered the level of criticality of the item. This level is used to rationalize the efficiency of the spare parts inventory policy.

A literature review on MCDM/A approaches in reliability and maintenance shows work conducted related to spare parts sizing (de Almeida et al. 2015).

8.2 Some Sizing Approaches for Spare Parts in Repair

This text highlights some approaches to the problem of sizing the need for spare parts (de Almeida and Souza 2001):

- An approach based on the risk of inventory shortages;
- An approach based on the risk of inventory shortages by using prior knowledge;
- An approach under the cost constraint;
- An approach according to an MCDM/A model.

8.2.1 Relevant Factors to Sizing Spare Parts

The system type, whether repairable or not repairable, will influence how to size the need for spare parts. For non-repairable systems, the desired lifecycle of the system should be considered as a variable time T (de Almeida 1996; de Almeida and Souza 2001). It should be noted, therefore, that the size of the stock is defined by the difficulty in acquiring spare parts (price, delivery time, availability of more than one supplier, etc.) and by issues directly to inventory management (available space, cost storage, etc.).

As to repairable systems, the variable T is equal to the time at which the item will be repaired, i.e., the system is restored when the defective item is replaced with a similar one that is already in stock. The number of spare parts in this case is equal to $N_S = N + 1$, since the defective item returns to stock after being repaired (de Almeida 1996; de Almeida and Souza 2001).

Another issue that will influence spare parts management is related to the behavior of the failure rate over time. As seen, the number of items available for spare parts is directly related to the number of failures, which in turn is directly related to the reliability of the equipment (system). Therefore, the problem is directly related to the behavior that the variable deemed the number of failures is a function of time. One should also consider the independence of failures among the items that make up the system.

Under the analysis of the bathtub curve, spare parts management, in the repair context, is usually dealt with only in the second life stage that matches the useful life or the operational phase of the equipment. At this stage, it is assumed that the failure rate $\lambda(t)$ has a constant behavior as a function of time (the reliability function is represented by an exponential probability function).

In the first phase of the bathtub curve, in which the predominant faults are classified as early failures, these are usually covered by the equipment manufacturer's warranty, with no need for the user to direct efforts to solve this problem, i.e., it is not necessary to have spare parts in stock to cover this period in the lifecycle of the equipment. In some specific kinds of contract, it is interesting to analyze the possibility of having spare parts. From the manufacturer point of view, the sizing decision for this stage has to be made and may follow the model presented in this section, with proper assumptions.

In the third phase of the bathtub curve, the equipment is at the end of its useful life. Therefore it might not make much sense to study the problem of dimensioning the need for spare parts, in the context of repair, because at this stage the failure rate is high due to wear and tear. The failure rate $\lambda(t)$ increases with time so that repair is not sufficient to change the behavior of degenerative equipment, so the equipment has reached its use limit at this stage. At this stage, what remains is to consider the policy for preventive maintenance, replacement, reconstruction or overhaul. If economically feasible, this period may be prolonged as necessary until the equipment is deemed obsolescent and can then be discarded.

The spare parts sizing in this different context should use the information collected for the maintenance decision in that particular context. For instance, if a preventive maintenance model, such as one of those in Chap. 5, is applied, then, the information from the decision model regarding to the amount of replacements necessary in the planning time horizon is related to the sizing of the spare parts.

Another relevant factor to be considered for sizing the need for spare parts is technological outdating, which can be a limiting factor in the lifetime of a piece of equipment (system) and thus, may well shorten its life expectancy. Therefore, whenever the equipment becomes technologically outdated before the end of its useful life, the spare parts for it that are in stock lose their functionality in short periods, there being an economic loss (obsolete inventory) that need to be written off.

Furthermore, what to do about perishable goods (spare parts subject to degradation while held in stock) should be made of the Wilson model defined in Rezg et al. (2008) and Ben-Daya et al. (2009). Gopalakrishnan and Banerji (2013) point out that perishable spares, with a short shelf life, must be identified, and the First in First out method must be practiced. Therefore, the optimal sizing of the total quantity of each spare part has to be determined, and must take into consideration the objectives of minimizing the cost to the system and wastage (loss of materials due to deterioration) as investment constraints (Padmanabhan and Vrat 1990).

Van Volkenburg et al. (2014) develop a model which addresses the effects of the shelf-life of spare parts (perishable items) on optimizing the stocking of spare parts because certain conditions exacerbate their deterioration, thereby affecting the reliability of the system being supported or the spare part being found to be unserviceable when required. This is especially evident in non-repairable components that are stored for extended periods.

8.2.2 Approach Based on the Risk of Inventory Shortages

This approach involves determining a number of spare parts *N*, for a given value of risk of stock shortages α within a particular time value *T*. Thus, cost is considered in an indirect way, because as the desire is to reduce the risk α, there is a resultant increase in cost and vice versa. So the cost is obtained at the instant that defines what level of risk to run (i.e., will be determined at the time of choosing the value of α).

The risk of stock shortages α means the probability that the number of spare parts in stock is less than the number of failures *x*, namely, $P(x > N)$ (Probability of Stockout of the spare parts (*PS*)). Thus, the Margin of Safety *(MOS)* is defined as $MOS = 1 - \alpha$, i.e. $MOS = 1 - PS$, which is a measure of the probability that the stock will not fall outside the range considered (de Almeida 1996; de Almeida and Souza 2001). Therefore,

$$MOS = 1 - \alpha = P(x \leq N) \tag{8.1}$$

where *N* is the number of spare parts kept in stock. Notice that the *MOS* corresponds to the cumulative probability distribution of the number of failures. Assuming a Poisson Process, for a system comprising n items:

$$MOS = P(x \le N) = \sum_{k=0}^{N} \frac{(n\lambda T)^k e^{-n\lambda T}}{k!} \tag{8.2}$$

As $\lambda_S = n\lambda$:

$$MOS = P(x \le N) = \sum_{k=0}^{N} \frac{(\lambda_S T)^k e^{-\lambda_S T}}{k!} \tag{8.3}$$

where N is the number of items held in stock; λ_S rate is the failure system and T is the time interval.

Finally, there is a procedure for calculating the number of spare parts of N, for some α risk of stockout or the *MOS*, so that, respectively:

$$P(x > N) < \alpha \tag{8.4}$$

or

$$P(x \le N) \ge MOS \tag{8.5}$$

Therefore, the procedure consists of finding each of the possible values of N, starting from $N = 0$ (no spare parts) until the first value of N is found that meets the condition of keeping the risk within the limit established.

8.2.3 Approach Based on the Risk of Inventory Shortages by using Prior Knowledge

There are practical situations where it is not possible to obtain the values for the parameters of reliability and/or maintainability of a system. This approach provides a procedure for sizing the need for spare parts where at least one of these parameters is not known (de Almeida 1996; de Almeida and Souza 2001).

In such cases, prior knowledge is used (as discussed in Chap. 3) with respect to the reliability and/or maintainability of the system. Therefore, the prior probability is applied to obtain the expected values of risk or *MOS* in order to determine what the appropriate number of spare parts to be held in stock should be.

For this study, three scenarios are considered:

- Lack of knowledge about the failure rate λ ;
- Lack of knowledge about the *MTTR* (Mean Time to Repair);
- Lack of knowledge about the parameter λ and *MTTR*.

In the first case, the absence of λ, one obtains the prior probability of λ: $\pi(\lambda)$; in the second case, one should obtain the prior probability on the *MTTR*: $\pi(MTTR)$. I.e., these two functions of prior probabilities are required. Therefore, to address the problem of sizing the need for spare parts in the absence of data, it is considered the expected value of *MOS* data as being derived from previously defined situations, respectively:

$$E_{\lambda}[MOS] = \int_{\lambda} (MOS)\pi(\lambda)d\lambda = \int_{\lambda} \left(\sum_{k=0}^{N} \frac{(n\lambda T)^k e^{-n\lambda T}}{k!} \right) \pi(\lambda)d\lambda \tag{8.6}$$

$$\begin{aligned} E_{MTTR}[MOS] &= \int_{MTTR} (MOS)\pi(MTTR)dMTTR \\ &= \int_{MTTR} \left(\sum_{k=0}^{N} \frac{(n\lambda(MTTR))^k e^{-n\lambda(MTTR)}}{k!} \right) \pi(MTTR)dMTTR \end{aligned} \tag{8.7}$$

$$E_{\lambda,MTTR}[MOS] = \int_{MTTR} \left[\int_{\lambda} (MOS)\pi(\lambda)d\lambda \right] \pi(MTTR)dMTTR \tag{8.8}$$

As shown in Chap. 3, the procedures for eliciting prior knowledge about parameters of interest, based on the Equiprobable Intervals Method (Raiffa 1968), is a very viable alternative.

8.2.4 Approach under the Cost Constraint

In this approach, the attribute of value is treated directly. The cost criterion is seen as a limiting factor, since for a given cost limit, one tries to minimize the risk of breakage of stock shortages, i.e., starting from the amount of (monetary) resources that have been allocated in order to determine the optimal number of inventory items that should be held (Goldman and Slatery 1977).

The decision process is to determine the threshold value of cost, which depends on the availability of resources. This determines what the number of spare parts of *N* is that minimizes the risk of stock shortages.

Therefore, an expression for calculating the number of spare parts of *N* needed, so that:

$$C_T \leq C_0 \tag{8.9}$$

where C_T is the final total cost and C_0 is the amount of resources available is:

$$C_T = N.C \tag{8.10}$$

where N is the number of spare parts and C is the unit cost of each item. The procedure consists of finding the value of N, starting from $N = 0$ (no spare parts), that minimizes the risk of inventory shortages that meets the condition previously established by the budget constraint.

In more complex situations, such as a modularized system that has equipment with a number of J different types of modules, where each module has its own λ_j failure rate. The final total cost is obtained by summing the total final costs of each module:

$$C_T = \sum_{j=0}^{J} N_j.C_j \tag{8.11}$$

where C_j is the individual cost of the module type j; and, N_j is the number of modules (items) of type j.

For this approach, a *MOS* in which $N = \left(N_1, \ldots, N_j\right)$ is considered to represent the probability that there will be no stock shortages of N, i.e., this is given by the product operator between the *MOS* modules:

$$MOS_N = \prod_{j=0}^{J} P\left(x_j \leq N_j\right) \tag{8.12}$$

Therefore the solution is to maximize MOS_N, such that the total cost is less than or equal to the initial cost imposed as a constraint, $C_T\left(N\right) \leq C_0$. For this, one needs to use non-linear optimization.

8.2.5 Use of MCDM/A Model

This approach comes from the perspective of multidimensionality (de Almeida 2001; de Almeida 1996). Multiple objectives can be aggregated to decision models such as by taking into account maximizing system revenues and minimizing the volume of total spares. However, the fact remains that when attempting to optimize any design aspect of an engineered system, the analyst is frequently faced with the demand of achieving several targets (e.g. low costs, high revenues, high reliability, low accident risks), some of which may very well be in conflict with each other. At the same time, several peculiar requirements (e.g. in spacecraft systems, maximum allowable weight, volume, etc.) should also be satisfied (Marseguerra et al. 2005).

Unlike a single objective approach, which often implies the poor performance of other desired objectives, the set identified by a multi-objective approach provides a spectrum of 'acceptable' solutions and attempts made to find a compromise. This is one of the advantages of working under the multidimensional aspect (multiobjective).

According to Jajimoggala et al. (2012), a systematic evaluation of the criticality of spare parts is the key to effective control of spare parts recorded by inventory systems. There are many factors to be considered which will measure the criticality of spare parts for maintenance activities, and these evaluation procedures involve several objectives and it is often necessary to make compromises among possibly conflicting tangible and intangible factors. In their study, Jajimoggala et al. (2012) use an MCDM approach to solve this kind of problem, namely by using a three-phase hybrid model: the first stage involves identifying the criteria; the second is to prioritize the different criteria using fuzzy ANP; and, finally in the third phase, the criticality of spare parts is ranked using fuzzy TOPSIS.

Molenaers et al. (2012) propose a spare parts classification method based on the criticality of items. Starting from a multicriteria analysis, the proposed model converts relevant criteria impacting the criticality of an item into a single score which thus represents the criticality level. This level is used to rationalize the efficiency of the spare parts inventory policy. They consider the following criteria:

- The criticality of equipment;
- The probability of an item failing;
- Replenishment time;
- The number of potential suppliers;
- The availability of technical specifications, and;
- Maintenance type.

de Almeida (2001) considers two criteria (risk and cost), which are combined through a multi-attribute utility function in a decision model for provisioning spares, i.e., spares provisioning can also be modelled by the multicriteria utility function (by the MAUT method) based on the need for spares and the risk of no supply.

MAUT has been rarely used for the spares provisioning problem. Several criteria such as: availability, risk, and cost are used to estimate the volume of spares needed. Risk is a common criterion used in Mickel and Heim (1990). Other models optimize a single criterion such as availability or risk subject to costs (Goldman and Slattery 1977; Barlow et al. 1996).

The combination of these two attributes will be made through the utility function (multiattribute function). The decision to be adopted in this approach is to determine values for the attributes of cost (C) and risk (α) in order to maximize the multi-attribute utility function of consequence $U(C,\alpha)$.

By concepts of Decision Theory, one has the space of action (a), which consists of the possible quantities of spare parts *N*, which is the element on which the decision maker can act in order to achieve the desired goal, in which case it is the maximization of the multiattribute utility function $U(C,\alpha)$.

Furthermore, the state of nature (θ), the reliability of the system and the maintainability of its structure need to be considered. They can be represented by the parameters of reliability and maintainability, which can be obtained by using a statistical procedure or the use of experts' prior knowledge, as previously mentioned (de Almeida 2001; de Almeida 1996).

The observation data (obtained by an analysis of likelihood) concerning the reliability and maintainability of the system under study allows some considerations about the behavior of the state of nature (θ). The state of nature has a direct influence on the results of the consequences of the decision made by the decision maker, but the decision maker does not have any control or influence over the state of nature.

The consequence space is given by the expected utility value $(E[u(C,\alpha)])$. The function $u(C,\alpha)$ is obtained using the procedure for eliciting a multi-attribute utility function (described in Keeney and Raiffa (1976)), which defines the DM's preference structure with respect to the values of cost and risk of stock shortages.

Finally, the goal of the approach is to determine the number of *N* spare parts, which maximizes $u(C,\alpha)$. The mathematical model is given by:

$$u(\theta,a)=E_{p|\theta,a}(u(p))=\int u(p)P(p\,|\,\theta,a)dp=\int u(C,\alpha)P(C,\alpha\,|\,\theta,a)dp \quad (8.13)$$

where $P(C,\alpha\,|\,\theta,a)$ is a consequence function given the decision-maker adopted an action a (defined by the combination of a certain value and risk α and cost *C*) and the state of nature (θ) that had occurred.

It is emphasized that the cost of spare parts depends exclusively on the action chosen to maximize the utility function, and there is not, for this case, dependence on the state of nature $C(\lambda,T)$.

On the other hand, the α risk depends on the state of nature (θ) and action (a) to be adopted by the decision maker, and thus there is no dependency between the attributes, thereby allowing the use of the conditional probability function $P(p\,|\,\theta,a)$ as follows:

$$P(p \mid \theta, a) = P(C, \alpha \mid \theta, a) = P(C \mid \theta, a) \cdot P(\alpha \mid \theta, a) \tag{8.14}$$

For every action (a_i) determined by the decision maker there is an associated cost, so having a deterministic view of the result of the cost function:

$$P(C_i \mid \theta, a) = 1 \;\; \textit{iff}, \; a = a_i \tag{8.15}$$

Hence,

$$P(p \mid \theta, a) = P(C \mid \theta, a) \cdot P(\alpha \mid \theta, a) = 1 \cdot P(\alpha \mid \theta, a) = P(\alpha \mid \theta, a) \tag{8.16}$$

For $P(\alpha \mid \theta, a)$, risk $\alpha = 1 - P(\lambda, N, T)$; therefore, with the values from λ, N and T, the value of risk α can be determined. Thus similarly:

$$\begin{cases} P(\alpha \mid \theta, a) = P(\alpha = 1 - MOS \mid \theta, a) = 1 \text{ iff, } MOS = \sum_{k=0}^{N} \frac{(n\lambda T)^k e^{-n\lambda T}}{k!} \\ P(\alpha \mid \theta, a) = P(\alpha = 1 - MOS \mid \theta, a) = 0 \text{ iff, } MOS \neq \sum_{k=0}^{N} \frac{(n\lambda T)^k e^{-n\lambda T}}{k!} \end{cases} \tag{8.17}$$

The behavior of the random number x to a system fault variable is represented by a Poisson probability distribution, due to the fact that the failure rate parameter shows a constant behavior in time function, given that the reliability function is represented by the function exponential probability. Likewise, in a deterministic view, $P(\alpha = 1 - MOS \mid \theta, a) = 1$.

Therefore, the maximization of $u(C, \alpha)$ is obtained through a deterministic approach, where $u(\theta, a_i) = u(p \mid \theta, a_i)$ that consists in determining the number of N spare parts to be available in stock.

Among the criteria of Decision Theory for maximizing a utility function one highlights the Bayesian method, which consists of choosing the action a_i, which is the number of spare parts available in stock in order to maximize expected utility, $u(\theta, a_i)$, depending on the prior probability $\pi(\theta)$ according to the following formulation:

$$\max_{a_i} \int_{\theta} u(a_i, \theta) \pi(\theta) d\theta \tag{8.18}$$

In this model, one considers that the state of nature has two dimensions: one that matches the reliability of the equipment comprising the system, represented by the rate of system failures (λ_s); and the second dimension is the maintainability of a repairable system, represented by the mean time to repair (*MTTR*). Therefore, the probability distribution of the state of nature $\pi(\theta)$ is defined as $\pi(MTTR)$. So the expected utility can be expressed by:

$$E_{p|\theta,a}[u(p)] = E_{C,\alpha|\theta,a}[u(C,\alpha)] = \int_{T_0}^{T_{\max}} \int_{\lambda_0}^{\lambda_{\max}} u(\lambda,T;N)\pi(\lambda)\pi(T)d\lambda dT \qquad (8.19)$$

Therefore, maximizing the multi-attribute utility function is obtained by maximizing the expected utility function, depending on the number of spare parts:

$$\max_N \left[E_{\lambda,T}[u(\lambda,T;N)]\right] \qquad (8.20)$$

Formulated mathematically as:

$$\max_N \left[\int_{T_0}^{T_{\max}} \int_{\lambda_0}^{\lambda_{\max}} u(\lambda,T;N)\pi(\lambda)\pi(T)d\lambda dT\right], \text{ iff,}$$
$$\alpha = 1 - \sum_{k=0}^{N} \frac{(n\lambda T)^k e^{-n\lambda T}}{k!} \text{ and } C = a_i \cdot C_i \qquad (8.21)$$

8.3 Multiple Spare Parts Sizing

Spare parts management usually considers issues of a single item and an independent decision problem of other system items. However, many items that require spare parts to be available compete for the same resources. For example, someone may evaluate the possibility of decreasing the number of spare parts of a given item to balance the increase in the number of spare parts of another item. It should be noted that these alternatives can have distinct global performance measures.

When considering the modeling of multiple spare parts simultaneously, the maintenance manager is not only interested in defining the optimal number of each item. In this case he is interested in finding an optimized allocation of resources, given that distributing a limited amount of resources among various items is considered a typical portfolio problem.

In this context, several papers address issues of spare parts policy using multi-objective genetic algorithms. Marseguerra et al. (2005) explore the possibility of using genetic algorithms to optimize the number of spare parts in a multicomponent system. The objectives considered are the maximization of system revenues and the minimization of the total volume of spares. A Monte Carlo simulation approach was defined to deal with system failure, repair and replacement stochastic processes. Ilgin and Tunali (2007) propose an approach using genetic algorithms to optimize preventive maintenance and spares policies of a manufacturing system operating in the automotive sector, while Lee et al. (2008) develop a framework that integrates a multi-objective evolutionary algorithm (MOEA) with a multi-objective computing budget allocation (MOCBA) method for the multi-objective simulation optimization problem of allocating spare parts for aircraft.

In general, the maintenance manager is interested in minimizing the total cost of spare parts and also minimizing the probability of stockout. In this section, a multi-objective genetic algorithm is proposed to tackle the multi spare parts problem. Firstly, it is important to point out that this model assumes the 'fixed' shape of the failure rate. Finkelstein and Cha (2013) state that this assumption is well founded for the spare parts setting. This feature of a failure rate makes sense only for spare parts used in corrective maintenance, rather than parts used in preventive maintenance for which consumption can be defined by a periodic replacement strategy. This setting justifies the use of the Poisson distribution for the computations of the probability of stockout of an item. It is also assumed that each item has a failure rate and purchase cost. Each item can be classified into two levels of importance to the system, in order to manage different levels of criticality of items to the system. There are critical and non-critical items which compete for the same resources of a limited budget.

A multi-objective model based on NSGA-II is developed to aid the management of multiple spare parts. The model was tested in an urban passenger bus transport company.

8.3.1 The Mathematical Model

The mathematical model proposed for the spare parts inventory problem was a multi-objective optimization model, where the objectives are to optimize the average of the probability of stockout, and the total cost of the spare parts purchased, which should be minimized.

The model aims to answer the main question inherent in any process of inventory management: what is the ideal inventory level for a spare part which can be obtained at minimum cost and provide maximum availability.

As shown by Kennedy et al. (2002) and Bevilacqua et al. (2008), the Poisson distribution is the most widely-used mathematical-statistical model in the literature for optimizing inventories of spare parts, and is premised on modelling the behavior of demand for the item by a probability distribution, which is widely used to describe rare random events. The Poisson distribution is represented by (8.22):

$$P_x(t) = \frac{(\lambda t)^x e^{-\lambda t}}{x!} \tag{8.22}$$

where x represents the consumption of replacement parts by time interval for which the wish is to estimate the probability; t is the time interval considered; λ is the historical consumption rate of the replacement parts by unit of time; and, $P_x(t)$ is the probability of there being x requests for replacement parts during time interval t.

The model can be represented as follows:

$$PS_i = P(x > N_i) = 1 - P(x \le N) = 1 - MOS = 1 - \sum_{i=0}^{N_i} \frac{(\lambda_i t)^i e^{-\lambda_i t}}{i!} \tag{8.23}$$

where PS_i is the probability of stockout of the i-th spare part associated with the quantity N_i; i represents each spare part (critical or non-critical); N_i is the amount in stock of each spare part; λ_i is the monthly rate of consumption of the i-th spare part; and, C_i is the unit cost of the i-th spare part.

Hence, the problem consists of minimizing the objective functions (8.24) and (8.25).

$$\min\left[\sum_{i=1}^{n} PS_i\right] \tag{8.24}$$

$$\min\left[\sum_{i=1}^{n} C_i\right] \tag{8.25}$$

The genetic algorithm used to solve the spare parts inventory problem was an adaptation of the elitist multi-objective genetic algorithm proposed by Deb et al. (2002), NSGA II. The algorithm is based on sorting the chromosomes based on non-dominance to find the Pareto front of multi-objective problems, as well as to maintain the good solutions during the evolutionary process, since it is an elitist algorithm. In the proposed algorithm, the length of the chromosome is equal to the number of different spare parts, where each gene represents the amount to be purchased.

The selection of parents for crossover operation to generate offspring is random, and, for each set of parents chosen, two descendants are generated using the genetic operators. The crossover operator is based on random selection of a position of the chromosome that is the cutoff point. The offspring 1 inherit the genes of parent 1 until the cutoff position and, from there on, they inherit the genes of parent 2. Additionally, the offspring 2, thus, inherit the genes of the parent 2 until the cutoff point, and, from that point on, inherit the genes of parent 1.

8.3.2 Case Study

For the case study, the procedure followed was: (1) to define the components of the replacement spare parts of the buses to be studied, such that 14 critical items used were defined in corrective maintenance actions alone, and (2) to define the parameters to be quantified, which in this case were the consumption rate λ, the unit price and the initial stock. These parameters are shown in Table 8.1. Items which cause transport service failure were determined as critical items.

The data were collected from an urban collective public transport company that has been operating buses for more than 25 years and is regarded as anonymous in this study. This company has a fleet of 83 buses, the average age of which is 5.69 years, which run 600,000 km per month.

Table 8.1 Initial data from the critical items

Part	λ	Unit Cost (C)	Initial Stock
P1	0.636	168.00	0
P2	0.364	660.00	0
P3	0.727	2,700.00	0
P4	1.0	1,843.00	0
P5	0.364	23.00	0
P6	2.273	882.00	0
P7	2.727	1,176.00	1
P8	1.364	136.00	0
P9	0.909	200.00	0
P10	18.0	1,180.00	13
P11	7.636	380.00	4
P12	85.455	14.49	74
P13	1.273	268.00	0
P14	3.0	30.00	1

Initially, the algorithm was run for the critical items. 99.9% was established as the upper limit of the average of probability of stockout, which ensures a high quality of service obtained by the purchase of critical items. As mentioned in the previous section, the initial solution of the genetic algorithm was generated as the result of applying the model based on cost benefit ratio (CB), which is obtained through the cost ratio by varying the level of service caused by the purchase of spares. The algorithm based on cost-benefit presented a total of 142 solutions in the Pareto front. The population size chosen for use in NSGA II was twice the amount of solutions obtained by the CB model, i.e., 284. The first 142 chromosomes of the initial solution are the same chromosomes obtained by the CB model, and the other half of the chromosomes is random generated, so that the diversity in the solutions is preserved.

After 250 iterations of the genetic algorithm, a total of 276 solutions on the Pareto front are obtained. Of this number, only 21 coincide with the solutions generated by the CB model, which shows that the genetic operators have diversified the initial solution a lot. If analyzed together, the two models generated a total of 397 different solutions, of which 363 are non-dominated. A comparative graph of the solutions of the model based on cost-benefit and NSGA II for critical items is shown in Fig. 8.1 and Fig 8.2.

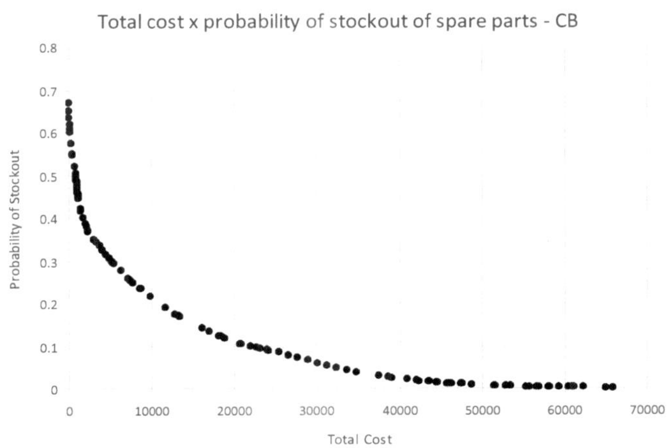

Fig. 8.1 Total cost versus probability of stockout for critical items (Cost-benefit)

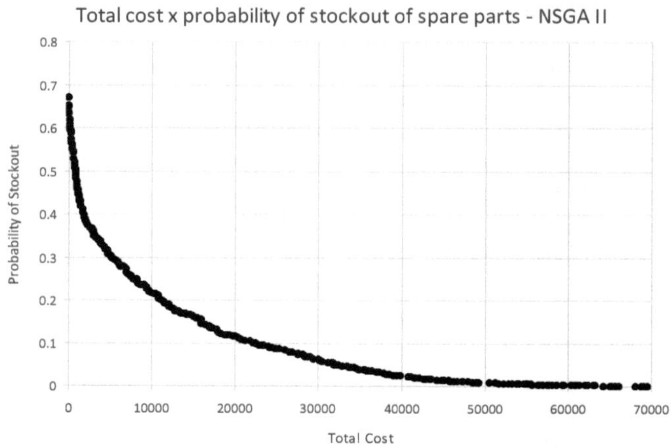

Fig. 8.2 Total cost versus probability of stockout for critical items (NSGA-II)

In Fig. 8.1 and Fig. 8.2, it can be realized that from an *PS* (Probability of Stockout) of 10% ahead, it is soon seen that there is a "saturation" in the curve, thus reversing the prevailing logic, i.e. there are then high investments for little return (low reduction in the probability of stockout), which clearly it is not worth the company's spending resources on, in this situation.

The option to deal separately with the critical items allows the manager to have greater flexibility in managing the contingency element of his/her budget, and certainly yields a better result for inventory management as it allows the logic of the program, based on the typical problem of a portfolio of assets, in which several items, within their group of criticality, compete for resources simultaneously, thus gaining the one that presents the lowest cost-benefit index, which brings a gain to the operation as a whole.

It can be concluded that the model developed and applied in a real situation reached its objective, as it allowed important parameters for controlling the inventory of replacement spare parts to be monitored efficiently, thus contributing to the management of an urban bus company. It is further understood that this model can be replicated in any other company which has replacement spare parts in its inventory and consumes them when carrying out corrective maintenance.

8.4 Spare Parts for CBM

Probability of failure, inspection period, holding cost and obsolescence are crucial factors in modeling spare parts inventories. In terms of the maintenance policy, one can argue that condition monitoring may well give a better forecast of the residual life of the system monitored and can support better decisions about acquiring spares, in the context of Condition-Based Maintenance (CBM). The demand for spare parts is commonly generated by the need for preventive maintenance actions and by failures. Besides, maintenance costs are influenced by the availability of spare parts. It needs to be borne in mind that penalties due to spare parts being unavailable usually consist of the cost of, for example, extended downtime and the high costs of acquiring spare parts in emergency situations. Technical advances in condition monitoring techniques have provided a means to ensure high availability and to reduce scheduled and unscheduled production shutdowns (Ferreira and Wang 2012; Wang 2012; Wang 2008).

Studies on spare parts dealing with failure based maintenance, age- or block-based replacement policies have been of interest to several researchers. A review of the literature on spare parts inventories was conducted by Kennedy et al. (2002). They set out how research directions were conducted on this theme, although no CBM model was found to be used in spare part inventory control.

In terms of age-based replacement, a comparative study between optimal stocking policy and the Barlow–Proschan age replacement policy shows the cost effectiveness of the former. Joint stocking and age-based replacement policy were studied by Zohrulb Kabir and Al-Olayan (1996). Barabadi et al. (2014) evaluated reliability models with covariates in the field of spare part predictions. Van Horenbeek et al. (2013) proposed a joint maintenance and inventory policy model based on predictive information in order to evaluate the added value of predictive information (RUL) for multi-component systems.

Rezg et al. (2008) proposed a joint optimal inventory control and preventive maintenance policy subject to a required minimum level of availability. Diallo et al. (2008) suggested a mathematical model which aims at maximizing the availability of a system under a budget constraint where the parameters for placing orders and the intervals of preventive maintenance are derived, based on the lifetime distribution of the system. Vaughan (2005) assumed that the demands for spare parts due to regularly scheduled preventive maintenance and the random failure of units in service are independent. Chang et al. (2005) proposed an inventory model for spare parts taking into account the criticality of the production equipment. Aronis et al. (2004) applied a Bayesian approach to forecast demand, based on prior distributions of the failure rates, where the number of spare parts is determined for a required level of service.

CBM strategies should be integrated with traditional models to indicate when and how many spares are needed. A hybrid of simulation and analytical models is proposed taking into account the residual life of equipment estimated by using condition monitoring techniques. The advantages of CBM include reducing the cost of the inventory, making better predictions of and planning for the volume of spares required, since the residual life can be better predicted by condition information, which can lead to better forecasting of the quantity of spare parts needed.

In CBM modelling, it is important recognize two fundamental classes of problems. Wang (2008) explains the concepts of direct and indirect monitoring. In direct monitoring, the actual condition of the item can be observed, and a critical level can be set up. While in indirect monitoring, one can only collect measurements related to the actual condition of the item monitored in a stochastic manner. Some enhancements to direct monitoring have been made. Rausch and Liao (2010) develop a model for joint production and spare part inventory based on CBM, where the condition monitored can be observed directly. Wang et al. (2009) present the concept of condition-based replacement and spares provisioning policy, and through the simulation method and the genetic algorithm, the decision variables were jointly optimized for minimizing the cost rate. Linear and exponential degradation models is evaluated by Elwany and Gebraeel (2008) in order to support the dynamic decisions of replacement and inventory based on the physical condition of the equipment. Ferreira et al. (2009) propose a multicriteria decision model to determine inspection intervals of condition monitoring based on delay time analysis.

Ferreira and Wang (2012) assume that there are a number of identical component items used in a system, which are condition monitored periodically. For example, there may be many critical and identical bearings installed on a paper machine and proper maintenance of these bearings should lead to better availability and lowering the operating costs of the machine as a result of having condition monitoring information. Having the appropriate volume of spare parts available at the right time is a relevant issue when managing maintenance activities.

Opportunities for maintenance actions such as condition monitoring and preventive maintenance times, likewise the order time and arrivals of acquisitions, are illustrated in Fig. 8.3 in order to represent the main features of the problem.

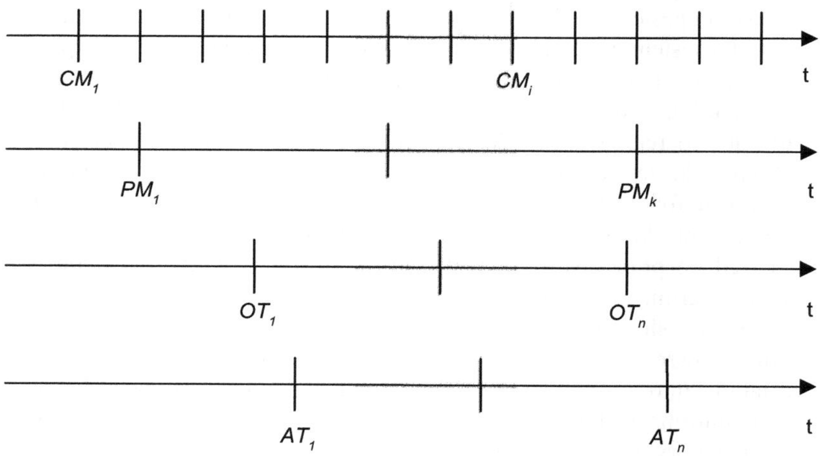

Fig. 8.3 Intervals of condition monitoring (*CM*), preventive maintenance (*PM*), order time (*OT*), order arrival (*AT*) and lead time (τ) of spare parts

Thus there is a decision problem at each replacement opportunity as shown in Fig. 8.4. Basically there are alternatives:

1. Alternative 1 – Replacements at present moment, subject to a stock level;
2. Alternative 2 – Replacements at the next condition monitoring opportunity (CM_i), subject to probability of failures and stock level;
3. Alternative 3 – Replacements at the next order arrival time opportunity (AT_n), subject to probability of failures and stock level;
4. Alternative 4 – Replacements at the next preventive maintenance opportunity (PM_k), subject to probability of failures and stock level;

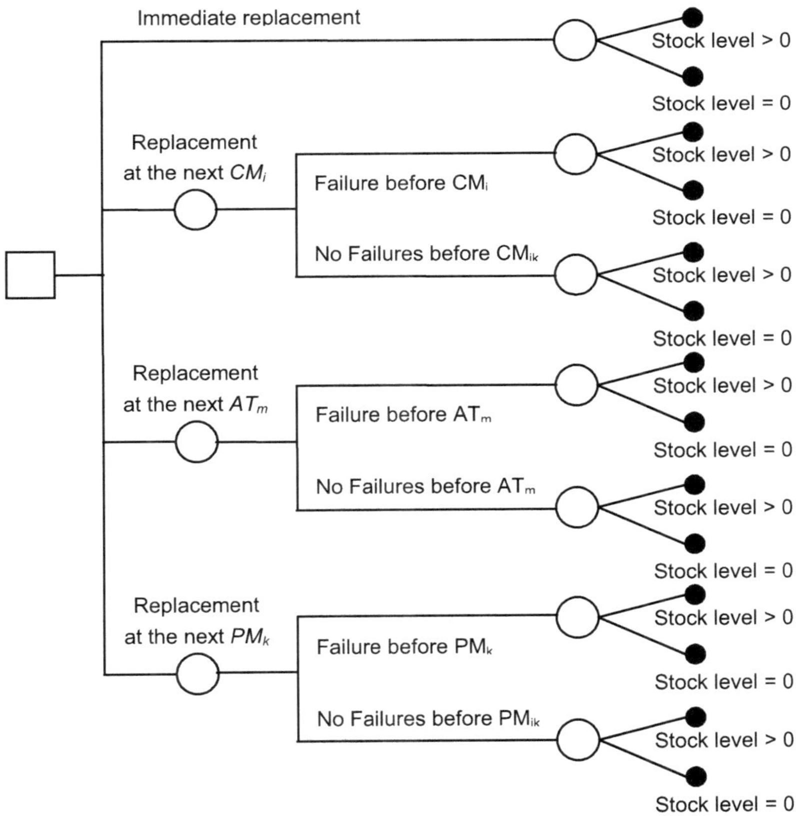

Fig. 8.4 Decision tree at each replacement opportunity

From the decision tree structure of Fig. 8.4, it is possible to evaluate dynamically the performance of a given maintenance policy by comparing the results and analyzing the replacement times. Based on the monitored information, the structure of the risk of stockout, costs and estimates of the residual life are derived. These estimates may vary which implies that the need for spare parts may change.

This section addresses a spare part problem by using condition monitoring information. CBM is a more cost effective maintenance policy than time-based maintenance since it can avoid premature maintenance or replacement while making better forecasts of the need for spare parts. In traditional CBM models, there is a strong assumption that spare parts are always available when needed, and in several practical situations this is not true.

References

Aronis K-P, Magou I, Dekker R, Tagaras G (2004) Inventory control of spare parts using a Bayesian approach: A case study. Eur J Oper Res 154:730–739

Barabadi A, Barabady J, Markeset T (2014) Application of reliability models with covariates in spare part prediction and optimization - A case study. Reliab Eng Syst Saf 123:1–7

Barlow RE, Proschan F (1965) Mathematical theory of reliability. John Wiley & Sons, New York

Ben-Daya M, Duffuaa SO, Raouf A, et al. (2009) Handbook of Maintenance Management and Engineering. Springer, London

Bevilacqua M, Ciarapica FE, Giacchetta G (2008) Spare parts inventory control for the maintenance of productive plants. 2008 IEEE Int. Conf. Ind. Eng. Eng. Manag. IEEM 2008. IEEE, Singapore, pp 1380–1384

Boylan JE, Syntetos AA (2010) Spare parts management: A review of forecasting research and extensions. IMA J Manag Math 21:227–237

Chang PL, Chou YC, Huang MG (2005) A (r,r,Q) inventory model for spare parts involving equipment criticality. Int J Prod Econ 97:66–74

de Almeida AT (1996) Multicriteria for spares provisioning using additive utility function. In: International Conference on Operational Research for Development, IFORS-ICORD II. Rio de Janeiro, RJ, pp 1414–1418.

de Almeida AT (2001) Multicriteria decision making on maintenance: Spares and contracts planning. Eur J Oper Res 129:235–241

de Almeida AT, Ferreira RJP, Cavalcante CAV (2015) A review of multicriteria and multi-objective models in maintenance and reliability problems. IMA Journal of Management Mathematics 26(3):249–271

de Almeida AT, Souza FMC (2001) Gestão da Manutenção: na Direção da Competitividade (Maintenance Management: Toward Competitiveness) Editora Universitária da UFPE. Recife

Deb K, Pratap A, Agarwal S, Meyarivan T (2002) A fast and elitist multiobjective genetic algorithm: NSGA-II. IEEE Trans Evol Comput 6:182–197

Diallo C, Ait-Kadi D, Chelbi A (2008) (s,Q) Spare parts provisioning strategy for periodically replaced systems. IEEE Trans Reliab 57:134–139

Duchessi P, Tayi GK, Levy JB (1988) A Conceptual Approach for Managing of Spare Parts. Int J Phys Distrib Logist Manag 18:8–15

Elwany AH, Gebraeel NZ (2008) Sensor-driven prognostic models for equipment replacement and spare parts inventory. IIE Trans 40:629–639

Ferreira RJP, de Almeida AT, Cavalcante CAV (2009) A multi-criteria decision model to determine inspection intervals of condition monitoring based on delay time analysis. Reliab Eng Syst Saf 94:905–912

Ferreira RJP, Wang W (2012) Spare parts optimisation subject to condition monitoring. In: 11th International Probabilistic Safety Assessment and Management. Conference and the Annual European Safety and Reliability Conference, Helsinki, Finland, 25-29 June 2012

Finkelstein M, Cha JH (2013) Stochastic Modeling for Reliability: Shocks, Burn-in and Heterogeneous populations. Springer, London

Goldman AS, Slattery TB (1977) Maintainability: a major element of system effectiveness. Robert E. Krieger Publishing Company, New York

Gopalakrishnan P, Banerji AK (2013) Maintenance and Spare Parts Management. PHI Learning, New Delhi

Ilgin MA, Tunali S (2007) Joint optimization of spare parts inventory and maintenance policies using genetic algorithms. Int J Adv Manuf Technol 34:594–604

Jajimoggala S, Rao VVSK, Beela S (2012) Spare parts criticality evaluation using hybrid multiple criteria decision making technique. Int J Inf Decis Sci 4:350

Kabir ABMZ, Al-Olayan AS (1996) A stocking policy for spare part provisioning under age based preventive replacement. Eur J Oper Res 90:171–181

Keeney RL, Raiffa H (1976) Decisions with multiple objectives: Preferences and Value Trade-Offs. Wiley Series in Probability and Mathematical Statistics. Wiley and Sons, New York

Kennedy WJ, Wayne Patterson J, Fredendall LD (2002) An overview of recent literature on spare parts inventories. Int J Prod Econ 76:201–215

Lee LH, Chew EP, Teng S, Chen Y (2008) Multi-objective simulation-based evolutionary algorithm for an aircraft spare parts allocation problem. Eur J Oper Res 189:476–491

Macchi M, Fumagalli L, Pinto R, Cavalieri S (2011) A Decision Making Framework for Managing Maintenance Spare Parts in Case of Lumpy Demand: Action Research in the Avionic Sector. In: Altay N, Litteral LA (eds) Serv. Parts Manag. pp 89–104

Marseguerra M, Zio E, Podofillini L (2005) Multiobjective spare part allocation by means of genetic algorithms and Monte Carlo simulation. Reliab Eng Syst Saf 87:325–335

Mickel LS, Heim RL (1990) The spares calculator: a visual aid to provisioning. Annu. Proc. Reliab. Maintainab. Symp. IEEE, Los Angeles, CA, pp 410 – 414

Molenaers A, Baets H, Pintelon L, Waeyenbergh G (2012) Criticality classification of spare parts: A case study. Int J Prod Econ 140:570–578

Padmanabhan G, Vrat P (1990) Analysis of multi-item inventory systems under resource constraints: A non-linear goal programming approach. Eng Costs Prod Econ 20:121–127

Porras E, Dekker R (2008) An inventory control system for spare parts at a refinery: An empirical comparison of different re-order point methods. Eur J Oper Res 184:101–132

Raiffa H (1968) Decision analysis: introductory lectures on choices under uncertainty. Addison-Wesley, London

Rausch M, Liao H (2010) Joint production and spare part inventory control strategy driven by condition based maintenance. IEEE Trans Reliab 59:507–516

Rezg N, Dellagi S, Chelbi A (2008) Joint optimal inventory control and preventive maintenance policy. Int J Prod Res 46:5349–5365

Roda I, Macchi M, Fumagalli L, Viveros P (2014) A review of multi-criteria classification of spare parts: From literature analysis to industrial evidences. J Manuf Technol Manag 25:528–549

Syntetos A, Keyes M, Babai M (2009) Demand categorisation in a European spare parts logistics network. Int J Oper Prod Manag 29:292–316

Van Horenbeek A, Scarf P, Cavalcante CAV, Pintelon L (2013) On the use of predictive information in a joint maintenance and inventory policy. In: Steenbergen RDJM, VanGelder PHAJM, Miraglia S, Vrouwenvelder ACWMT (eds) 22nd Annual Conference on European Safety and Reliability (ESREL), Amsterdam, 2013. Safety, Reliability and Risk Analysis: Beyond the Horizon. Taylor & Francis Group, London, UK, p 758

Van Volkenburg C, Montgomery N, Banjevic D, Jardine A (2014) The effect of deterioration on spare parts holding. 2014 Reliab. Maintainab. Symp. IEEE, Colorado Springs, CO, pp 1–6

Vaughan TS (2005) Failure replacement and preventive maintenance spare parts ordering policy. Eur J Oper Res 161:183–190

Wang L, Chu J, Mao W (2009) A condition-based replacement and spare provisioning policy for deteriorating systems with uncertain deterioration to failure. Eur J Oper Res 194:184–205

Wang W (2008) Condition-based maintenance modelling. In: Kobbacy KAH, Prabhakar Murthy DN (eds) Complex Syst. Maint. Handb. Springer London, pp 111–131

Wang W (2012) A stochastic model for joint spare parts inventory and planned maintenance optimisation. Eur J Oper Res 216:127–139

Chapter 9
Decision on Redundancy Allocation

Abstract: Redundancy allocation is a decision that involves assessing and choosing where to locate additional components or subassemblies, above the minimum required for an existing system to operate, in order to promote the system's reliability. Specifically, the field of multi-objective redundancy allocation has received several contributions since the 1970s and the combinatorial complexity of these problems has mainly encouraged researchers to develop search algorithms focused on the Pareto front definition, the most frequent approach in this literature. Finding a set of non-dominated solutions based on heuristics is a step that demands much computational effort to solve the problem. Despite these difficulties, the DM's preferences should be evaluated in order to recommend a solution that represents the best compromise among the criteria considered, such as reliability, cost and weight. This chapter covers redundancy allocation problems from a multi-criteria perspective. Therefore, basic concepts related to the typical criteria and tradeoff in redundancy allocation problems are presented and a brief review of the literature on MCDM/A redundancy allocation is given. To illustrate the MCDM/A approach for redundancy allocation, a decision model considering a standby system based on Multi-attribute Utility Theory (MAUT) is presented including the DM's behavior to risk (prone, neutral and risk averse). The problem approached in this chapter involves a question about how to select a suitable maintenance strategy in order to evaluate the tradeoff between a system's availability and cost, including experts' prior knowledge to deal with the uncertainty of failure and repair rate parameters.

9.1 Introduction

Redundancy allocation is a decision that involves assessing and choosing where to locate additional components or subassemblies, above the minimum required for an existing system to operate, in order to promote system reliability. This theme is one of the classic issues in reliability theory, in which the system design seeks to balance fundamental factors such as reliability, cost and weight. The balance amongst these factors has been the subject of research since the classic publications on reliability theory, such as that by Barlow and Prochan (1965).

A.T. de Almeida et al., *Multicriteria and Multiobjective Models for Risk, Reliability and Maintenance Decision Analysis*, International Series in Operations Research & Management Science 231, DOI 10.1007/978-3-319-17969-8_9

Barlow and Prochan (1965) presented models involving redundancy and revealed how to allocate redundancy among the various subsystems under linear constraints on weight, volume and cost, in order to maximize system reliability. The problem of maximizing the reliability of a series system subject to one or more constraints on total cost, weight, volume, has a number of variations depending on whether the redundancy is parallel or standby. In parallel systems, redundant units operate simultaneously, and they are subject to failure. In standby redundancy, redundant units are place on stand-by as spares and used successively for replacement, and they are not subject to failure while in standby condition. In some active parallel redundant configurations, it may be required that k out of the n units must be working for the system to function. The reliability of a k out of n system, with n independent components in which all the unit reliabilities are equal, is expressed by the binomial reliability function (O'Connor and Kleyner 2012; Kuo and Zuo 2003).

Even though Barlow and Prochan (1965) did not use the terminology of an MCDM/A problem, they indicated a class of problems in which no specific set of constraints is provided. In this case, one may wish to generate a family of non-dominated allocations in terms of reliability and cost both in parallel as well as standby redundancy.

The concept of non-dominated solutions is defined by Barlow and Prochan (1965) as: $x^0 = (x_1^0, x_2^0, \ldots, x_n^0)$ is non-dominated if $R(x)>R(x^0)$ implies $c_j(x)>c_j(x^0)$ for some j, whereas $R(x) = R(x^0)$ implies either $c_j(x)>c_j(x^0)$ for some j or $c_j(x)=c_j(x^0)$ for j, where $c_j(x) = \sum_{i=1}^{n} c_{ij} x_i$.

This property is the same as the classical definition of the Pareto Front used in multiobjective formulations presented in Chap. 2. Barlow and Prochan (1965) stated that if the set consisting of all non-dominated redundancy allocations is obtained (the complete family of non-dominated redundancy allocations), then the solution of a redundancy allocation problem with a set of constraints must be a member of this family. In other words, Barlow and Prochan (1965) realized that the mono-objective formulation is a particular case of the MCDM/A formulation and the solution of a mono-objective case is one from the Pareto Front set.

They presented a procedure that is able to generate an incomplete family of non-dominated allocations in a single cost factor. The procedure is based on the principle of adding the most reliability obtained per dollar spent in each iteration, starting with no redundancy in the system. For a multiple cost factor case, a simple weighted function of reliability is proposed and arbitrarily chosen values of weights are recommended. They also suggested a procedure to find a complete family of non-dominated allocations based on the dynamic programming algorithm of Kettelle Jr (1962).

A literature review of the optimal redundancy allocation models was carried out by Tillman et al. (1977). They classified early references in the field in terms of optimization techniques and system configurations. Among the optimization techniques, no MCDM/A approach was cited.

Kuo and Prasad (2000) updated the literature review of Tillman et al. (1977) and this included identifying if an MCDM/A approach had been analyzed as a way to help optimize system-reliability. They found twelve papers within this scope. They stated that an MCDM/A approach was an important but not widely studied problem in reliability optimization. Although some exact methods can be used to solve redundancy allocation problems, heuristics used include: ant colony optimization method; hybrid genetic algorithm and tabu search.

Kuo and Wan (2007) cited multiobjective optimization as a recent topic and indicated eleven references to this, which have published since 2000 in this field. They defined four problem structures, namely: 1) The traditional reliability-redundancy allocation problem; 2) The percentile life optimization problem; 3) Multi-state system optimization; 4) Multiobjective optimization.

Some kinds of system configuration are defined as shown in Fig. 9.1 and Fig. 9.2.

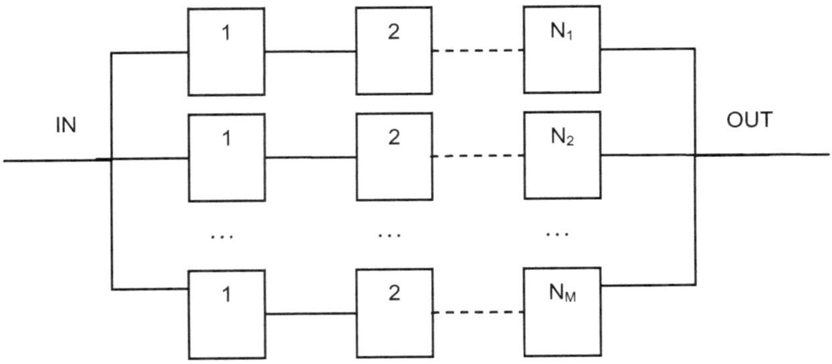

Fig. 9.1 Mixed series-parallel system, *N* components are connected in series, and *M* such series connections are connected in parallel to form the system

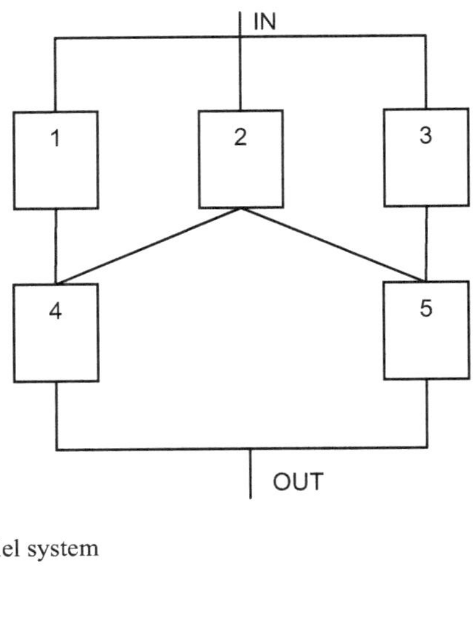

Fig. 9.2 Non series-parallel system

A simple version of the redundancy allocation problem is shown in Fig. 9.3. It is a series system which regards system reliability as an objective function $R(x_j)$. The system has n Stages in series with $x_j + 1$ independent identical distributed units in parallel in Stage j. x_j is the number of parallel redundant components in Stage j. c_{ij} is the cost of type i of each component in Stage j. The cost types include monetary values, weight and volume. p_j is the reliability of each component in Stage j. c_j is some specified cost of each component in Stage j. r is the number of cost types considered. It is assumed that all units fail independently.

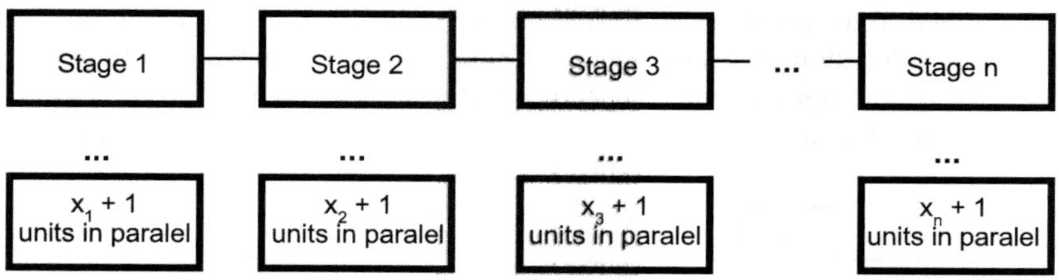

Fig. 9.3 Structure of a simple redundancy allocation problem

The problem is mathematically stated as follows in (9.1):

$$
\begin{aligned}
&\text{maximize}\, R(x_j) = \prod_{j=1}^{n} (1-(1-p_j)^{x_j+1}) \\
&\text{subject to} \sum_{i=1}^{n} c_{ij} x_j \le c_j, \quad j = 1,\ldots,r \\
&\qquad 0 \le x_j \le u_j, \quad x_j \,\text{integer}, \; j = 1,\ldots,n
\end{aligned}
\tag{9.1}
$$

In mono-objective formulations, it is possible to maximize the reliability of a system subject to the constraints on the amount of available resources or to minimize the cost of some resource subject to the constraint that the reliability of the system must meet a specified reliability target. Cost, weight and volume can be limited by constraints.

Kuo and Zuo (2003) presented some measures for the importance of a component, such as structural importance, reliability importance, criticality importance and relative criticality. These factors can be useful in order to compare components in terms of their importance to a system.

Kuo and Zhu (2012) defined three types of standby redundancy: hot standby, warm standby, and cold standby. A hot standby has the same failure rate as the active component. A cold standby has a zero failure rate. Warm standby implies that inactive components have a failure rate that is between zero and the failure rate of active components. A warm standby and a hot standby may fail while in the standby condition, but a cold standby will not fail.

Some limitations of redundancy allocation problems are important to be considered. For example, redundant components can be subjected to the same external loads and common failures modes that limit the effectiveness of the redundancy (Paté-Cornell et al. 2004).

In terms of classifying models, there are redundancy models which assume that only two component states are possible: the operating and failed states. But there are some models assuming more than two component states. These are called multi-state systems.

According to a literature review (de Almeida et al. 2015) on reliability and maintenance models based on MCDM/A approaches, 18.8% of the of publications are related to redundancy allocation. A set of relevant publications of the MCDM/A redundancy allocation problems is presented in Table 9.1. Most articles use Reliability, Cost and Weight as optimization objectives. Among the techniques for finding solutions, a diverse range of proposals has been suggested.

Table 9.1 A list of publications on MCDM/A redundancy allocation problems

References	Reliability	Cost	Weight	Other Criteria	Search method
Khalili-Damghani et al. (2013)	X	X	X		Multiobjective particle swarm optimization
Garg and Sharma (2013)	X	X			Fuzzy multiobjective particle swarm optimization
Cao et al. (2013)	X	X	X		Decomposition-based approach
Sahoo et al. (2012)	X	X			Tchebycheff; Lexicographic; Genetic Algorithms
Safari (2012)	X	X			NSGA-II
Okafor and Sun (2012)	X	X			Genetic Pareto set identification algorithm
Khalili-Damghani and Amiri (2012)	X	X	X		ε-constraint method and data envelopment analysis
Zio and Bazzo (2011a); Zio and Bazzo (2011b)		X	X	Availability	Clustering procedure; Level Diagrams and MOGA
Li et al. (2009)	X	X	X		NSGA-II and data envelopment analysis
Kumar et al. (2009)	X	X			Multiobjective hierarchical genetic algorithm; SPEA2 and NSGA-II
Tian et al. (2008)		X		System utility	Physical programming; Genetic algorithms

(continued)

Table 9.1 (continued)

Limbourg and Kochs (2008)		X		Life distribution	Feature models; NSGA-II
Taboada et al. (2008)		X	X	Availability	Multiobjective multi-state genetic algorithm
Zhao et al. (2007)		X	X		Multiobjective ant colony system;
Taboada et al. (2007)	X	X	X		NSGA
Liang et al. (2007)	X	X			Variable neighbourhood search
Chiang and Chen (2007)		X		Availability and net profit	Simulated annealing and genetic algorithms
Tian and Zuo (2006)		X	X	System performance utility	Physical programming; genetic algorithms and fuzzy theory
Salazar et al. (2006)	X	X			NSGA-II
Coit and Konak (2006)				Subsystem reliability	Multiple weighted objective heuristic; linear programming
Marseguerra et al. (2005)	X			Reliability estimated variance	Genetic algorithms and Monte Carlo simulation
Coit et al. (2004)	X			Reliability estimated variance	Weighted sum;
Elegbede and Adjallah (2003)		X		Availability	Weighted sum; Genetic algorithms
Huang (1997)	X	X			Fuzzy and multiobjective optimization;
de Almeida and Souza (1993); de Almeida and Bohoris (1996)		X		Interruption time	Multi-attribute utility theory
Gen et al. (1993)	X	X	X		Fuzzy goal programming model
Dhingra (1992); Rao and Dhingra (1992)	X	X	X		Fuzzy goal-programming and goal-attainment
Misra and Sharma (1991)	X	X	X		Efficient search multiobjective programming; min-max concept
Sakawa (1980); Sakawa (1981)	X	X	X	Volume	Surrogate Worth Trade-off method and dual decomposition method
Sakawa (1978)	X	X			Surrogate Worth Trade-off method
Inagaki et al. (1978)	X	X	X		Interactive Optimization

From this set of 35 publications listed in Table 9.1, 23 used Reliability as objective function (65.7%); 32 used Cost (91.4%) and 17 used Weight (48.6%) on multiobjective redundancy allocation problems.

Redundancy allocation problems are complex by nature. Chern (1992) evaluated the computational complexity of allocating reliability redundancy in a series system and proved that some reliability redundancy optimization problems are Non-deterministic Polynomial-time hard (NP-hard).

Due to the complexity of the problem, there is a focus of research with emphasis on use of heuristic methods to find solutions of Pareto fronts. However, an absence of a preference modeling is relevant shortcoming in the selection process of alternatives. There is a real need on the part of DM in choosing which of the set of the Pareto solutions provides the best balance for a given preferences structure.

9.2 An MCDM/A Model for a 2-Unit Redundant Standby System

In this section, a decision model (de Almeida and Souza 1993) for a standby system based on the MAUT is presented. This model addresses the waiting time to call a repair facility when the first piece of equipment of a 2-unit standby system fails. The first failure implies only a reliability reduction, not system failure, since the other unit is still operating. This scheme of waiting-time when the first fault occurs avoids overtime costs in the repair facility. An expert prior knowledge approach is lead in order to deal with the uncertainty of the parameters failure and repair rates. Another decision model (de Almeida and Bohoris 1996) extends this first model, introducing a Gamma distribution to the repair time. The possible states for a 2-unit redundant standby system are shown in Fig. 9.4.

The problem involves a question about how to select a suitable maintenance strategy in order to combine system availability and cost preferences. There is an assumption that the capacity of repair is limited, and instantaneous repair is not applicable. An MCDM/A approach can solve the conflicting requirements of system availability and cost through a multi-attribute utility function taking into account DM's preferences over these requirements. In this way, MAUT can also deal with uncertainty of the consequences.

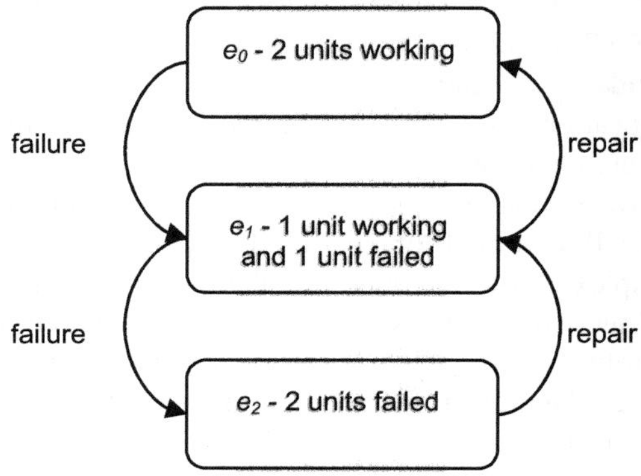

Fig. 9.4 States for a 2-unit redundant standby system

It is noteworthy that several redundancy allocation models assume that the system configuration is fixed for a given time horizon, which reflects an emphasis on design aspects and system reliability, corresponding to a planning stage, prior to system operation. Moreover, the maintenance actions define a strategy that would have a balanced way in terms of cost and availability on the system operation phase. Assuming the design phase the system was planned with redundant units operating in standby, the time limit in which a repairman perform the repair or replacement of a failed unit needs to be established. Clearly, there is a conflict between the cost of maintenance and system availability. The parameters of the model are given in Table 9.2 (de Almeida and Souza 1993).

Table 9.2 Model parameters

Parameters	Description
λ	Failure rate of the equipment
μ	Repair rate of the equipment
a	An action, element of the action space, representing the maintenance strategy
e_0, e_1, e_2	State of the system when [0, 1, 2] of the units failed
T_a	Decision variable representing the repair delay corresponding to a
T_0	Time at which the first failure occurs
T_1	Time at which the second failure occurs
T_2	Time at which the first-failed unit resumes operation, which could be returning the system to e_0
TTR	T_2 - T_a
$\pi_1(\lambda)$	Prior knowledge distribution about λ
$\pi_2(\mu)$	Prior knowledge distribution about μ
A_i	Scale parameter of π_i

(continued)

Table 9.2 (continued)

B_i	Shape parameter of π_i
C_i	Cost for a_i
FC_i	Fixed cost for a_i
CR_i	Repair cost-rate for a_i
MCR_i	Mean CR_i
$U\{TI,C\}$	Multi-attribute utility function for interruption time and cost

The assumptions of the model (de Almeida and Souza 1993) are:

1. The probability distribution of failure of two units are identically distributed;
2. Each unit has two states: good and failed;
3. The system is down when no unit is available for operation;
4. There is one repair facility;
5. Failure rate (λ) is constant and the number of failures follows a Poisson distribution;
6. A unit repaired becomes as good as new;
7. Repair rate is constant (μ);
8. If during a repair of a failed unit, the other unit also fails, the latter unit waits for repair until the first unit is repaired;
9. There is prior knowledge about λ and μ represented as prior probability distributions over these parameters;
10. Failure and repair states are *s*-independent;
11. The DM has a structure of preferences over the consequence space (*TI*,*C*) according to the axiomatic preferences of the utility theory;
12. *C* and *TI* are *s*-independent;
13. The objective function is to maximize the multi-attribute utility function $U\{TI,C\}$

The decision model building was based on the context of a telecommunication system of an electric power company with a 2-unit standby redundant system. The DM's preference elicitation over consequences (interruption time and cost) produces a multi-attribute utility function, which is introduced into the decision model, according to (9.5). The expected utility of alternatives is given by (9.2).

$$E_{(\lambda,\mu)}\{U\{(\lambda,\mu),a_i\}\} = \int_{\lambda_0}^{\lambda_m}\int_{\mu_0}^{\mu_m} \pi_1(\lambda)\cdot\pi_2(\mu)\cdot U\{(\lambda,\mu),a_i\}d\lambda d\mu \qquad (9.2)$$

where:

$$\pi_1(\lambda) = (B_1/A_1)\cdot(\lambda/A_1)^{B_1-1}\cdot\exp[-(\lambda/A_1)^{B_1}] \qquad (9.3)$$

$$\pi_2(\mu) = (B_2 / A_2) \cdot (\mu / A_2)^{B_2 - 1} \cdot \exp[-(\mu / A_2)^{B_2}] \tag{9.4}$$

$$U\{TI, C\} = K_t \cdot \exp(-K_{kt} \cdot TI) + Kc \cdot U\{C_i\} \tag{9.5}$$

$$C_i = FC_i + (\lambda / \mu) \cdot CR_i \tag{9.6}$$

$$U\{(\lambda, \mu), a_i\} = K_c \cdot U\{C_i\} + \left(\frac{K_t \cdot \mu}{K_{kt} + \mu} \right) \cdot \left[1 + \left(\frac{K_{kt}}{\lambda + \mu} \right) \cdot \exp(-\lambda \cdot T_{ai}) \right]. \tag{9.7}$$

The problem is solved by applying (9.2) into (9.8).

$$Max_{a_i} (E_{(\lambda, \mu)} \{U\{(\lambda, \mu), a_i\}\}) \tag{9.8}$$

Prior knowledge about the states of nature can be obtained from prior distributions of λ and μ. There are several prior probability elicitation procedures available in the literature, such as that given by Winkler (1967). The elicitation procedure applied is based on equal probable intervals. Based on experts on the equipment and the system maintainability, respectively, these $\pi_1(\lambda)$ and $\pi_2(\mu)$ were obtained according to (9.3) and (9.4), which are illustrated in Fig. 9.5 and Fig. 9.6.

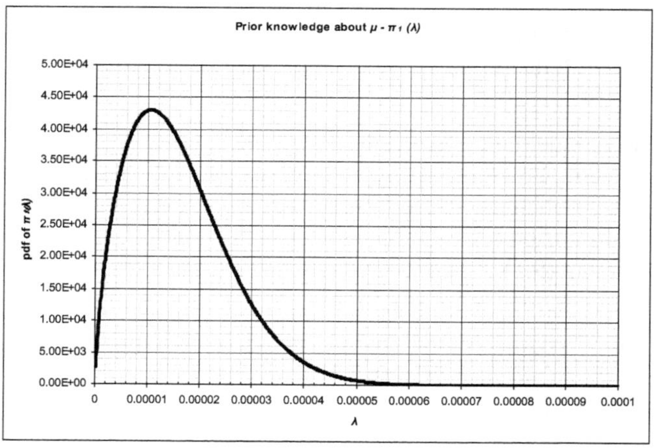

Fig. 9.5 Prior knowledge about λ, $\pi_1(\lambda)$ with $A_1 = 18.06 \cdot 10^{-6}$ and $B_1 = 1.68$

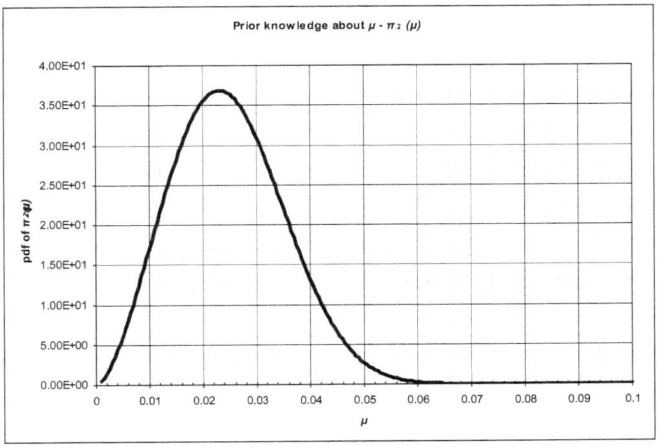

Fig. 9.6 Prior knowledge about μ, $\pi_2(\mu)$ with $A_2 = 0.028$ and $B_2 = 2.57$

There are three possible situations in the state e_1 (de Almeida and Souza 1993):

- $T_1 > T_a$, and $T_1 > T_2$; therefore $TI = 0$;
- $T_1 > T_a$, and $T_1 < T_2$; therefore $TI = T_2 - T_1 > 0$;
- Otherwise, there is an emergency, and the repair facility is called immediately, so that T_a is set equal to T_1.

Then, (9.9) represents the interruption time formulation.

$$TI = \max(0, \min(T_a + TTR - T_1, TTR)) \tag{9.9}$$

The set of alternatives for this problem is represented by maintenance strategies in terms of repair delay, as follows:

- a_1 - There is no repair delay, then $T_{a1} = 0$. It is assumed that maintenance department have infra-structure and resources to repair a unit immediately upon a failure.
- a_2 - There is zero repair-delay only during usual work hours and there is repair-delay during non-usual work hours, then T_{a2} is a random variable between 0 and 14 hours. It is assumed that the expected value of T_{a2} is equal to 7 hours.
- a_3 - Zero repair-delay is only in the usual work hours, with a cheaper structure, but the accessibility is lower, then T_{a3} is a random variable between 0 and 62 hours. It is assumed that the expected value of T_{a3} is equal to 31 hours.
- a_4 - A repair delay is allowed so that the resources are shared with other tasks. Thus, $T_{a4} = 360$ hours.

Fixed cost and repair cost rate (FC_i and CR_i) were obtained from the company for these four alternatives and the alternative a_3 got the best performance of the multi-attribute utility function.

References

Barlow RE, Proschan F (1965) Mathematical theory of reliability. John Wiley & Sons, New York

Cao D, Murat A, Chinnam RB (2013) Efficient exact optimization of multi-objective redundancy allocation problems in series-parallel systems. Reliab Eng Syst Saf 111:154–163

Chiang C-H, Chen L-H (2007) Availability allocation and multi-objective optimization for parallel–series systems. Eur J Oper Res 180:1231–1244

Coit DW, Jin T, Wattanapongsakorn N (2004) System optimization with component reliability estimation uncertainty: a multi-criteria approach. Reliab IEEE Trans 53:369–380

Coit DW, Konak A (2006) Multiple weighted objectives heuristic for the redundancy allocation problem. IEEE Trans Reliab 55:551–558

de Almeida AT, Bohoris GA (1996) Decision theory in maintenance strategy of standby system with gamma-distribution repair-time. Reliab IEEE Trans 45:216–219

de Almeida AT, Ferreira RJP, Cavalcante CAV (2015) A review of multicriteria and multi-objective models in maintenance and reliability problems. IMA Journal of Management Mathematics 26(3):249–271

de Almeida AT, Souza FMC (1993) Decision theory in maintenance strategy for a 2-unit redundant standby system. Reliab IEEE Trans 42:401–407

Dhingra AK (1992) Optimal apportionment of reliability and redundancy in series systems under multiple objectives. Reliab IEEE Trans 41:576–582

Elegbede C, Adjallah K (2003) Availability allocation to repairable systems with genetic algorithms: a multi-objective formulation. Reliab Eng Syst Saf 82:319–330

Garg H, Sharma SP (2013) Multi-objective reliability-redundancy allocation problem using particle swarm optimization. Comput Ind Eng 64:247–255

Gen M, Ida K, Tsujimura Y, Kim CE (1993) Large-scale 0–1 fuzzy goal programming and its application to reliability optimization problem. Comput Ind Eng 24:539–549

Huang H-Z (1997) Fuzzy multi-objective optimization decision-making of reliability of series system. Microelectron Reliab 37:447–449

Inagaki T, Inoue K, Akashi H (1978) Interactive Optimization of System Reliability Under Multiple Objectives. Reliab IEEE Trans R-27:264–267

Kettelle Jr JD (1962) Least-Cost Allocations of Reliability Investment. Oper Res 10:249–265

Khalili-Damghani K, Abtahi A-R, Tavana M (2013) A new multi-objective particle swarm optimization method for solving reliability redundancy allocation problems. Reliab Eng Syst Saf 111:58–75

Khalili-Damghani K, Amiri M (2012) Solving binary-state multi-objective reliability redundancy allocation series-parallel problem using efficient epsilon-constraint, multi-start partial bound enumeration algorithm, and DEA. Reliab Eng Syst Saf 103:35–44

Kumar R, Izui K, Yoshimura M, Nishiwaki S (2009) Multi-objective hierarchical genetic algorithms for multilevel redundancy allocation optimization. Reliab Eng Syst Saf 94:891–904

Kuo W, Prasad VR (2000) An annotated overview of system-reliability optimization. Reliab IEEE Trans 49:176–187

Kuo W, Wan R (2007) Recent Advances in Optimal Reliability Allocation. In: Levitin G (ed) Comput. Intell. Reliab. Eng. SE - 1. Springer Berlin Heidelberg, pp 1–36

Kuo W, Zhu X (2012) Importance measures in reliability, risk, and optimization: principles and applications. John Wiley & Sons, New York

Kuo W, Zuo MJ (2003) Optimal reliability modeling: principles and applications. John Wiley & Sons, New York

Li Z, Liao H, Coit DW (2009) A two-stage approach for multi-objective decision making with applications to system reliability optimization. Reliab Eng Syst Saf 94:1585–1592

Liang Y-C, Lo M-H, Chen Y-C (2007) Variable neighbourhood search for redundancy allocation problems. IMA J Manag Math 18:135–155

Limbourg P, Kochs H-D (2008) Multi-objective optimization of generalized reliability design problems using feature models - A concept for early design stages. Reliab Eng Syst Saf 93:815–828

Marseguerra M, Zio E, Podofillini L, Coit DW (2005) Optimal design of reliable network systems in presence of uncertainty. Reliab IEEE Trans 54:243–253

Misra KB, Sharma U (1991) An efficient approach for multiple criteria redundancy optimization problems. Microelectron Reliab 31:303–321

O'Connor P, Kleyner A (2012) Practical reliability engineering. John Wiley & Sons, Chichester

Okafor EG, Sun Y-C (2012) Multi-objective optimization of a series–parallel system using GPSIA. Reliab Eng Syst Saf 103:61–71

Paté-Cornell ME, Dillon RL, Guikema SD (2004) On the Limitations of Redundancies in the Improvement of System Reliability. Risk Anal 24(6):1423–1436

Rao SS, Dhingra AK (1992) Reliability and redundancy apportionment using crisp and fuzzy multiobjective optimization approaches. Reliab Eng Syst Saf 37:253–261

Safari J (2012) Multi-objective reliability optimization of series-parallel systems with a choice of redundancy strategies. Reliab Eng Syst Saf 108:10–20

Sahoo L, Bhunia AK, Kapur PK (2012) Genetic algorithm based multi-objective reliability optimization in interval environment. Comput Ind Eng 62:152–160

Sakawa M (1978) Multiobjective reliability and redundancy optimization of a series-parallel system by the Surrogate Worth Trade-off method. Microelectron Reliab 17:465–467

Sakawa M (1980) Reliability design of a standby system by a large-scale multiobjective optimization method. Microelectron Reliab 20:191–204

Sakawa M (1981) Optimal Reliability-Design of a Series-Parallel System by a Large-Scale Multiobjective Optimization Method. Reliab IEEE Trans R-30:173–174

Salazar D, Rocco CM, Galván BJ (2006) Optimization of constrained multiple-objective reliability problems using evolutionary algorithms. Reliab Eng Syst Saf 91:1057–1070

Taboada HA, Baheranwala F, Coit DW, Wattanapongsakorn N (2007) Practical solutions for multi-objective optimization: An application to system reliability design problems. Reliab Eng Syst Saf 92:314–322

Taboada HA, Espiritu JF, Coit DW (2008) MOMS-GA: A Multi-Objective Multi-State Genetic Algorithm for System Reliability Optimization Design Problems. Reliab IEEE Trans 57:182–191

Tian Z, Zuo MJ (2006) Redundancy allocation for multi-state systems using physical programming and genetic algorithms. Reliab Eng Syst Saf 91:1049–1056

Tian Z, Zuo MJ, Huang H (2008) Reliability-Redundancy Allocation for Multi-State Series-Parallel Systems. Reliab IEEE Trans 57:303–310

Tillman FA, Hwang C-L, Kuo W (1977) Optimization Techniques for System Reliability with Redundancy: A Review. Reliab IEEE Trans R-26:148–155

Winkler RL (1967) The Assessment of Prior Distributions in Bayesian Analysis. J Am Stat Assoc 62:776–800

Zhao J-H, Liu Z, Dao M-T (2007) Reliability optimization using multiobjective ant colony system approaches. Reliab Eng Syst Saf 92:109–120

Zio E, Bazzo R (2011) Level Diagrams analysis of Pareto Front for multiobjective system redundancy allocation. Reliab Eng Syst Saf 96:569–580

Zio E, Bazzo R (2011a) A clustering procedure for reducing the number of representative solutions in the Pareto Front of multiobjective optimization problems. Eur J Oper Res 210:624–634

Chapter 10
Design Selection Decisions

Abstract: The design selection problem in the RRM context considers long-term performance, and represents higher additional costs if unforeseen features that should have been included during the project design phase had to be implemented afterwards. Design decision involves multiple aspects and may be more critical depending on the kind of item, such as consumer appliances, industrial equipment or projects that have to consider safety aspects (airplanes or facilities). Reliability has an essential role for design selection although other aspects have to be considered such as maintainability and risk depending on the specific design problem. Therefore, a multidimensional approach is usually required. In this chapter, all these aspects are discussed in order to illustrate the importance of a broader perspective when facing design decision problems. The fundamental requirements are to consider reliability, maintainability and risk aspects so as to establish features in the design project, including the definition of material, redundancies, control systems and safety barriers. To illustrate these decisions, aspects such as reliability (e.g. MTBF), maintainability (e.g. MTTR), safety, cost, service life, efficiency, are discussed as criteria for these problems. Multi-attribute utility theory (MAUT) is applied in this chapter to illustrate how reliability, maintainability and risk aspects are included in an MCDM/A model for design selection incorporating states of nature. The decision regarding the selection of which features to include in a design project may be considered as an MCDM/A portfolio problem. Finally, an introductory view is given of how the redesign problem arises in the maintenance context with multicriteria approaches.

10.1 Introduction

The term design may have different meanings, such as project, and aesthetical conception. The main issue in this chapter is related to the former, although the latter is specifically applied as one of the criteria decision for a car project, subsequently presented.

Decisions about the design of a product are determinant to its reliability. Errors in the design process can increase considerably the costs during the product development cycle. In this way, reliability is highly connected with problems of engineering design (O'Connor and Kleyner 2012).

A.T. de Almeida et al., *Multicriteria and Multiobjective Models for Risk, Reliability and Maintenance Decision Analysis*, International Series in Operations Research & Management Science 231, DOI 10.1007/978-3-319-17969-8_10

In this field, performance capability and cost are two of the most important factors of the design. While performance capability means the adequacy of a design to perform required functions, cost means how much these performance requirements of a design in monetary units are. To deal with a three-way trade-off among performance capability, cost and reliability, into the design process, the designer (the DM) have to examine how reliability requirements are set to build reliability into a design (Lewis 1987). A literature review on MCDM/A approaches in reliability and maintenance, shows that 16.7% of the work conducted is related to design (de Almeida et al. 2015). It is important to note that this decision has different implications depending on the mode and probability of failures. Reliability requirements can be defined by several ways, such as by the designer, by the buyer of the product and by government agencies.

Products that are more reliable imply higher capital costs and lower costs for maintenance and repair. For the classical optimization approach, it is possible to describe a function that represents the total cost including capital and repair costs. Thus, an optimal solution can be found when this function is minimized. However, DM can have a preference for other solutions instead of the optimal that minimizes the single objective cost function. This occurs in several practical situations when the DM has some additional considerations about the trade-off between capital and maintenance costs. The DM's preferences are illustrated for three examples: a mobile phone, an automated industrial machine and an aircraft turbine.

A mobile phone is a consumer appliance in which reliability requirements rely on consumer expectations. In this case, the reliability of this product can increase the prices, reduce repair costs, provide elements to offer a longer warranty than a competitor and boost sales. While some negative consequences are due to excessive reliability such as lower sales, some inconveniences are due to the opposite, such as company reputation for poor design. Design decisions in this context are made by the manufacturer taking into account public's preferences about price and reliability through market surveys.

In a design of an automated industrial machine, or other equipment directed to large organizations, the DM's preference structure is completely different. In this case, the trade-off between capital cost and production lost through breakdowns should be evaluated. Thus, design decisions have to consider aspects of reliability and maintainability.

Finally, in a design of an aircraft turbine, failure consequences are so severe that a higher level of reliability is required. It means that the increase of a turbine cost can be justifiable due to the level of reliability required to this product. Additionally, aspects of delays and safety in airline maintenance are also relevant (Sachon and Paté-Cornell 2000). In this case, insurance underwriters and government agencies are responsible to define reliability specifications and risk analysis performs an important role in the design.

10.1.1 The Reliability Role in System Design

Reliability estimation of a new product, before it is manufactured, is attractive for designers. This information can allow accurate forecasts of support costs, spares requirements, warranty costs and marketability (O'Connor and Kleyner 2012). Reliability is a key aspect in the design selection. Different standards establish a comprehensive design specification, general requirements and descriptions of activities in order to guide the industry and government for developing reliable products and systems (IEEE 1998; US MIL-STD-785B 1980; IEC 61160 2005; BS 5760-0 2014).

According to Ren and Bechta Dugan (1998), the design requirements typically consider reliability, cost, weight, power consumption, physical size, and other system attributes. In order to meet these requirements, the DM should observe the whole system as a set of components, which each design component can be chosen from a set of design alternatives. Fu and Frangopol (1990) deal with the problem of optimal structural design from a multi-objective perspective that taking into account weight, system reliability and system redundancy.

Designers are experts in the creative process to provide reliable products. They must know the types and extent of the loading, and the range of environmental conditions under which the product operates. Additionally, they must know the physics of the potential failure modes to ensure the required level of reliability. Design margins, redundancy allocation and protection against strength degradation are frameworks that can help designers to enhance reliability (Lewis 1987; O'Connor and Kleyner 2012).

Design margin is a framework that considers that the reliability can be increased by ratio of capacity of components and loads applied to them. In Fig. 10.1, the probabilistic mechanism of the failure function is illustrated for different levels of loading, $load_i$, where $load_1 < load_2$. That is, the failure rate decreases as the component load is reduced for a given operating features. For instance, a pavement design for a road network can reduce drastically the probability of failures and maintenance costs by means of this analysis.

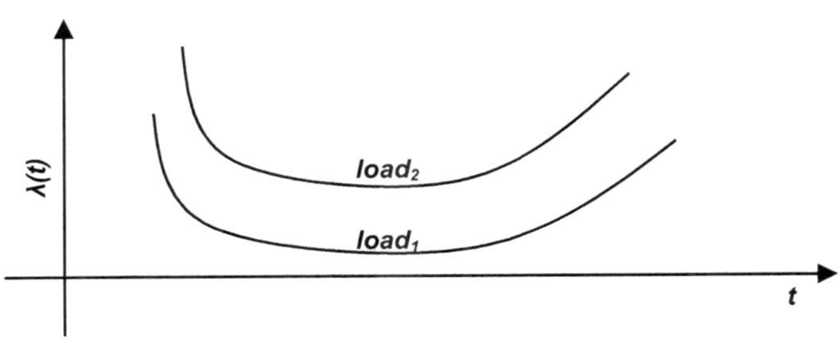

Fig. 10.1 Failure rate function at different levels of loading

In other words, a product is designed to have a reliability performance in excess of that stated in its specifications. This often leads to an overdesign, early in the development phase of a design. Hurd (1966) presents some examples of this overdesign, for instance in structural design, a factor of 10 is used. A structure is designed to hold 10,000 lb, while the maximum specification is 1,000 lb. A similar situation is illustrated in an electronic context, when electronic parts are designed to be used to 10 or 15 per cent of their rated capabilities.

Redundancy allocation in design allows the increasing of system reliability by means of addition of components in parallel. It means that one or more components can fail without result a system failure. Design decisions of multi-state weighted k-out-of-n systems is a example of a relevant redundancy allocation problem (Li and Zuo 2008). Decisions on redundancy allocation problems are discussed in Chap. 9.

Strength degradation includes several complex mechanisms such as fatigue in metals, corrosion and wear. Based on these mechanisms, the designer can specify a fatigue limit of operation. Tests can provide the required data by generating failures under known loading conditions, and reliability estimation can be carried out. The designer must specify maintenance procedures for inspection, lubrication or scheduled replacement when a suitable protection is not possible to be determined by the design (O'Connor and Kleyner 2012). Inspections and maintenance need to be planned at the design stage when fatigue failures are present. Economic criteria to minimize lifecycle maintenance costs but satisfying a minimum reliability level must be considered in design decisions. Reliability-based maintenance strategies can help designers to deal with the conflict of minimizing maintenance costs and maximizing reliability levels, specifically when some fatigue mechanisms need to be inspected (Guedes Soares and Garbatov 1996; Garbatov and Guedes Soares 2001).

According to Sahoo et al. (2012) while most reliability optimization problems have been formulated into the single objective optimization approach, it can be recognized that most real-world design problems involving reliability optimization require a broader perspective, and that is done by optimizing simultaneously more than one objective function.

10.1.2 The Maintainability Role in System Design

A contribution to the problems associated with maintainability comes from the engineering design specification. It should specify the engineering requirements, including reliability and maintainability issues. Designer should comprehend the international standards and their specific content. There are international standards that deal with maintainability issues in design phase (IEC 60706-2 2006; BS EN 60706-2 2006).

As stated in the last section, the reliability has a central role in system design. Additionally, the overall performance of a plant is also associated with maintainability. The availability of a plant depends on the frequency and the downtime of interruptions. Maintainability is a design characteristic that reflects the probability of an item be restored in a given period to specified conditions by maintenances actions that meet some procedures and resources requirements (Goldman and Slattery 1977).

The fundamental function of equipment depends on its availability or readiness. Thus, some trade-offs between reliability and maintainability could be made in order to meet the availability requirements. Different levels of reliability $R(t)$ and maintainability $M(t)$ can result an availability $A(t)$ level.

As stated by Goldman and Slattery (1977), maintainability is a concept related to different aspects from the basic physical characteristic of the design to the strategic level of the maintenance function. Amongst the different issues comprised on the spectrum of aspects related to maintainability, it is possible to highlight design requirements, selection of maintenance strategy, logistic provisioning, and so on.

Despite these different issues, this section focus on the role of that maintainability plays, providing guidelines on the design specification, in order to provide to the overall equipment some features related with effectiveness and lifetime support cost.

In a design selection problem, while the designer has to develop the system, taking into account various aspects, including the maintainability, and different constraints imposed by budget project and standards; the user has to accept the final configuration of the design and handle with the challenges related to effective operational use of this built design. From the difference between user and designer perspectives, sometimes, feedbacks have to be considered in both directions: from the operation to design and from the design to operation.

Design development is complex and non-failure-free task. Thus, besides the problems that could exceed the design phase, some other problems could arise related to the discordance between designer and user views. Therefore, analysis of the design process is essential not only as an alternative to handle with the views discordances, but also to reduce the number of failures that comes from the project phase. In this process of improvement, the feedbacks are essential. A simple reason is that some problems come up after the launch of the product. As any item is subject to failure, maintenance could be a quite frequent activity. In this case, the maintenance should be done in order to restore the system for the operational state, as quickly as possible, so that interruptions resulting from failures do not affect the production targets.

In order to act as quick as possible, maintenance teams have to deal with some barriers related with design selection issue considering the difficulty associated to the maintenance activity. In some cases, the maintainability attributes might be too restrictive resulting in difficulties to reduce maintenance times. For example, the downtime reflects relations among three main groups of elements: design decisions, maintenance policies and technician requirements. Therefore, after the phase of design, the decisions related with the other elements taking into account the constraints of the project (Goldman and Slattery 1977).

The maintainability is not only a criterion at the design selection problem, but it should be considered in other problems such as downtime reduction or availability increase. In some design problems, the maintainability formulated as state of nature. The downtime is a non-controllable variable by a configuration of the design, once there are human aspects such as motivation and ability that have great impact on the downtime (time to repair).

10.1.3 The Risk Role in System Design

According to Lewis (1987), a range of consequences can be produced from a failure of a system, affecting people and the environment. With the intention of reduce this possibility, several levels of the acceptability of risk can be established, based on specific characteristics such as procedures, resolutions and standards, type of industry, industry or design location, etc.

Two risk aspects are involved in systems' design, the first is the level of acceptability defined in different sectors (such as civil, energy and chemical engineering) can be affected by a specific accident, for instance, the catastrophe at Chernobyl in 1986. The second is how technological levels impact directly in the occurrence of a hazard situation (Vrijling et al. 1998).

Thus, in the context of systems' design, an effective risk management (see Chap. 3) is necessary to mitigate or prevent the risk occurrence, such a way that a minimum risk level is reached. The observance of procedures, resolutions and standards is a fundamental question to minimize the chance of the risk occurrence. In Sect. 10.3, it is illustrated the use of standards as input to design selection.

A conflicting question in the context of system or equipment design should be observed considering the following perspectives: reliability and safety. From the safety perspective, if a hazardous event occurs, the risk to the public should be minimized by the plant shutdown. From the reliability point of view, the plant should stand in operation waiting for a failure occurrence before a shut down take place or, as a last resort; the repair of the plant should be performed if the shut down it is not possible. Thus, the challenge of an effective risk management in system's design context is to reduce the possibility of an accident, reducing the probability to very low levels through design and safety detailed analysis (Lewis 1987).

Another interesting aspect with regard to safety and design is the increase in the requirements of safety barriers, once each safety feature added to the project increases the project costs, it also reduces the facility profitability. Nuclear power plants faced such situation after the Fukushima accident in 2011. After the accident, safety requirements has been revised, compromising the economic viability of specific project designs that would have to raise safety in order to meet risk acceptance levels more conservative.

10.2 An MCDM/A Model for the Design Selection for a Car

In this section is presented an illustrative example of the design selection for a car according to the MCDM/A approach presented in Chap. 2.

According to the first step discussed in Chap. 2, it is important to make some remarks and observations to characterize a DM involved in such problem. In this illustrative example, the DM is the senior engineer responsible for defining which car project should be selected as the best design from the product development stage.

At this point is important to notice that all standards and requirements have already been achieved by each project, thus, the decision regards on evaluating these alternatives considering the factors or objectives related to this decision, which recalls the second step pointed in Chap. 2.

As the objectives for this kind of decision, one may list the following objectives: Maintainability, Reliability, Safety, Cost, Service life, Efficiency, Aesthetic.

For simplification purposes this illustrative example will not consider the last three objectives. Focusing on the first four objectives allows to explore more deeply the RRM perspective of such problem.

Other objectives may be more emphasized during different phases of the product development, taking the initial list of objectives as a reference or by the addition of other objectives depending on specific aspects related to the problem context.

The definition of criteria related to each objective refers to the third step given in Chap. 2 with regard to establishing criteria.

In order to measure maintainability, the concepts given in previous sections and in Chap. 3 shall be used, thus as the maintainability concept relates to the time spent during repairs.

The maintainability function for any probability distribution may be represented as in (10.1) (Dhillon 1999; Stapelberg 2009), where t represents time and $f_r(t)$ is the pdf of the repair time.

$$m(t) = \int_0^t f_r(t)dt \tag{10.1}$$

An Exponential, Lognormal, Weibull, Normal, and others may represent the repair time depending on the equipment considered.

A few indices are applied for analyzing maintainability in a deterministic context. For instance, Dhillon (1999) presents some measures for maintainability:

- Mean time to repair (MTTR).
- Mean active preventive maintenance time and median active corrective maintenance time.
- Maximum corrective maintenance time.
- Mean maintenance downtime.

In a probabilistic context, considering the step 5 of the procedure for building a MCDM/A model (see Chap. 2), the pdf of the maintainability is introduced into the model, as a state of nature, such as in the decision model of Chap. 7 and Chap. 9. The MTTR represents the expected value of t, given $f_r(t)$. In other models, the maintainability may be modeled as a consequence, such as in this subsection.

Simplifying with deterministic indices, it could be applied for the maintainability either: a quantile of the maintainability distribution or use the standard deviation with the MTTR, as explained in Chap. 2.

As previously observed in Chap. 3, the objective reliability is related to a probabilistic concept with regard to the time whenever the equipment shall fail.

In order to provide an easier scale for measuring the reliability of each car project for a DM, the mean time between failures (MTBF) concept may be used for evaluating each alternative's reliability, as pointed by O'Connor and Kleyner (2012) as an alternative for measuring reliability of repairable items. The definition of MTBF is given by (10.2), from the reliability function (Stapelberg 2009).

$$MTBF = \int_0^\infty R(t)dt \tag{10.2}$$

In order to standardize the reliability measurements of each car project, one may consider it for each car project measured by the number of hours of highway driving to evaluate its operation.

There are many procedures for assessing a car project safety. Most of these procedures involve a crash test. Thus, there is an evaluation of the entire project and the outcomes of safety items such as seatbelts, airbags, anti-intrusion bars (side protection), laminated windshields, crumple zones, cargo barriers, safety cell, and others.

For each new car project there is a study based in crash tests to assess its safety through a New Car Assessment Program (NCAP). Depending on the region, the Euro or US NCAP, for example, may assess a new car project.

Considering the US NCAP, the crash tests includes the evaluation of a frontal crash, a side crash, and the risk of rollover in a five star safety rating, from *1* to *5* stars, with *5* being the best rating. Therefore, the criteria adopted for measuring safety shall be the lowest estimate rating of the car project in the frontal crash and side crash test.

The objective cost measures and evaluates all aspects that may be converted into a monetary value scale. It is important to emphasize that MCDM/A approaches provide methodological support to understand and value alternatives among different objectives scales and provide a global evaluation that includes all objectives.

Thus, when measuring cost, all factors that may be represented by monetary scale shall be summed in order to evaluate the respective cost of each car project. In this example, it shall be assumed that all these factors have been considered for estimating each car project cost.

The elements of the set of alternatives are the car projects. The problematic relies on the choice of the best car project according to the corresponding criteria.

The states of nature in this illustrative problem corresponds to factors that are not under the DM's control and are subject to uncertainty, influencing the decision outcomes, such as when a failure may occur or the time for a repair service. Such uncertainties may be represented by probability distributions that provide the performance estimates for each car project.

Assuming that the DM preferences fit the axiomatic structure required by MAUT, the next step regards the intra criterion evaluation. Establishing intra criterion evaluation consists in the definition of a utility function for each criterion by assessing its shape and parameters from the DM's evaluation of the outcomes in each criterion through lotteries or other elicitation procedure.

For a DM with additive independence condition, the additive utility function is represented by (10.3):

$$u_a = k_m E_a[u_m(m)] + k_r E_a[u_r(r)] + k_s E_a[u_s(s)] + k_c E_a[u_c(c)], \qquad (10.3)$$

where u_a is the expected utility of a, based on the additive function over the expected utility of the attributes maintainability (m), reliability (r), safety (s) and cost (c); k_m, k_r, k_s and k_c are the respective scale constants.

The scale constants in MAUT are elicited through lotteries as given in Chap. 2. For this illustrative problem, consider k_m, k_r and k_s equal to *0.3* and k_c equals to *0.1*.

The respective one-dimensional utility functions are $u_m(m)$, $u_r(r)$, $u_s(s)$ and $u_c(c)$. For maintainability (m) and reliability (r), the random variable repair time (m) and time to failure (r) are considered consequences, with its respective pdfs $f_a(m)$, $f_a(r)$, for each alternative a. Their expected utilities are respectively given as follows by (10.4) and (10.5).

$$E_a[u_m(m)] = \int_0^\infty u_m(m) f_a(m) dm \qquad (10.4)$$

$$E_a[u_r(r)] = \int_0^\infty u_r(r) f_a(r) dr \qquad (10.5)$$

Similar formulation are given to safety (s) and cost (c).

Considering seven non-dominated car project alternatives for this design selection problem, named as Alt1, Alt2, Alt3,..., Alt7, Table 10.1 presents each alternative expected utility over the considered attributes and its correspondent additive utility value.

In a single attribute perspective, alternatives 4, 6, 3 and 5 gives the best performance for maintainability, reliability, safety and cost, respectively. Alternatives 1, 2 and 7 provide a more distributed performance among the attributes, as a result, an MCDM/A approached is required to evaluate tradeoffs and value the overall value of these alternatives in order to provide a recommendation for the selection problem.

Table 10.1 Car project alternatives evaluation for a design selection problem

	$u(m)$	$u(r)$	$u(s)$	$u(c)$	u_a
Alt1	0.558	0.812	0.126	0.453	0.494
Alt2	0.419	0.750	0.106	0.818	0.464
Alt3	0.626	0.586	0.976	0.600	0.716
Alt4	0.892	0.761	0.760	0.200	0.744
Alt5	0.139	0.508	0.091	0.941	0.315
Alt6	0.739	0.881	0.626	0.105	0.685
Alt7	0.861	0.563	0.765	0.606	0.717

From the results given by Table 10.1 the best car project for such DM would be the alternative 4, which achieved the highest value in the additive utility function. By evaluating alternative 4 individual utilities is interesting to observe that it has the best performance on the maintainability attribute, although presenting one of the worst values for the attribute cost, only better than alternative 6.

Thus, for a DM with such preferences, the alternative 4 values for maintainability, reliability and safety are compensating its bad outcome for the cost attribute. A different DM may have different preferences and make different tradeoffs, leading to the selection of a different alternative. As pointed in Chap. 2, the sensitivity analysis is an important step in order to evaluate the robustness of the preliminary recommendation, allowing to increase the accuracy of the elicitation process when it is required.

For a DM with a non-compensatory rationale, the selected alternative could be other due to a different kind of preference structure. Such DM with a different preference structure, would set different model parameters and establish different comparisons, such that in many times would be reflected as the selection of a different alternative. The use of the MCDM/A approach enriches the decision process by allowing to incorporate these particularities of the DM preferences with accuracy, by incorporating it in the decision model.

10.3 Risk Evaluation for Design Selection

During the design phase, it is possible to improve system reliability and risk barriers in order to reduce risks and avoid unnecessary costs to adjust and match the project to the required risk standards. During this phase, accident rates can be influenced when deciding on what material and components to use in the project.

Many studies consider that risk evaluation follows a trend, which reflects probabilities separately from consequences. As a result of this trend, many decisions are taken which consider only probabilities or consequences, without aggregating these two important factors when evaluating the overall risk.

This may occur due to the difficulties in estimating and/or simulating these processes to quantify probabilities and the magnitude of the consequences. However, these two measures can be considered together with human judgment if utility functions are used, as do Baron and Paté-Cornell (1999), Brito and de Almeida (2009), Brito et al. (2010), Almeida-Filho and de Almeida (2010) and Garcez and de Almeida (2014).

The concept of ALARP has been questioned by many authors in the literature. Melchers and Stewart (1993) shows that each individual can have a different level that he/she finds acceptable for different types of risk and also that this can change also from one culture to another.

Aven and Vinnem (2005) presented a different risk analysis regime that is not based on risk acceptance criteria at all. They argue that a rule based on cost-effectiveness should do better than pre-defined risk acceptance limits. In some situations, it is possible to achieve risks below ALARP levels. Thus a methodology is required that can consider a DM's tradeoffs among costs and other loss dimensions such as environmental and potential losses of life.

Aven and Kristensen (2005) presented a discussion on several perspectives of risk, establishing a common basis for the different perspectives, emphasizing how important it is to consider all possible consequences associated with their uncertainties.

When considering the context of risk analysis in the literature on the oil and gas industry, there are two main models for risk management with a specific focus on risk evaluation and risk reduction that can guide the selection and design decisions by evaluating risk levels. Besides these models there are other models in the literature that can be used after start-up at the facility, such as the framework proposed by Øien (2001) for structuring risk indicators for risk control during operation.

Khan et al. (2002) present an example of design selection in order to implement safety measures. Khan et al. (2002) present an offshore oil and gas facility and design alternatives that may reduce risk.

The approach of inherent safety design was presented initially by Kletz (1985), and detailed later in Kletz (1998). Khan and Amyotte (2002) presented a study showing that safety measures should be a concern from the design stage for a facility in order to reduce costs throughout its life span.

10.3.1 Risk Assessment Standards

One of the main models for risk evaluation can be found in ISO/IEC Guide 51: 2014 and another in the NORSOK Standard Z-013. The model described in ISO/IEC Guide 51: 2014 updates the 1999 version regarding this subject. It can also be used combined with IEC Guide 73: 2009, which refers to the vocabulary and meanings regarding risk management, so as to consolidate terminologies.

The NORSOK Standard Z-013 is a standard edited by the NPD (Norwegian Petroleum Directorate), which is a Norwegian agency in charge of regulating oil industry activities in the North Sea.

Brandsæter (2002) describes the implementation and uses of risk analysis using quantitative and qualitative methodologies for the oil and gas offshore industry with contributions to the EC-JRC International Workshop on "Promotion of Technical Harmonisation on Risk-Based Decision Making (2000)", which is formatted as if it were a response to a set of questions prepared by workshop organizers, and in which both models mentioned are discussed.

In ISO/IEC Guide 51: 2014, risk evaluation is defined as a wide process of estimation and analysis. For ISO/IEC Guide 51: 2014, risk analysis terminology is defined as the systematic use of information to identify hazards and estimate risk, and risk estimation is defined as a procedure to determine if the risks are tolerable or not.

Thus the ISO/IEC Guide 51: 2014 model is represented by an iterative process to evaluate and reduce risks that can be applied to qualitative and quantitative risk evaluations. In ISO/IEC Guide 51: 2014, it is clear that some tolerance criteria have to be defined, as in ALARP. However, it does not suggest any procedure to deal with the situation where the limits have already been satisfied (or not) even how to choose between non-dominated alternatives considering multiple risk dimensions.

This iterative process considers that each hazard must be considered and must satisfy a tolerable risk level. According to ISO/IEC Guide 51, it is necessary to identify each hazardous situation and event by anticipating stages and conditions for the system, including installation, operation, maintenance, repair and destruction/ disposal. This iterative process considers an entire process of risk assessment.

The ISO/IEC Guide 51: 2014 presents a "three-step method" from the design phase and additional measures at the use phase. The risk reduction process starts from the design of the installation, beginning with inherently safe design as a way to start the risk reduction process.

Additional risk reduction alternatives have to be implemented after the design stage such as training and procedures that will reduce residual risks after all protective measures have been deployed.

The NORSOK Standard Z-013 model for the process of assessing risk and emergency preparedness describes this in a similar way to that of the ISO/IEC, although NORSOK includes an assessment of preparedness for emergencies in its process.

The previous version of the process for this standard already gave more emphasis to a typical quantitative risk analysis methodology when considering the risk assessment process. It emphasized the importance of the estimation, analysis and evaluation approach for typical quantitative risk analysis methodologies when applied to offshore oil and gas structures. It defines risk as a probability or an expected frequency, and requires a risk acceptance criterion to be defined, which should consider the probability or frequency of an associated consequence, thereby establishing a risk index and an acceptable risk limit. For a human dimension, for example, the risk for an individual can be used and/ or the FAR (Fatality Accident Rate) which should be compared with acceptable limits by the adopted standards in order to establish the risk picture.

10.3.2 MCDM Framework for Risk Evaluation in Design Problems

Throughout the risk analysis process there is no specific framework about how to aggregate preferences amongst multiple risk dimensions, especially when there is a decision problem where some of the alternatives have already reached the acceptable risk levels defined in standards.

Thus, all dimensions are considered in terms of constraints that must be respected and considered in terms of cost benefit evaluation. Nevertheless, during the design process, there are opportunities to improve safety and prioritize safety alternatives.

To address this issue, this section presents a framework with a numerical application that aggregates preferences by using metrics, which consider probability, give values to human judgment on consequences and their behavior with regard to risk (prone, neutral and risk averse). These metrics are provided by Utility Theory.

Based on the literature, this procedure supports the structuring of decision problems in the context of evaluating multidimensional risk, based on the framework given in Chap. 2.

The approach given in this section addresses decision problems regarding risk reduction and safety improvements for the design of hazardous facilities. These kinds of decision problems can refer to a choice, ranking, sorting or a portfolio decision problem, and depending on the type of problematic (Roy 1996) a specific methodology should be used to aggregate the DM's preferences and doing so should be amongst its objectives.

While due consideration should be given to the models for risk estimation, analysis and evaluation from both standards (NORSOK and ISO/IEC), this MCDM/A procedure can be used as a framework to evaluate risk reducing alternatives. This can be done even if some of these risks have already achieved the acceptable risk levels and assuming there are still safety improvements that could be implemented. Also decisions can be taken about which risk reduction measures should be implemented, according to the priority of each action and the risk involved in different parts of the process.

This kind of decision has two main actors who, in general terms, can be identified as a DM and an analyst who will give methodological support to the DM. Both of these actors will exert influence during the decision process, in which the former, a DM, will influence the decision because of cognitive aspects and his/her preference structure. The analyst will influence it in such way that he may bring bias to the process due to his/her own opinion about the subject and/or because of the use of his/her preferred methodological approaches (Almeida-Filho and de Almeida 2010).

Figure 10.2 presents the steps to consider an MCDM/A approach in order to choose or prioritize design alternatives for risk reduction based on the procedure for building MCDM/A model presented in Chap. 2.

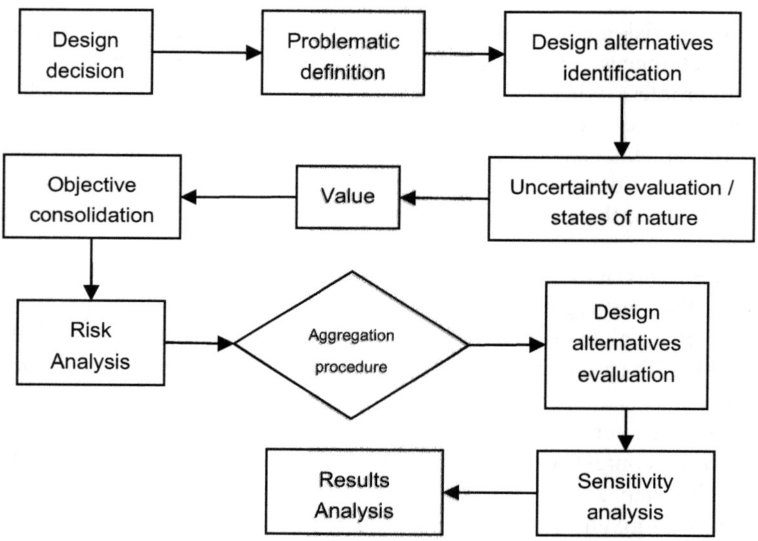

Fig. 10.2 Multi dimensional risk evaluation in design problems

It starts with a Decision Situation that represents the phase when a decision problem has been identified or has appeared and needs to be outlined. This includes determining the risk level achieved and the risk reduction alternatives to be considered in design. The step ahead is to define the kind of problematic (Roy 1996; Vincke 1992) that addresses the problem itself.

The process of identifying alternatives should be extensive and exhaustive with a view to this set comprising as many alternatives as possible, except for those, which can be previously considered as dominated alternatives. This is a very important step since it seeks to avoid a situation where good alternatives are neglected.

After the set of alternatives is well defined, it is necessary to evaluate uncertainties regarding each alternative (action) and their possible states of nature. For this stage, the same QRA techniques can be applied for its estimation as suggested by NORSOK Standard Z-013 or the ISO/IEC Guide 51: 2014, including simulation and estimation models for damage radii (DR) of different propensities, for example. To estimate probability, the same QRA methods discussed in Chap. 3 can be applied such as fault trees analysis, event tree analysis and expert's knowledge elicitation amongst other techniques.

Afterwards it is necessary to establish the DM's preferences, in order to evaluate alternatives. The use of utility theory to evaluate risk in each criterion enables the probability and the consequence value for the DM of each possible outcome to be considered together, thereby providing a metric for each risk dimension and also the DM's behavior (probe, neutral and risk averse) in each risk dimension. This often considers financial aspects, human potential losses and environmental damages as the three dimensions usually considered. Thus, what is required is to elicit the DM's utility for consequences in each risk dimension. (Brito and de Almeida 2009; Brito et al. 2010; Alencar et al. 2010; Lopes et al. 2010; Garcez et al. 2010; Almeida-Filho and de Almeida 2010; Garcez and de Almeida 2014). These consequences result from the combination of alternatives and the possible states of nature, as shown in Chap. 2 (Table 2.1).

To aggregate all risk dimensions evaluations, an aggregation method must be considered. To make this choice some aspects will have to be taken into account, as which kind of preferences structure the DM has and his/her preferences should be modeled to determine if he/she has a compensatory or non-compensatory rationality.

As pointed in Chap 2, the rationality behind the DM's preferences guides the choice of a compatible aggregation method. As to non-compensatory methods there are, for instance, the ELECTRE family of methods (Roy 1996) and the PROMETHEE family of methods (Brans and Mareschal 2002).

As to compensatory approach, there are several methods that can be used, of which MAUT (Keekey and Raiffa 1976) is amongst the most used methods that considers the risk evaluation structure.

The evaluation of alternatives is the phase where the MCDM/A method chosen is applied and its parameters should be obtained through an elicitation process that may change according to the nature of each kind of aggregation methodology.

These steps are detailed in Chap. 2, thereby it is possible to consider multiple risk dimensions and aggregate them from the perspective of the DM's preferences in order to reduce risks and improve safety conditions of a hazardous installation by considering different design alternatives.

This MCDM/A framework allows the DM not only to use acceptance levels as references but also to evaluate them according to his/her preferences and risk behavior (prone, neutral, averse).

10.3.3 Illustrative Example of Risk Evaluation in a Design Problem

In this section, an illustrative example is presented based on a realistic problem of implementing a safety project to illustrate an application of MCDM/A for risk evaluation in facilities design. Thus, there is a set of safety projects that can be implemented and a DM has to define which sub-set of safety projects to implement, with regard to an offshore oil and gas platform, specifically in the primary process (Khan et al. 2002).

Oil and gas are well-known hazardous materials and when they are extracted, there are several sources of hazard, one of which is the primary process where the crude oil from the wellhead (a mixture of oil, gases and water) is separated before it is processed.

In a general way, the primary process on an offshore oil and gas platform consists of a first separator to separate the crude oil from the gases and water, which it then sends to a transportation line. A second separator is used to separate the residual water from the gas and send it to other subsequent units to separate the wet gas and the dry gas. The other units comprise two compressor units, a flash drum unit and a drier unit.

The decision situation consists of improving safety throughout the primary process on an offshore oil and gas platform by choosing whether or not to implement the design of some safety features. Therefore, the set of alternatives may be globalized or fragmented. The former consists that each alternative exclude the others, while the latter considers the combinations of the set of alternatives (Vincke 1992). Thus, if considering different safety features, the set of the alternatives may be the combination of all features that may be considered for the design project.

The problematic involved in this decision situation is about choosing to implement one or more features. In this particular case, the model could use a choice or a portfolio problematic. Although the decision relies in maximizing safety and minimizing costs simultaneously, while technical aspects are formulated as constraints into the model, for instance, acceptable risk limits. Therefore, a knapsack problem (Martello and Toth 1990) would be a model representation considering to objectives and technical aspects to define the design as presented in the model given by (10.6).

$$\begin{aligned} &\max E\left[U\left(x_1, x_2, x_3, \ldots, x_i, \ldots, x_{n-1}, x_n\right)\right] \\ &s.t. \quad r_j\left(x_1, x_2, x_3, \ldots, x_i, \ldots, x_{n-1}, x_n\right) \le AL_j \text{ for each } j = 1 \text{ to } m. \end{aligned} \tag{10.6}$$

By a knapsack problem representation the set of alternatives would be fragmented and each design project alternative would be represented by the vector $(x_1, x_2, x_3,\ldots, x_i,\ldots, x_{n-1}, x_n)$ which is all the possible combinations of the n safety features considered for the design problem. By maximizing the expected MAU function value, $E[U(x_1, x_2, x_3,\ldots, x_i,\ldots, x_{n-1}, x_n)]$, subjected to the technical constraints, represented in the model given by (10.6) only by the m risk acceptance levels (AL_j) for simplification purposes as other technical aspects may be included in this formulation. Thus, the design project alternative recommended by this model is in compliance with the technical aspect considered. For this illustrative example, the alternative risk level $r_j(x_1, x_2, x_3,\ldots, x_i,\ldots, x_{n-1}, x_n)$ in dimension j has to be lower than AL_j.

For this application, five features were considered that could be implemented in the design of the facility in order to improve safety throughout the primary process on the offshore oil and gas platform. The first feature (A) introduces improvements in the first separator; the second feature (B) introduces improvements in the second separator; the third feature (C) introduces improvements in the compressor units; the fourth feature (D) introduces improvements in the flash drum unit; and the fifth feature (E) introduces improvements in the drier unit.

If all these features were implemented, they would improve safety throughout the process by reducing risk. In other words, they would reduce the probabilities of events that would generate different accident scenarios. This could be to substitute some kinds of materials for stronger and more reliable ones or it could also be to implement different control procedures and automation throughout the unit, for example.

Given the structure of MAUT, when considering an MCDM/A portfolio analysis there are some issues that must be observed. There are effects associated with the different utility scales on the results of an MCDM/A portfolio, especially non linearity as occurs in the utility scale for evaluating the consequences (de Almeida et al. 2014). When such aspects are involved, a different approach has to be used in order to avoid the bias due to utility scales issues into the aggregation procedure for the MCDM/A portfolio analysis. Thus, to avoid misleading results, a complete enumeration approach for the portfolio problem given the five safety features is used to avoid possible bias effects associated with the utility scales. Thus, it allows illustrating this portfolio problem as a choice given all possible design project combinations considering the safety features. Enumeration schemes are an alternative approach to solve knapsack problems (Yanasse and Soma 1987; Martello and Toth 1990).

Therefore, a choice problematic is used for modeling, and thus all possible alternatives are enumerated for considering all five safety features combinations in order to provide all the design projects that represent the set of alternatives. So, from the choice problematic definition given in Chap. 2, a DM can choose a subset of this, which, in this case, is one of the design projects.

The identification of alternatives considers the existence of features A, B, C, D and E only, and that the DM can choose all of them if he/she thinks that this is worthwhile. Thus, the set of alternatives consists of all the combinations of implementing (1) or not (0) each project. This can be summarized in 32 alternatives as shown in Table 10.2.

Table 10.2 Set of alternatives

Alternative	Action	A	B	C	D	E
1	No feature implemented	0	0	0	0	0
2	Implement E	0	0	0	0	1
3	Implement D	0	0	0	1	0
4	Implement D and E	0	0	0	1	1
5	Implement C	0	0	1	0	0
6	Implement C and E	0	0	1	0	1
7	Implement C and D	0	0	1	1	0
8	Implement C, D and E	0	0	1	1	1
9	Implement B	0	1	0	0	0
10	Implement B and E	0	1	0	0	1
11	Implement B and D	0	1	0	1	0
12	Implement B, D and E	0	1	0	1	1
13	Implement B and C	0	1	1	0	0
14	Implement B, C and E	0	1	1	0	1
15	Implement B, C and D	0	1	1	1	0
16	Implement B, C, D and E	0	1	1	1	1
17	Implement A	1	0	0	0	0
18	Implement A and E	1	0	0	0	1
19	Implement A and D	1	0	0	1	0
20	Implement A, D and E	1	0	0	1	1
21	Implement A and C	1	0	1	0	0
22	Implement A, C and E	1	0	1	0	1
23	Implement A, C and D	1	0	1	1	0
24	Implement A, C, D and E	1	0	1	1	1
25	Implement A and B	1	1	0	0	0
26	Implement A, B and E	1	1	0	0	1
27	Implement A, B and D	1	1	0	1	0
28	Implement A, B, D and E	1	1	0	1	1
29	Implement A, B and C	1	1	1	0	0
30	Implement A, B, C and E	1	1	1	0	1
31	Implement A, B, C and D	1	1	1	1	0
32	Implement all features	1	1	1	1	1

The consequence evaluation considered the most credible scenarios for these units (Khan et al. 2002) are summarized in Table 10.3.

Table 10.3 Consequences for the most credible scenarios

Units	DR 100% Fatality / Damage (m)	DR 50% Fatality / Damage (m)	DR 100% 3rd. Degree of Burn (m)	DR 50% 3rd. Degree of Burn (m)	Possibility of Spills
First Separator	230	288	333	428	yes
Second Separator	53	74	69	78	yes
Compressor Units	24	35	44	57	no
Flash Drum	25	42	56	77	yes
Drier	73	92	106	136	yes

The most credible scenario for the first separator is a BLEVE followed by fire; for the second separator it is VCE followed by fire; for the compressor units it is a gas release possibly turning into a jet fire; for the Flash Drum unit it is a VCE followed by fire; and for the Drier unit the most credible scenario is a BLEVE followed by fire.

These scenarios, which are the most credible ones, have a higher probability in the present situation and a lower probability after implementing safety features, for each scenario (Khan et al. 2002). These probabilities are illustrated in Table 10.4.

Table 10.4 Probabilities considering design implementation

Scenario	Probability in present situation	Probability after implementing the design
Normality	0.9990804690	0.9999998688
Accident in First Separator	0.0000107000	0.0000000179
Accident in Second Separator	0.0009474000	0.0000000155
Accident in Compressor Unit	0.0136400000	0.0000013110
Accident in Flash Drum Unit	0.0009060000	0.0000000786
Accident in Drier Unit	0.0000028310	0.0000000347

The first objective to be considered by a DM would be the potential number of lives that could be saved by simply choosing a specific alternative for the facility design. Another concern, of a DM in this situation, would be the environmental dimension, which would be affected if an accident scenario occurs. There are also many monetary or financial aspects to be evaluated, such as property losses, downtime in production and several financial compensations and fines, which would have to be paid, and also the costs of any safety improvement. The aggregation procedure can be more extensive depending on the methodology used to model and elicit DM's preference structure, as given in Chap. 2, it may also consider a value focused thinking approach (Keeney 1992) for structuring DM's objectives in order to inspire design features and so, creating design project alternatives.

With regard to the criteria or objectives for this problem, these can be summarized by a human objective, which implies minimizing loss of human life; an environmental objective, which implies minimizing environmental losses; and a last objective which is a financial objective, that of minimizing any expected financial loss and also minimizing the costs of implementing safety improvement (actions) considering that these costs will occur if these actions are chosen.

For each dimension, it is necessary to elicit the conditional utility functions. These multiple risk dimensions are aggregated considering MAUT as an MCDM/A approach. Thus, the alternatives are evaluated by using a MAUT to provide a complete rank of all alternatives considered using an additive MAU function, such as (10.7), where k_h, k_e and k_f are the scale constants of the additive utility function, which represents the trade-offs amongst these objectives, i.e. human (h), environmental (e) and financial (f).

$$u(h,e,f) = k_h u_{h1}(h) + k_e u_e(e) + k_f u_f(f) \tag{10.7}$$

The elicitation of these scale constants considers the range (variability) in consequences and the importance of each criterion, so this measure represents these two figures. As to the values for scale constants, these are considered as 0.5 for the Human dimension, 0.49 for the Environmental dimension and 0.01 for the Monetary or Financial dimension. These values reflect the difference between the range of best and worst consequences in each dimension and the importance relation for a DM of changes in values among dimensions. This also reflects a DM who would be more inclined to spend by some proportion if this will reduce the probabilities of injures to people or to the environment.

After evaluating the alternatives, it is possible to provide a complete rank of all the alternatives considered. The ranking of 10 alternatives is presented in Table 10.5, with the main result in order to compare these alternatives, which is the difference ratio between them.

Since the utility scale is highly affected by the huge difference between normality and any of the accident scenarios, the analysis of the differences ratio of the alternatives shows more information than the utility scale itself. This measure is used due to the nature of the utility measure, which is based in an interval scale, so what really matters is the size of the utility difference ratio instead of absolute differences between them.

Table 10.5 The ranking of the design alternatives

Rank	Alternative	Difference Ratio $(u_i - u_{i+1} / u_{i+1} - u_{i+2})$
1	5	0.65
2	7	1.65
3	6	12.07
4	13	0.08
5	8	2.45
6	15	1.70
7	1	0.28
8	21	5.09
9	3	2.25
10	14	0.07

Thus the last column of Table 10.5 shows that the difference between alternative 5 and 7 is 65.39% bigger than the difference between alternative 7 and 6, and that the difference between 8 and 15 is 245.48% bigger than the difference between 15 and 1 (implement none of safety features). This measure gives to DMs a clear idea about the difference of these alternatives considering probabilities, consequences, their individual preferences and their behavior towards risk.

By conducting a sensitivity analysis, it was possible to observe that the first positions of the ranking would not change if the values chosen for the scale constants were changed. It is also interesting to highlight that according to the results, safety will be improved, whereas the alternative that represents no investment in safety appears only in 7th position in the ranking.

10.4 Redesign Required by Maintenance

From the perspective of the maintenance function, redesign is the action that is done when the *status quo* is not acceptable. Some reasons for that are: higher performance requirements due to competitiveness; more conservative standards related to environment and safety; and more severe degradation not covered by the initial design.

According to Moubray (1997), when a failure of a device implies safety and environmental losses and, there is no effective maintenance activity to reduce these consequences, the redesign may be undertaken with at least one objective, such as: Reducing the probability of failure modes; Mitigating the consequences of failures; Reducing the downtime.

The probability of critical failure modes can be reduced by increase the quality of the components or making changes that affect specifically the reliability. For the second objective, mitigating the consequence of failures usually is made by addition of protective devices that reduce the chance of serious consequences happen. Finally, the third objective can be achieved by design changes that make the maintenance actions faster.

In this way, the selection of the parts or equipment that need to be redesigned, can be defined as an MCDM/A problem, in which the set of alternatives is composed by equipment and criteria are related to the attributes of maintainability, and others associated with the cost of the redesign and possible consequences of failures due to the permanence of the *status quo*.

Efforts to redesign should be planned based on the potential gain in reducing or increasing the frequency of the occurrence of specific operating systems. Thus, for a plant with distinct redesign demands, a ranking of these demands based on this expected gains can be useful in order to manage resources efficiently (Heins and Roling 1995).

The redesign process is usually an expensive process and the probability that it will not solve the performance problem can be high. When the design provides opportunity of improvement, maintenance actions should help to achieve the desirable performance. However, when the desired performance is beyond what the design could provide, maintenance actions are ineffective.

References

Alencar MH, Cavalcante CAV, de Almeida AT, Silva Neto CE (2010) Priorities assignment for actions in a transport system based on a multicriteria decision model. In: Bris R, Soares CG, Martorell S (eds) European safety and reliability conference, Prague, September 2009. Reliability, Risk, and Safety: Theory and Applications, Vol. 1-3. 2009. Taylor and Francis, London, UK, p 2480

Almeida-Filho AT de, de Almeida AT (2010) Multiple dimension risk evaluation framework. In: Bris R, Soares CG, Martorell S (eds) European safety and reliability conference, Prague, September 2009. Reliability, Risk, and Safety: Theory and Applications, Vol. 1-3. 2009. Taylor and Francis, London, UK, p 2480

Aven T, Kristensen V (2005) Perspectives on risk: review and discussion of the basis for establishing a unified and holistic approach. Reliab Eng Syst Saf 90:1–14

Aven T, Vinnem JE (2005) On the use of risk acceptance criteria in the offshore oil and gas industry. Reliab Eng Syst Saf 90:15–24

Baron MM, Paté-Cornell ME (1999) Designing risk-management strategies for critical engineering systems. Eng Manag IEEE Trans 46:87–100

Brandsæter A (2002) Risk assessment in the offshore industry. Saf Sci 40:231–269

Brans JP, Mareschal B (2002) Prométhée-Gaia: une méthodologie d'aide à la décision en présence de critères multiples. Éditions de l'Université de Bruxelles
Brito AJ, de Almeida AT (2009) Multi-attribute risk assessment for risk ranking of natural gas pipelines. Reliab Eng Syst Saf 94(2):187–198
Brito AJ, de Almeida AT, Miranda CMG (2010) A Multi-Criteria Model for Risk Sorting of Natural Gas Pipelines Based on ELECTRE TRI integrating Utility Theory. Eur J Oper Res, 200:812-821
BS 5760-0 (2014) Reliability of systems, equipment and components. Guide to reliability and maintainability. British Standard.
BS EN 60706-2 (2006) Maintainability of equipment- Part 2: Maintainability requirements and studies during the design and development phase, British Standards Institution
de Almeida AT, Vetschera R, de Almeida JA (2014) Scaling Issues in Additive Multicriteria Portfolio Analysis. In: Dargam F, Hernández JE, Zaraté P, et al. (eds) Decis. Support Syst. III - Impact Decis. Support Syst. Glob. Environ. SE - 12. Springer International Publishing, pp 131–140
de Almeida AT, Ferreira RJP, Cavalcante CAV (2015) A review of multicriteria and multi-objective models in maintenance and reliability problems. IMA Journal of Management Mathematics 26(3):249–271
Dhillon BS (1999) Engineering Maintainability: How to Design for Reliability and Easy Maintenance. Gulf Professional Publishing
Fu G, Frangopol DM (1990) Balancing weight, system reliability and redundancy in a multi-objective optimization framework. Struct Saf 7:165–175
Garbatov Y, Guedes Soares C (2001) Cost and reliability based strategies for fatigue maintenance planning of floating structures. Reliab Eng Syst Saf 73(3):293–301
Garcez TV, Almeida-Filho AT de, de Almeida AT, Alencar MH (2010) Multicriteria risk analysis application in a distribution gas pipeline system in Sergipe. In: Bris R, Soares CG, Martorell S (eds) Reliability, risk and safety: theory and applications vols 1-3. European safety and reliability conference (ESREL 2009), Prague, September 2009. Taylor and Francis, 1043-1047
Garcez TV, de Almeida AT (2014) Multidimensional Risk Assessment of Manhole Events as a Decision Tool for Ranking the Vaults of an Underground Electricity Distribution System. Power Deliv IEEE Trans 29(2):624–632
Goldman AS, Slattery TB (1977) Maintainability: a major element of system effectiveness. Robert E. Krieger Publishing Company, New York
Guedes Soares C, Garbatov Y (1996) Fatigue reliability of the ship hull girder accounting for inspection and repair. Reliab Eng Syst Saf 51(3):341–351
Hurd Jr W (1966) Engineering design and development for reliable systems. In: Ireson W (ed) Reliab. Handb. McGraw-Hill, New York, pp 10–33
IEC 60706-2 (2006) Maintainability of equipment - Part 2: Maintainability requirements and studies during the design and development phase, International Electrotechnical Commission
IEC 61160 (2005) Design review. International Electrotechnical Commission.
IEEE (1998) Standard Reliability Program for the Development and Production of Electronic Systems and Equipment. IEEE Std 1332-1998:i.
ISO/IEC (2014) Guide 51: Safety aspects – Guideline for their inclusion in standards. ISO/IEC
Keeney RL (1992) Value-focused thinking: a path to creative decisionmaking. Harvard University Press, London
Keeney RL, Raiffa H (1976) Decisions with multiple objectives: Preferences and Value Trade-Offs. Wiley Series in Probability and Mathematical Statistics. Wiley and Sons, New York
Khan FI, Amyotte PR (2002) Inherent safety in offshore oil and gas activities: a review of the present status and future directions. J Loss Prev Process Ind 15:279–289
Khan FI, Sadiq R, Husain T (2002) Risk-based process safety assessment and control measures design for offshore process facilities. J Hazard Mater 94:1–36
Kletz TA (1985) Inherently safer plants. Plant/Operations Prog 4:164–167

Kletz TA (1998) Process plants: A handbook of inherently safer design. 2nd ed, Taylor & Francis, Philadelphia

Lewis EE (1987) Introduction to reliability engineering. Wiley, New York

Li W, Zuo MJ (2008) Optimal design of multi-state weighted k-out-of-n systems based on component design. Reliab Eng Syst Saf 93(11):1673–1681

Lopes YG, de Almeida AT, Alencar MH, Wolmer Filho LAF, Siqueira GBA (2010) A Decision Support System to Evaluate Gas Pipeline Risk in Multiple Dimensions. In: Bris R, Soares CG, Martorell S (eds) European Safety and Reliability Conference (ESREL), Prague, Czech Republic, 2009. Reliability, Risk and Safety: Theory and Applications. CRC Press-Taylor & Francis Group, p 1043

Martello S, Toth P (1990) Knapsack problems: algorithms and computer implementations. John Wiley & Sons, Chichester

Melchers RE, Stewart MG. (1993) Probabilistic risk and hazard assessment. Balkema, Rotterdam

Moubray J (1997) Reliability-centered maintenance. Industrial Press Inc., New York

NORSOK (2010) NORSOK Z-013: Risk and emergency preparedness analysis. Rev. 2, Norwegian Technology Centre

O'Connor P, Kleyner A (2012) Practical reliability engineering. John Wiley & Sons, Chichester

Øien K (2001) A framework for the establishment of organizational risk indicators. Reliab Eng Syst Saf 74:147–167

Polovko AM, Pierce WH (1968) Fundamentals of reliability theory. Academic press New York

Rathod V, Yadav OP, Rathore A, Jain R (2013) Optimizing reliability-based robust design model using multi-objective genetic algorithm. Comput Ind Eng 66:301–310

Ren Y, Bechta Dugan J (1998) Design of reliable systems using static and dynamic fault trees. Reliab IEEE Trans 47:234–244

Roy B (1996) Multicriteria Methodology for Decision Aiding. Springer US

Sachon M, Paté-Cornell E (2000) Delays and safety in airline maintenance. Reliab Eng Syst Saf 67(3):301–309

Sahoo L, Bhunia AK, Kapur PK (2012) Genetic algorithm based multi-objective reliability optimization in interval environment. Comput Ind Eng 62:152–160

Stapelberg RF (2009) Handbook of reliability, availability, maintainability and safety in engineering design. Springer-Verlag, London

US MIL-STD-785B (1980) Reliability Program For System and Equipment Development and Production, US Military Standard

Vanem E, Endresen Ø, Skjong R (2008) Cost-effectiveness criteria for marine oil spill preventive measures. Reliab Eng Syst Saf 93:1354–1368

Vincke P (1992) Multicriteria Decision-Aid. John Wiley & Sons, New York

Vrijling JK, van Hengel W, Houben RJ (1998) Acceptable risk as a basis for design. Reliab Eng Syst Saf 59:141–150

Yanasse HH, Soma NY (1987) A new enumeration scheme for the knapsack problem. Discret Appl Math 18:235–245

Chapter 11
Decisions on Priority Assignment for Maintenance Planning

Abstract: This chapter presents multicriteria (MCDM/A) models to classify and assign maintenance priorities in order to allow maintenance planning to be more effective. From traditional maintenance planning techniques such as RCM (Reliability Centered Maintenance), TPM (Total Productive Maintenance) and others, a common aspect of these techniques is the definition of maintenance priorities, based on a criticality classification for RCM, for example. As maintenance planning has to satisfy multiple objectives, such as availability, maintainability, detectability, safety and reliability besides cost, the maintenance manager is a decision maker (DM) who has to establish tradeoff amongst multiple criteria. This chapter presents an MCDM/A model integrated with the RCM structure using Utility Theory principles to include states of nature and the DM's behavior to risk (prone, neutral and averse) in a decision model based on Multi-attribute utility theory (MAUT). To illustrate situations when a DM has a non-compensatory rationality, and requires an outranking method, a decision model based on ELECTRE TRI is applied. In addition, TPM aspects are discussed in order to emphasize potential MCDM/A problems that may be approached.

11.1 Introduction

Among the decisions regarding maintenance planning, one of the most important decisions is related to define which kind of maintenance actions are more appropriated. This decision involves subjective and technical aspects in order to evaluate the consequences of failures. This chapter presents MCDM/A models considering the assignment of priorities before establishing a maintenance plan.

A maintenance plan can be defined with different approaches. Selective maintenance is an example for building a maintenance plan. This approach includes the specification of each action that should be done for each item in a multicomponent system, for an interval longer than one cycle, observing the constraints for optimizing a single objective. Originally, this problem was formulated considering a fixed time window (Lust et al. 2009).

A.T. de Almeida et al., *Multicriteria and Multiobjective Models for Risk, Reliability and Maintenance Decision Analysis*, International Series in Operations Research & Management Science 231, DOI 10.1007/978-3-319-17969-8_11

In practice, this problem requires the observation of multiple aspects and an MCDM/A approach enhances the solution by considering multiple aspects, such as: system performance, costs, total time spent at maintenance, number of repaired components and availability of spare parts.

Note that, the selective maintenance approach can also be used to build the annual maintenance plan for items without considering time window constraints. In the other hand, as the selective maintenance approach requires accurate information about the changes that occur in each cycle, it may demand too many information, turning the maintenance plan definition a complex task. Thus, due to the challenge that is building an annual maintenance plan, it is not revised as much as it should be.

There are other approaches for building a maintenance plan, which are based on the definition of the maintenance strategy (Bashiri et al. 2011). It considers that there is a most appropriate action for each component in order to optimize a specific criterion. Some authors considered MCDM/A approaches for defining a maintenance strategy in order to build a maintenance plan (Gómez de León Hijes and Cartagena 2006; Zaeri et al. 2007; Bevilacqua and Braglia 2000). A literature review considers MCDM/A models in maintenance (de Almeida et al. 2015) and points out the increasing number of research dealing with these models.

The maintenance literature considers that while in the selective maintenance approach it can be difficult to establish a maintenance plan due to its information requirements, the maintenance strategy selection problem may be too simplistic and inconsistent with some realities, especially when there is no interest in updating the strategy, and consequently the maintenance plan defined. Therefore, practical situations show that maintenance plans should be updated and revised continuously (Berrade et al. 2013; Berrade et al. 2012; Scarf and Cavalcante 2012).

From this perspective, priority assignment is an important step before establishing, updating or revising maintenance plans. The definition of which systems (subsystems, items) or failure modes are more critical for the producing system mission is an important decision when considering approaches such as RCM (Reliability Centered Maintenance), used to assist maintenance planning. As a result, a maintenance plan or strategy is only defined after the defining the critical systems (subsystems, items) or failure modes in the system. Considering this principle, the maintenance will be more effective and maintenance plans/ strategy shall be more accurate.

Based on this perspective, this chapter presents different MCDM/A models to establish criticality before building a maintenance plan. In addition, considerations are given to traditional approaches, such as RCM and Total Productive Maintenance (TPM) structure and its integration with MCDM/A models.

11.2 An MCDM/A Model for the RCM Approach

In this section, a quantitative MCDM/A model for evaluating the consequences of failure is presented (Alencar and de Almeida 2011). The procedure for resolution of problems and building MCDM/A models presented in Chap. 2 is referenced in some stages of the model. The model enhance the RCM approach features by providing a structured decision making process, taking into account uncertainties and the DM's preferences.

11.2.1 Traditional RCM Consequence Evaluation

Two important aspects of the traditional RCM approach are presented in this subsection in order to provide a better understanding of the MCDM/A model built: the procedure steps and the evaluation of failure consequences.

From the twelve steps introduced in Chap. 3, Moubray (1997) emphasizes these following steps:

- Establish the functions of each asset within the operating context considering the associated desired standards of performance;
- Define failures that may occur in the physical asset;
- Identify the failure modes;
- The fourth step involves checking the effects of failure;
- The fifth step is to verify and analyze the consequences of failure;
- Finally, the last step is to establish maintenance actions that could be verified by applying two techniques: proactive tasks and default actions.

Additionally, RCM approach classifies the consequences of failure into four categories (Moubray 1997):

- Hidden failure consequences: when it does not present a direct impact, but can expose the organization to multiple failures with serious consequences (including catastrophic);
- Safety and environmental consequences: when it presents safety consequences considering the possibility of injury or death. The environmental consequences might mean that an organization has violated a national or international environmental standard;
- Operational consequences: failures which affects only production;
- Non-operational consequences: failures in this category do not affect either production or safety, involving only the direct cost of repair.

According to the facilities analyzed, a failure could produce irrelevant consequences or compromise essential systems for the organization or society or safety. In RCM approach, the consequences are evaluated by verifying the impacts of the effects of a failure mode on system operation, physical security, the environment, and the economy of the process. Clemente et al. (2012) state that RCM, when used with other approaches can offer a more complete understanding of the operational context, providing financial and management information for decision making.

11.2.2 RCM Based on MCDM/A Approach

Some terms are relevant to build the model such as: the observed context, the availability of information and its degree of accuracy, the rationality required the DM's preference structure and the problematic. An important aspect is the rationality for the DM in the problem under study that involves a non-compensatory or compensatory approach. In this sense, the decision model presented in this subsection sets out to improve the RCM approach by incorporating contributions from MAUT.

According to de Almeida (2007), MAUT considers that the DM's preferences are modeled for computing the MAU function in which the aggregation of uni-dimensional utility functions must respect MAUT axiomatic structure. Additionally, Brito and de Almeida (2009) state that MAUT can be applied to aggregate valued preferences and uncertain consequences related with multiple criteria, providing results that can be used as input in the process of maintenance management.

The stages of this RCM MCDM/A model are shown in Fig. 11.1. Traditional RCM steps that remains are: define the functions of assets; identify functional failures; define failure modes; identify the effects of failure; and, establish maintenance actions. Therefore, this subsection focuses mainly on the MCDM/A model built for evaluating the consequences of failures.

For each objective defined, a dimension of consequences is proposed, representing the objectives of this decision model. Thus, a set of consequence dimensions are established.

The consequences of failures are evaluated based on five categories defined as the dimensions of the consequences, in which some of the characteristics differ from those established by the traditional RCM approach, as follows:

- Human dimension (*h*): considers the damage with respect to people affected by the consequences of failures;
- Environmental dimension (*e*): considers the area affected due to a failure;
- Financial dimension (*f*): considers the financial losses due to a failure;
- Operational dimension:
  - Operational dimension I (*o'*): considers failures that do not interrupt the producing system operation;
  - Operational dimension II (*o''*): considers failures that interrupt the producing system operation.

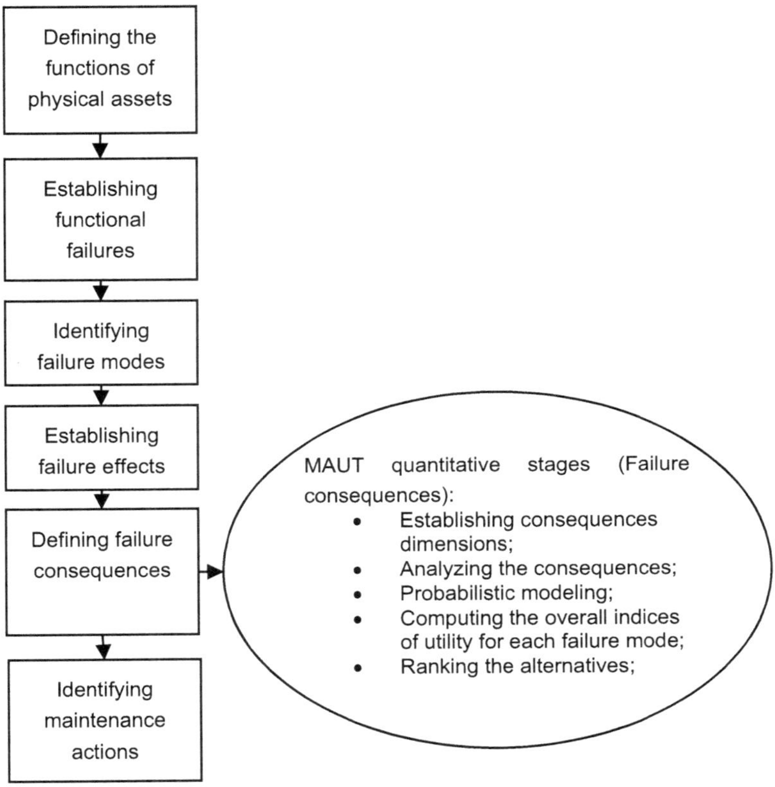

Fig. 11.1 Stages of an MCDM/A model

The identification of state of nature is based on the step 5 of procedure for resolution of problems and building MCDM/A models presented in Chap. 2.

For evaluating the consequences, elements of decision theory are applied, in which θ is established as the state of nature. It is used to express the uncertainty related with the problem. The consequences are represented by c and the set of all actions under study is represented by A.

A probabilistic approach is applied to incorporate the associated uncertainties in A considering a probability distribution over consequences and by eliciting utility functions for these consequences. The probability of each state of nature is defined as $\pi(\theta)$. $U(\theta,a_i)$ is the utility when θ and action a_i are considered (Berger 1985).

The utility values are defined in an interval scale between $[0, 1]$, where 0 is associated to the least preferred while the extreme 1 is related to the most preferred (Keeney and Raiffa 1976). The utility function of these consequences is shown by (11.1) when the set of consequences is discrete.

$$U(\theta,a_i)=\sum_c P(c|\theta,a_i)U(c) \tag{11.1}$$

Finally, (11.2) shows the utility function of these consequences for continuous cases.

$$U(\theta,a_i)=\int_c P(c|\theta,a_i)U(c)dc \tag{11.2}$$

Following the step 6, the preference modeling is structured (Vincke 1992; Keeney 1992). Step 7 includes intra-criteria evaluation, mandatory to define the functions of the consequences considered.

Step 8 consists of the inter-criteria evaluation, for establishing each criterion scale constant, k_i, and the overall utility function (Keeney and Raiffa 1976).

Assuming an additive MAU function, (11.3) represents the overall utility.

$$U(h,e,f,o',o'')=k_1U(h)+k_2U(e)+k_3U(f)+k_4U(o')+k_5U(o'') \tag{11.3}$$

where:
k_i is a scale constant that represents the value of the tradeoff;

When the DM's preferences require limiting tradeoff effects, a model that considers veto can be incorporated (de Almeida 2013).

The final results are presented by the obtained ranking, established by multi-attribute utility values found for each failure mode.

The interval scale of utility function allows the incremental value to be compared to the failure modes Keeney and Raiffa (1976). Applying the interval scale, it may be affirmed that the difference $U(MF_x)_{\beta x} - U(MF_y)_{\beta x+1}$ is M times greater than the difference $U(MF_y)_{\beta x+1} - U(MF_z)_{\beta x+2}$. This can be seen from the increment ratio IR of these differences since $IR = (U(MF_x)_{\beta x} - U(MF_y)_{\beta x+1})/(U(MF_y)_{\beta x+1}) - U(MF_z)_{\beta x+2})$.

11.2.3 Illustrative Example

For this illustrative example, 16 failure modes are considered, FM_x, $x=1,2,\ldots,16$. Each FM_x is associated with human (*h*), environmental (*e*), financial (*f*), operational I (*o'*) and operational II (*o''*) consequence dimensions (Alencar and de Almeida 2011).

There is a prior probability $\pi(\theta_x)$ associated to each FM_x, as can be observed from Table 11.1.

Table 11.1 A prior probability of failure modes

Component	Failure Mode	Prior Probability
X1	FM_1	0.0766
X2	FM_2	0.0256
X3	FM_3	0.0578
X4	FM_4	0.0333
X5	FM_5	0.0835
X6	FM_6	0.0259
X7	FM_7	0.0768
X8	FM_8	0.0493
X9	FM_9	0.0876
X10	FM_{10}	0.0087
X11	FM_{11}	0.07
X12	FM_{12}	0.0563
X13	FM_{13}	0.0367
X14	FM_{14}	0.0154
X15	FM_{15}	0.0958
X16	FM_{16}	0.0757

The scale constants $k_1=0.19$, $k_2=0.13$, $k_3=0.27$, $k_4=0.11$, $k_5=0.30$, are elicited from the DM adopting structured protocols (Keeney and Raiffa 1976).

The interval scale of the utility function allows comparison of the differences in utility among failure modes. These differences are verified in Table 11.2 (fourth column).

Table 11.2 Comparisons of differences in utility among failure modes

Ranking position (β_x)	Failure Mode FM_i	$U(FM_x)_{\beta x}$	$U(FM_x)_{\beta x}$ - $U(FM_y)_{\beta x+1}$	Difference ratio
β_{01}	FM_{15}	0	0.08788	0.51200
β_{02}	FM_9	0.08788	0.17164	4.15593
β_{03}	FM_1	0.25952	0.0413	0.40249
β_{04}	FM_3	0.30082	0.10261	1.46064
β_{05}	FM_5	0.40343	0.07025	0.85598
β_{06}	FM_7	0.47368	0.08207	0.70308
β_{07}	FM_{14}	0.55575	0.11673	14.70151
β_{08}	FM_8	0.67248	0.00794	0.41966
β_{09}	FM_{11}	0.68042	0.01892	1.09745
β_{10}	FM_4	0.69934	0.01724	0.19334

(continued)

Table 11.2 (continued)

β_{11}	FM_{13}	0.71658	0.08917	2.21540
β_{12}	FM_{02}	0.80575	0.04025	0.99187
β_{13}	FM_{12}	0.84600	0.04058	1.33399
β_{14}	FM_{16}	0.88658	0.03042	0.36651
β_{15}	FM_{6}	0.91700	0.083	-
β_{16}	FM_{10}	1	-	-

The values presented in Table 11.2 provide important information for the DM. The difference between the values of the utilities associated with the failure modes FM_{14} and FM_8 is 0.11673; and the difference between the values of the utilities associated with the failure modes FM_8 and FM_{11} is 0.00794. The ratio among differences (fifth column) allows the DM to understand the relative difference among each FM quantified by the utility scale.

This measure allows to state that the relative difference between FM_{14} and FM_8 is approximately 15 times greater than the difference between FM_8 and FM_{11}. It is important to highlight that these values presented in Table 11.2 reflects the DM's preferences among four consequence dimensions. From the numerical example given, it is possible to observe how undesirable is the differences among such failure modes for DM.

11.3 An MCDM/A Vision for the TPM Approach

There is no doubt of how the quality of maintenance activities affects the performance of a producing system. In some cases, system failures occurrence is affected predominantly from the influence imposed by personnel, whereas the ageing is a secondary failure mechanism (Levitin and Lisnianski 2000; Wang and Pham 2006; Scarf and Cavalcante 2012).

Furthermore, the implementation of any model, technique or procedure developed to support the maintenance effectiveness relies on the maintenance personnel effort. Thus, tools which get personnel involved and with high commitment become essential in the operational level.

Total Productive Maintenance (TPM) is a technique in which one of the main goals is to keep people engaged and motivated to participate in process of improvements related to maintenance issues. A TPM principle is to bring the attention of the operator for the signals of non-regular operations, in order to find and fix some small problems in the system. On the absence of a sophisticated monitoring system, this is a way to provide the continuous inspection. Therefore, personnel become a kind of monitoring system.

The Japan Institute of Plant Maintenance (JIPM) created TPM in the 1970s, during the Japanese quality improvement movement. It considers basic pillars that are divided by topics. Furthermore, TPM has an evolutionary structure, which makes this technique more flexible to be implemented. Therefore, the attention can be focused in one phase per time.

Although the importance of the TPM, there are few researches conducted using an MCDM/A approach in maintenance problems under the TPM framework (de Almeida et al. 2015). Thus, there still a niche to be explored addressing MCDM/A decision problems that arise under a TPM program.

Some potential MCDM/A models to explore decision problems in the TPM context include:

- Maturity evaluation of specific TPM pillars, taking into account multiple aspects involved for assessing the maintenance program in the organization;
- Priorities assessment of TPM pillars, for the budget allocation and team effort to improve the potential results from TPM;
- Overall Equipment Efficiency measure, including multiple dimensions through an MCDM/A approach in order to consider DM's preferences;

These problems can be addressed using the framework described in Chap. 2 to build an MCDM/A decision model.

As an example, consider the problem related with assigning priorities among TPM pillars, the focus is to point out which pillars should receive more attention during the implementation in order to maximize the chances of the TPM to succeed. During different moments, the organizations are subjected to an environment and constraints that would require improvements in different directions. Similar dynamic environment is considered when considering Goldratt's Theory of Constraints, thus specific efforts are deployed to achieve the goals required in the actual state of the system, in other words, the actual constraint that represents a bottleneck.

Therefore, the set of alternatives would be related to different pillars combinations, which will lead to a portfolio of actions to be prioritized considering the resources available for the maintenance function. TPM literature recommends a top-down implementation, which means that the DM would represent the board of the company.

The criteria considered for such problem includes each one of the pillars, with their strategic considerations, highlighting the gaps between the status quo and the organization goals.

11.4 Modeling a Problem for Identifying Critical Devices

This section presents a model for identifying critical devices, which classifies the items from an industrial plant into predetermined category of criticality. Using the general procedure proposed in Chap. 2, an MCDM/A model is introduced. For simplification purposes, only some steps of the Chap. 2 procedure are highlighted in the model presentation.

Maintenance planning requires a thorough understanding of the system and of the goals, as well as of the consequence dimensions associated with the failure of the item. Deciding "what to do" may be based on technical, environmental and financial aspects. Most of the models try to establish a severity index for representing a measure in different dimensions in order to support DM for deciding "what to do". The weakness of these approaches is that DM's preferences are not considered when building such measures.

Facing a large set of piece of equipment, the DM seeks to organize this set into classes of criticalities. This classification helps the DM to specify the most appropriated set of actions for each class, considering the adequate resources to be deployed in a more effective way.

For example, in a power distribution network there are several similar items, however their location in the network, despite its similarity adds the specific branch characteristics resulting in different criticality levels, which will point for different maintenance actions or polices depending on this specific item. Depending on the item which fails, multidimensional consequence will arise. The specific location of the item can characterize the number of affected customers, public services, and result in losses for business supplied in the distribution branch affected. Depending on the kind of failure, safety aspects also may arise. If such failure occurs in an underground distribution network, for example, and has the potential to cause explosions in a high density area, such as in large cities (Garcez and de Almeida 2014a, Garcez and de Almeida 2014b).

Considering the specific problem presented in this section for assigning devices into priority classes, by means of an MCDM/A sorting model, based on ELECTRE TRI, using information about the characteristics of device and the multidimensional consequences associated with its failures.

ELECTRE TRI was designed to describe actions in ordered categories, it enables a set of alternatives pre-defined and ranked categories based on multiple criteria to be classified, as illustrated in Fig. 11.2, where each device x_i ($x = 1 \ldots n$) is classified according to the device's criticality.

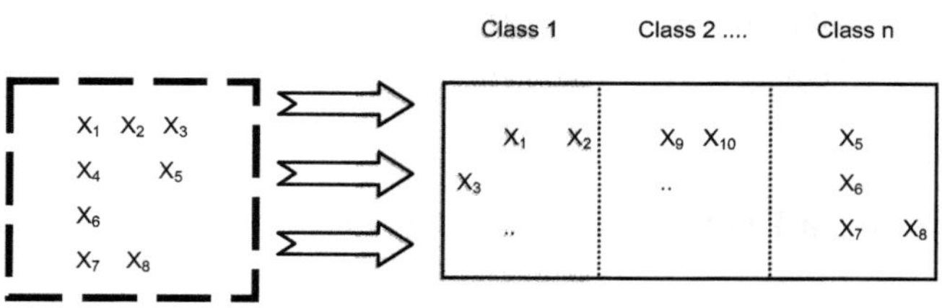

Fig. 11.2 Criticality classification of devices

This MCDM/A classification process can be replicated in different levels, starting from the equipment, device, component and failure modes level, successively, as soon as the need information becomes available.

Thus, the MCDM/A model sorts each device into priority classes that supports the maintenance management, giving an initial filter before addressing the elaboration of a maintenance plan for the entire plant. Therefore, an application is presented with an illustrative example of MCDM/A model.

The criteria considered are:

- Safety and environment losses (g_1): it refers to the possibility of someone being injured; or an environmental damage caused by the device failure;
- Financial losses (g_2): it considers monetary losses resulting from a device failure, including repair costs and other costs from downtime;
- Frequency of the device faults (g_3);
- Delay-time (g_4): expected time elapsed since the arrival of the defect until a device failure;
- Detectability (g_5): it represents the level of difficulty of the fault detection.

The set of alternatives is formed by 10 generic devices $\{x_1, x_2, x_3, ...x_{10}\}$.

All these criteria are measured considering a semantic scale from 1 to 5. These scales are detailed for the evaluation of the performance for each device according to the Tables 11.3-7, respectively for each criterion.

Table 11.3 Safety and environment damage scale

Description	Scale
Catastrophic consequence	5
Major consequence	4
Severe Consequence	3
Minor Consequence	2
Trivial Consequence	1

Table 11.4 Financial losses scale

Description	Scale
Loss of more than 20,000 monetary unities	5
Loss of 15,001 a 20,000 monetary unities	4
Loss of 10,001 a 15,000 monetary unities	3
Loss of 5,001 a 10,000 monetary unities	2
Loss of 0 a 5,000 monetary unities	1

Table 11.5 Frequency of the device faults scale

Description	Scale
Failed more than 15 times on interval	5
Failed from 12 to 15 times on interval	4
Failed from 8 to 11 times on interval	3
Failed from 4 to 7 times on interval	2
Failed from 0 to 3 times on interval	1

Table 11.6 Delay-time scale

Description	Scale
Mean Delay-time of 0 a 10 time unities	5
Mean Delay-time of 11 a 20 time unities	4
Mean Delay-time of 21 a 30 time unities	3
Mean Delay-time of 31 a 40 time unities	2
Mean Delay-time bigger than 40 time unities	1

Table 11.7 Detectability scale

Description	Scale
Almost Impossible detection	5
Difficult detection	4
Moderate detection	3
Easy detection	2
Immediate detection	1

The matrix of consequences should be elicited from a multidisciplinary team, including experts. It is shown in Table 11.8.

Table 11.8 Matrix of consequences

Alternatives\Criterion	g_1	g_2	g_3	g_4	g_5
x_1	1	1	2	1	3
x_2	4	5	1	3	4
x_3	3	2	3	4	2
x_4	3	4	3	4	1
x_5	5	5	1	5	1
x_6	4	3	2	4	3
x_7	1	2	5	2	2

(continued)

Table 11.8 (continued)

x_8	2	3	5	3	5
x_9	1	1	3	2	2
x_{10}	2	3	4	5	3

Besides the scale defined for each criterion, the definition of the parameters of the preference functions in order for the ELECTRE TRI. It was defined that the values of the limits of preference and indifference are both equal to 0. Disregarding the veto threshold and setting $\lambda = 0.5$.

Related to the intra-criteria evaluation, the weights defined by the DM are given by Table 11.9.

Table 11.9 The weights of the criteria

Criterion	Weight
g_1	0.25
g_2	0.35
g_3	0.18
g_4	0.12
g_5	0.1

In this study, five categories are considered, ordered according to their degree of importance concerning the priority of planning and conduct of maintenance actions. The classes considered are:

- Highly critical device: the occurrence of a fault in any device belonging to this class will bring serious damage to the organization;
- High priority devices;
- Intermediate priority devices;
- Low-priority devices;
- Extremely low-priority devices: i.e., one can, in a way, neglect the maintenance of equipment belonging to this class so as to direct more concentrated efforts to the most critical equipment;

The equivalence classes serve as standards by which the devices will be classified. The equivalence classes adopted for this study are defined by lower and upper ("Profiles") limits, as shown in Table 11.10.

Table 11.10 Classes of equivalence and their lower and upper limits

Class	Lower Limit	Upper Limit
C_1	4.5	-
C_2	3.5	4.5

(continued)

Table 11.10 (continued)

C_3	2.5	3.5
C_4	1.5	2.5
C_5	-	1.5

The results are presented in Table 11.11.

Table 11.11 Results

Equipment	Pessimistic	Optimist
x_1	C_5	C_5
x_2	C_2	C_2
x_3	C_3	C_3
x_4	C_3	C_3
x_5	C_1	C_1
x_6	C_3	C_3
x_7	C_4	C_4
x_8	C_3	C_3
x_9	C_5	C_5
x_{10}	C_3	C_3

Analyzing the results from Table 11.11, only one device was sorted as extremely critical (x_5). This is due to the fact that, if a failure occurs in this device, there will be catastrophic losses for the company, with regard to matters relating to the financial, human and environmental dimensions; and besides, this device has a short delay-time, which deserves careful attention.

The simulation is therefore useful since it enables the manager of maintenance to drive the maintenance actions in such a way as to focus on the most critical devices, while it reinforces that equipment considered less important can be neglected.

The impact from the application of an MCDM/A approach on the maintenance management process may be reflected in the improved operating performance of the device, due to more efficient maintenance planning for each class of device having been adopted.

References

Alencar MH, de Almeida AT (2011) Applying a Multicriteria Decision Model So as to Analyse the Consequences of Failures Observed in RCM Methodology. In: Takahashi RC, Deb K, Wanner E, Greco S (eds) Evol. Multi-Criterion Optim. SE - 41. Springer Berlin Heidelberg, pp 594–607

Bashiri M, Badri H, Hejazi TH (2011) Selecting optimum maintenance strategy by fuzzy interactive linear assignment method. Appl Math Model 35:152–164

Berger JO (1985) Statistical decision theory and Bayesian analysis. Springer Science & Business Media, New York

Berrade MD, Cavalcante CAV, Scarf PA (2012) Maintenance scheduling of a protection system subject to imperfect inspection and replacement. Eur J Oper Res 218:716–725

Berrade MD, Scarf PA, Cavalcante CAV, Dwight RA (2013) Imperfect inspection and replacement of a system with a defective state: A cost and reliability analysis. Reliab Eng Syst Saf 120:80–87

Bevilacqua M, Braglia M (2000) The analytic hierarchy process applied to maintenance strategy selection. Reliab Eng Syst Saf 70:71–83

Brito AJ, de Almeida AT (2009) Multi-attribute risk assessment for risk ranking of natural gas pipelines. Reliab Eng Syst Saf 94(2):187–198

Clemente T, Almeida-Filho AT de, Alencar MH, Cavalcante CAV (2013) A Decision Support System Based on RCM Approach to Define Maintenance Strategies. In: Poels G (ed) Enterp. Inf. Syst. Futur. SE - 9. Springer Berlin Heidelberg, pp 122–133

de Almeida AT (2007) Multicriteria decision model for outsourcing contracts selection based on utility function and ELECTRE method. Comput Oper Res 34(12):3569–3574

de Almeida AT (2013) Additive-veto models for choice and ranking multicriteria decision problems. Asia-Pacific J Oper Res 30(6):1-20

de Almeida AT, Ferreira RJP, Cavalcante CAV (2015) A review of multicriteria and multi-objective models in maintenance and reliability problems. IMA Journal of Management Mathematics 26(3):249–271

Garcez TV, de Almeida AT (2014a) A risk measurement tool for an underground electricity distribution system considering the consequences and uncertainties of manhole events. Reliab Eng Syst Saf 124:68–80

Garcez TV, de Almeida AT (2014b) Multidimensional Risk Assessment of Manhole Events as a Decision Tool for Ranking the Vaults of an Underground Electricity Distribution System. Power Deliv IEEE Trans 29(2):624–632

Gómez de León Hijes FC, Cartagena JJR (2006) Maintenance strategy based on a multicriterion classification of equipments. Reliab Eng Syst Saf 91(4):444–451

Keeney RL (1992) Value-focused thinking: a path to creative decision making. Harvard University Press, London

Keeney RL, Raiffa H (1976) Decisions with multiple objectives: Preferences and Value Trade-Offs. Wiley Series in Probability and Mathematical Statistics. Wiley and Sons, New York

Levitin G, Lisnianski A (2000) Optimization of imperfect preventive maintenance for multi-state systems. Reliab Eng Syst Saf 67:193–203

Lust T, Roux O, Riane F (2009) Exact and heuristic methods for the selective maintenance problem. Eur J Oper Res 197:1166–1177

Moubray J (1997) Reliability-centered maintenance. Industrial Press Inc., New York

Scarf PA, Cavalcante CAV (2012) Modelling quality in replacement and inspection maintenance. Int J Prod Econ 135(1):372–381

Vincke P (1992) Multicriteria Decision-Aid. John Wiley & Sons, New York

Zaeri MS, Shahrabi J, Pariazar M, Morabbi A (2007) A combined multivariate technique and multi criteria decision making to maintenance strategy selection. Ind. Eng. Eng. Manag. 2007 IEEE Int. Conf. IEEE, Singapore, pp 621–625

Chapter 12
Other Risk, Reliability and Maintenance Decision Problems

Abstract: In this chapter, specific problems in risk, reliability and maintenance context are described, such as location of backup units, sequencing of maintenance activities, natural disasters, operation planning of a power system network, integrated production and maintenance scheduling, maintenance team sizing and reliability acceptance tests. This chapter presents a multicriteria decision model with an illustrative application for most of these problems. Amongst the MCDM/A approaches considered for the illustrative applications in this chapter are: Multi-attribute utility theory (MAUT), PROMETHEE II, NSGA-II. Regarding the reliability acceptance test an MCDM/A Bayesian approach is presented. For these problems, several aspects have been considered such as: size of population, degree of industrialization, the extent of health services (location of backup units); degree of damage, consumption, electric load, special clients, healthcare services, SAIDI and SAIFI (sequencing of maintenance activities); human, environmental, financial and infrastructure concerns (natural disasters); expected tardiness and maintenance costs (integrated production and maintenance scheduling); waiting time and cost of personnel (maintenance team sizing); probability of accepting equipment not in accordance with the reliability specified by the manufacturer; and delaying the project conclusion (reliability acceptance test). Finally, some aspects of multiobjective optimization are discussed.

12.1 Introduction

A literature review found 186 papers related to maintenance and reliability problems based on MCDM/A published between 1978 and 2013. Studies from various countries contributed to this subject. In fact, more than 30 countries were identified (de Almeida el al. 2015). This spread around the world is shown in Fig. 12.1, in which the size of circles indicates the number of such studies found per country relative to each other.

A.T. de Almeida et al., *Multicriteria and Multiobjective Models for Risk, Reliability and Maintenance Decision Analysis*, International Series in Operations Research & Management Science 231, DOI 10.1007/978-3-319-17969-8_12

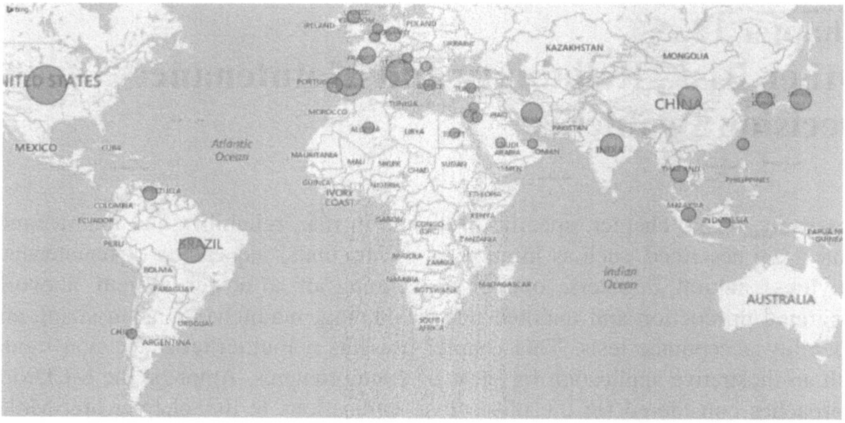

Fig. 12.1 World map of publications on the use of MCDM/A in maintenance and reliability research

Figure 12.2 shows that there has been an exponential trend in citations which explains the growth of publications on this subject. The ever increasing amount of publications indicates the relevance of the topic and the perspectives in the area.

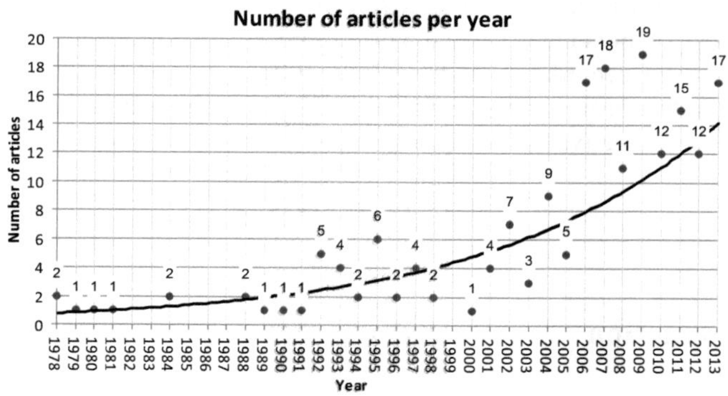

Fig. 12.2 Number of articles per year on MCDM/A in maintenance and reliability research

Furthermore, the 170 articles considered until 2012 had received 4,306 citations from 1996 to 2013 according to the Scopus database, which represents an average of 25.33 citations per paper. Fig. 12.3 reflects the impact of this research area, measured by citations per year since 1996 (de Almeida el al. 2015). For instance, the articles received 831 citations in 2012.

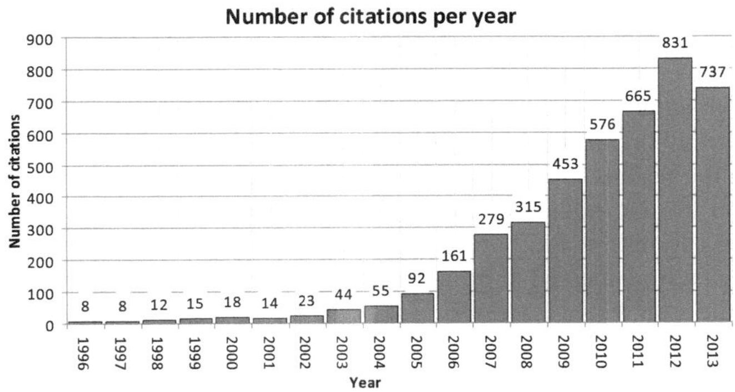

Fig. 12.3 Number of citations per year on MCDM/A in maintenance and reliability research

From this perspective, this chapter presents RRM problems that arise in different particular contexts not mentioned in previous chapters.

Amongst the many specific problems that require an MCDM/A approach, this chapter illustrates RRM problems that require appropriate modeling in order to allow a DM to consider multidimensional consequences.

Thus, this chapter covers topics related to the following problems:

- Location of backup units;
- Sequencing of maintenance activities;
- Natural disasters;
- Reliability in power systems;
- Integrated production and maintenance scheduling;
- Maintenance team sizing;
- Reliability acceptance testing.

12.2 Location of Backup Units in an Electric System

One of the main objectives of the maintenance function is to minimize the occurrence of failure, i.e., reduce its frequency. This can be achieved by design improvements, proper use of assets, preventive maintenance, and condition monitoring. There is also an interest in minimizing the time spent on corrective actions when failures occur in order to maximize system availability.

Two significant portions of time are usually considered in corrective actions. Each has a different impact on the total time spent on maintenance actions. First, there is the time needed for the logistics i.e. from identifying the failure, placing the work order, obtaining and preparing the resources needed to perform the maintenance such as tools, labor, parts; and dislocating the maintenance staff to the place of the service. When the maintenance team is ready to conduct the service, there is an elapse of time from service startup to completion, and then the asset returns to operational status.

In many practical applications, is possible to observe that the maintenance time needed for a repair has been considered one of the most relevant. On the other hand, when the resources available to perform maintenance are scarce, some other factors need to be addressed to minimize the elapsed time for corrective maintenance actions. In addition to this scarcity, the location of some assets may be geographically dispersed and may have a strong influence on the time needed to conduct the maintenance actions.

The example given in this section for location of backup units considers the context of an electric power distribution company. For this kind of companies, there are several geographically dispersed systems throughout the distribution network. Given the geographically dispersion, the maintenance function has to overcome the logistics obstacles to cope with the required performance standards that require the availability of equipment located in each of the electric power substations distributed along the network.

The equipment considered includes high-tension power transformers, which are heavy, expensive assets and have a high useful life. Such equipment costs millions of dollars and has a useful life around 30 years. Moreover, the lead-time for ordering such equipment can take several months, besides the time referring the logistics of installing the equipment.

Although expensive equipment with a low failure rate cannot justify investment in a high amount of redundancy for electric power substations, the best use of resources should be evaluated to deal with an emergency, especially when the impact on system unavailability is high.

In the case of power transformers, it is known that a limited number of backup units are available to electric power substations for possible replacements due to a failure. The decision problem is to define the locations of backup transformers to minimize the overall consequences of a failure and the need for emergency replacement.

The consequences of these equipment failures can be characterized as in Chap. 1, when considering service producing systems. In this particular case, the number of users affected can escalate from thousands to millions, depending on the consequences of the failure and the blackout effects. Such consequences vary depending on the equipment location, similar to the example given in Chap. 11 regarding the implications to the priority assignment for maintenance planning decisions.

To prevent and minimize such consequences, it is essential to plan the location of backup units in order that it can quickly restore the system in case of failure.

Therefore, the location of backup transformers involves many factors, which directly influence the operation of a power distribution system. Such process involves objectives beyond costs related to service interruptions prolonged due to the absence of a backup unit. These factors have a direct influence on system's availability and maintainability (de Almeida et al. 2006; Ferreira et al. 2010; Ferreira and Ferreira 2012).

Thus, the decision model for this problem seeks to ensure that customers are minimum affected by the inconvenience of service interruption and the losses associated. Although a similar problem structure is addressed in classical facility location (Drezner and Hamacher 2004), different objectives are considered.

A failure in each particular location affects multidimensional consequences. For this particular problem, three dimensions are considered: number of customers, health services and local economy affected.

Brandeau and Chiu (1989) give an overview of location problems, which have been previously studied, with the emphasis on models that were developed in the field of operations research, formulated as optimization problems, such as the p-median. The p-median is a classical problem in the field of combinatorial optimization problem. An algorithm with re-optimization procedures for multi-objective combinatorial optimization problems is proposed by Bornstein et al. (2012).

The decision model consists of an MCDM/A p-median model based on three criteria:

- The size of the population (pop_i);
- Degree of industrialization (ind_i);
- The extent of health services (hs_i).

The p-median model was adapted to minimize three criteria as shown in (12.1). The distance factor (d_{ij}) represents the distance amongst the electric power substations. The distance factor works as a multiplier weight in relation to *pop*, *ind* and *hs*.

$$
\begin{aligned}
Max \quad & \sum_{i=1}^{ns}\sum_{j=1}^{ns}\begin{bmatrix} K_1 \cdot U(pop)_{ij} + K_2 \cdot U(ind)_{ij} + K_3 \cdot U(hs)_{ij} \\ + K \cdot U(pop)_{ij} \cdot U(hs)_{ij} \end{bmatrix} \cdot x_{ij} \\
s.t. \quad & \sum_{i=1}^{ns} x_{ij} = 1; \quad j \in N \\
& \sum_{j=1}^{ns} x_{jj} = nb \\
& x_{ij} \le x_{jj}; \quad i, j \in N \\
& x_{ij} \in \{0,1\}; \quad i, j \in N
\end{aligned}
, \qquad (12.1)
$$

where:
K_1, K_2, K_3, and K are scale constants related to the respective attributes;
N is a set of electric power substations, $N = \{1, \ldots, ns\}$;
nb is the number of back-up transformers;
pop_i is the size of population served by the substation i;
ind_i is the degree of industrialization served by the substation i;
hs_i is the extent of health services served by the substation i;

x_{ij} is a decision matrix variable, where $x_{ij} = 1$ if the backup transformer of the substation i is allocated to substation j, and $x_{ij} = 0$, otherwise. And x_{jj}=1 if the substation j is allocated to store a back-up transformer (a median) and x_{jj}=0, otherwise.

For the specific application, regarding the inter-criteria evaluation all attributes are found to be preference independent, except health service and population criteria. Thus, DM's preferences over health are affected when varying values of the size of the population. The model, which represents these conditions, corresponds to a multi-linear model expressed in the MAU function, given by (12.1).

The objective function represents the utility for substation k, if the backup is located in substation j. In order to have an indicator for the location of the backup in substation j, the sums of the utilities of location for all the electric power substations k should be calculated.

By calculating the maximum utility for each substation, the recommendation will be to locate a backup transformer in the power substation for which this transformer will provide the highest maximum utility.

The considered company has to decide which substation should host a backup transformer from 19 options considered.

As to the multi-attribute p-median model, the results are shown in Fig. 12.4. The maximum value of the multi-attribute function for nb = 6 is 17.071. The constant scales represent the preferences of the manager of the current project. The parameters K_1 = 0.2, K_2 = 0.5, K_3 =0.2, K = 0.1 were obtained using a structured process suggested by Keeney and Raiffa (1976). The electric power substations chosen are S = {3, 6, 11, 8, 13, 19}.

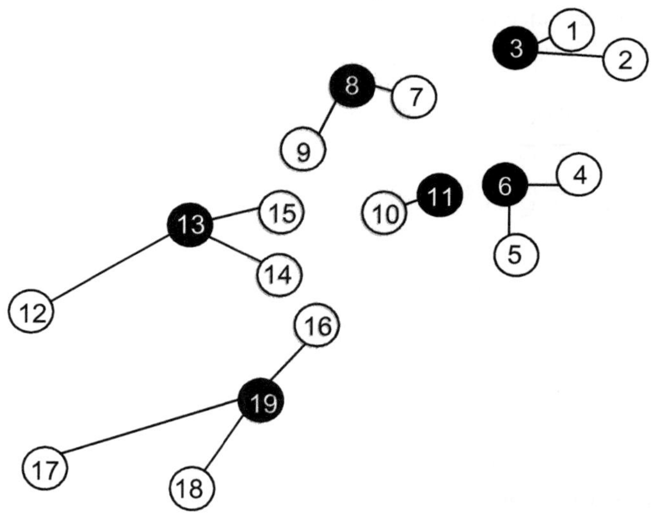

Fig. 12.4 Example of a solution of the multi-attribute p-median model (S = {3, 6, 11, 8, 13, 19})

The robustness of this model is verified by a sensitivity analysis of the scale constants K_1, K_2, K_3 and K.

This model recommendation suggests the best alternative in terms of the trade-off between the multidimensional consequences and the logistics for restoring the system availability by using a backup unit.

12.3 The Sequencing of Maintenance Activities

The sequencing of maintenance activities is an important problem although not necessarily an issue for most maintenance systems. Depending on the size of the system, the strategies and priorities that have been established are sufficient to define the sequence of maintenance activities, and so too, of course, are the technical constraints.

If considering large systems such as an electric power distribution network, a rail network or a water supply network for example, there are several maintenance services such as inspection and repairs that must be performed if personnel are available to do so.

If compared with the sequencing and scheduling of manufacturing orders, there is a different set of criteria besides cost that shall be considered, such as system's availability, quality, service dependability, degradation effects on product quality and other factors that should be considered due to the nature of the system.

This section describes an MCDM/A decision model according to the general procedure for building an MCDM/A given in Chap. 2. This MCDM/A model deals with the planning of maintenance activities by establishing the most appropriate sequence among a large number of maintenance services. This model for sequencing maintenance activities has been applied in an electrical power distributor assisted by a Decision Support System (DSS) (Almeida-Filho et al. 2013).

As the model considers a real situation, the contextual factors related to this situation have been taken into account when formulating the problem and defining the model. Considering the size an electrical power distribution network, the number of repair and inspection services to be performed represents a large sequencing problem.

This model for sequencing maintenance activity was built from data taken from a specific Brazilian electrical power distribution network, which extends over 128,412.5 km in order to supply almost two hundred towns, in an area of about 98,500 square kilometers. It has almost 3.1 million customers who consume 12,266,246 MWh per year.

Throughout this network, there are several components such as voltage transformers, isolators and so forth, which are exposed to severe weather conditions that degrade these pieces of equipment and age components more quickly.

The maintenance database for this power distribution network is updated from data from inspections scheduled on a calendar that covers the entire power distribution network over a period of ten years, with periodical activities that take place at one, two, five and then ten year intervals.

The maintenance strategy adopted is that of immediately restoring the system when a failure that disrupts the service supply occurs or even if this failure is reported as not disrupting the service but it does expose the population to risk.

The maintenance culture adopted in this context is similar to that described by Moubray (1997), who set out three typical states related to an equipment mode: the normal state, a defect and a failure. Depending on the effect of a failure on the functioning of the item, the failure is classified either as:

- A potential failure, which is an observable condition, which implies there will be a functional failure if no preventive action is taken (Moubray 1997);
- A functional failure, which means the inability of an item of equipment to perform a specific function within desirable operational limits (Moubray 1997).

The focus of this problem on sequencing maintenance activities is to do with the potential failures, which are identified by inspections included in the calendar and recorded on the maintenance information system. Thus, these potential failures are prioritized to avoid a disruption to the service and its consequences for strategic and operational objectives. The sequence of maintenance services is based on MCDM/A, which define the order among services to be performed.

In the particular problem addressed, there were about 25 thousand potential failures identified by inspections. Given the capacity of the maintenance division workforce and the annual budget set aside for preventive maintenance, only four thousand potential failures can be tackled per year. The practical meaning is that potential failures are sequenced taking their priority into account and are corrected over the year, which leaves a set of potential failures to be reevaluated by maintenance services in the following year together with other new potential failures identified in inspections undertaken as per the inspection calendar. This preventive maintenance budget is usually defined in the sense that the potential failures identified and pending repairs are still at a tolerable level in the light of the organizational targets. There are several models regarding the definition of preventive maintenance time interval (Jiang and Li 2002; Shafiee and Finkelstein 2015), although there are practical situations when the DM has to consider also the resources available and the production scheduling to define the exact time of a maintenance repair or replacement. Sect. 12.6 illustrates this problem with a decision model.

Some of these organizational targets are defined in order to meet regulatory aspects, such as those defined by ANEEL, which is the Brazilian government agency responsible for regulating the generation of electrical power, its transportation and the distribution companies involved. This agency defines operational and service levels for these companies, and has the power to levy fines in accordance with the regulatory rules. Another important issue is that the electrical power tariff is proportional to the service quality provided. Thus, improving the service level reflects directly on the company′s revenue. There are two main measures of quality of service considered by ANEEL: DEC and FEC. DEC is related to the duration of service disruptions whenever these occur and FEC considers the frequency of disruption to the service (ANEEL 2012).

The reliability indices used by ANEEL are similar to those defined by IEEE (2012), where DEC corresponds to the System Average Interruption Duration Index (SAIDI) and FEC to the System Average Interruption Frequency Index (SAIFI).

Given the MCDM/A nature of this problem, the decision model requires a method that allows the preference among criteria to be elicited in order to find the most adequate sequence of maintenance repairs to be performed. For this specific decision model, the PROMETHEE II method was used. PROMETHEE II is one of the methods of the PROMETHEE family which have been evolving since 1982 (Brans and Mareschal 1984; Brans and Mareschal 2002).

The choice of this method is justified as it can provide a complete ranking order that considers a wide range of value functions and has an easy-to-understand elicitation procedure to assess the DM's preferences. Thus, an important factor for choosing this MCDM/A method is related to the simplicity with which it elicits and requires parameters. This is important as it consolidates the DM's readiness to understand the recommendations provided from the decision model.

Another important issue is the calculation process. Given that there are about 25 thousand alternatives and that this number may grow, an MCDM/A method needs to be able to give a response within an appropriate interval of time so DMs may build scenarios and conjectures and use sensitivity analysis.

With regard to PROMETHEE II, the literature raises questions regarding rank reversal when new alternatives are added to the sets of alternatives, which is a frequent issue when using methods based on a pair-wise comparison process. Mareschal et al. (2008) presented conditions when this situation may occur, which is restricted to very limited situations. This is one of the reasons for choosing this method rather than other outranking methods for the decision model.

The PROMETHEE II method allows the DM to choose between six different value functions, namely, defining each criterion as the usual criterion; a u-shape criterion; a v-shape criterion; a level criterion; a v-shape with an indifference criterion; or a Gaussian criterion (Brans and Mareschal, 1984; Brans and Mareschal, 2002).

PROMETHEE II uses pairwise comparisons throughout its process to aggregate preference indices and outranking flows. Equation (12.2) represents the preference indices, and expresses to what degree a is preferred to b over all the criteria, where $W = \sum_{j=1}^{k} w_j$, and w_j represents the weight of criterion j, $w_j \geq 0$.

$$\pi(a,b) = \frac{1}{W} \cdot \sum_{j=1}^{k} w_j P_j(a,b) \tag{12.2}$$

Equation (12.3) represents the net outranking flow, which consists of the difference between the positive and the negative flow of an alternative a. Based on the net outranking flow, a complete pre-order is provided that ranks all alternatives.

$$\phi(a) = \frac{1}{n-1}\left[\sum_{\substack{b=1 \\ b\neq a}}^{n} \pi(a,b) - \sum_{\substack{b=1 \\ b\neq a}}^{n} \pi(b,a)\right] \tag{12.3}$$

The criteria to be considered in the MCDM/A model were assessed together with a DM when structuring the problem. Thus, each type of potential failure had specific characteristics due to the location of the equipment. This has to be considered since similar failures, which are located in different segments along the distribution network, would cause different consequences and damages.

Thus, the set of criteria identified for this decision model is:

- Degree of Damage (to installation and people, a verbal scale is used);
- Average Affected Consumption;
- Electric Load;
- Percentage of Regional Network Electric Load (considering the network branch);
- Special Clients Affected (subjected to regulatory special rules);
- Healthcare Services;
- Slack on DEC/SAIDI (difference between branch DEC/SAIDI and Aneel target for DEC/SAIDI);
- Slack on FEC/SAIFI (difference between branch FEC/SAIFI and Aneel target for FEC/SAIFI);
- Political Consequences of a Failure.

The main screen of the DSS is presented in Fig. 12.5. MCDM/A concepts and the company's maintenance culture have been combined into a DSS. Therefore, the DSS draws on Moubray's RCM critical levels and uses verbal scales to determine the level of degradation of the equipment.

The DSS allows scenario and application notes to be recorded, so information regarding the decision process for the maintenance activities planning, including personnel involved may be retrieved later, and also to be compared with different scenarios when such evaluation is required. The top section of the screen shown in Fig. 12.5 represents such input information.

The lower section of the screen in Fig. 12.5 shows the interface between the parameters of the MCDM model. On the left, there is the list of criteria, followed by the type of preference function chosen for each criterion and a scroll button for setting the parameters of the preference function. On the right, the input of weights is displayed numerically and graphically.

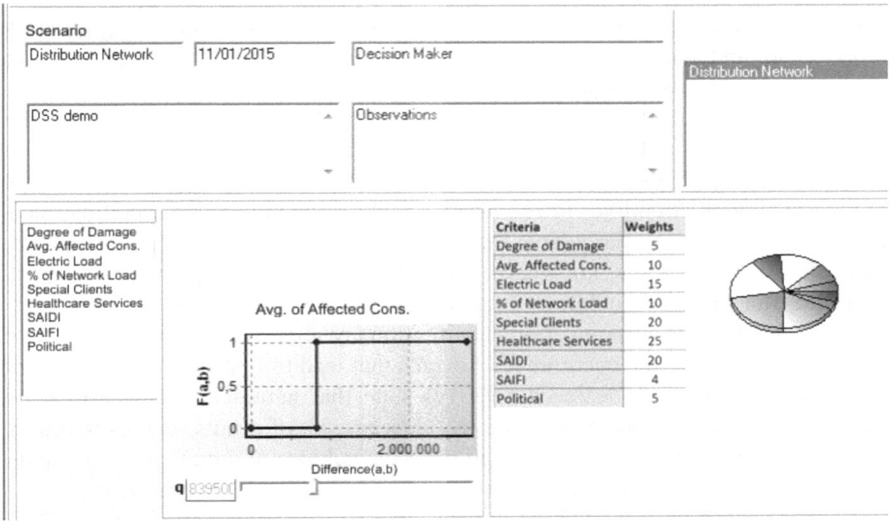

Fig. 12.5 DSS decision model parameters- main screen

After inputting the MCDM/A model parameters into the required fields for the decision model and performing the PROMETHEE II method, the sequencing of the maintenance orders is obtained by considering their priority according to the set of criteria under which all preventive maintenance orders were evaluated.

The DM can also generate reports on preventive maintenance orders which takes account of the budgetary constraints and of an analysis of the sensitivity analysis report. The DSS also enables the DM to perform a scenario analysis supported by graphs so that he/she can compare the effectiveness of each action while considering costs and the managerial objectives (DEC and FEC), and can evaluate the cost levels incurred to carry out preventive maintenance orders as against losses from potential failures (such as, in revenue and because of fines), these being the consequences if the failure became a functional one.

It is interesting to observe that some prioritized maintenance actions may not prove to be financially effective. However, they do prevent losses in other dimensions, such as service quality, which is monitored by the regulatory agency (ANEEL) or any special clients affected, for example. This illustrates the importance of considering the MCDM/A nature of such problems; if this is not done, these factors would not be given appropriate consideration.

12.4 Natural Disasters

It is well-known that there has been an increase in research studies on natural hazards and the relationship of the latter with the climatic changes that have been occurring around the world in recent years. Additionally, human migration to urban areas and the consequent growth in and/or density of the population in urban areas increases the impact of natural disasters significantly.

Urban settlement worldwide is becoming more and more evident. Half of the world's population now resides in urban areas. There is an expectation that these numbers will increase in the coming decades (Linnekamp et al. 2011). A large part of the urban population lives in coastal areas, where the impacts of specific effects of climate change potentially have the most critical consequences. In this context, Li et al. (2014) states that knowledge about future extreme events is important to support actions in order to define more appropriate safety levels for the society.

Thus, the increase of population density in many regions and cities has a direct impact by provoking the occurrence of events that lead to financial losses (billions of dollars). Keller and DeVecchio (2012) state that natural disasters affect the lives of millions of people around the world as a result of events such as flooding, earthquakes and hurricanes, which lead to an annual loss of around 80,000 human lives, besides economic losses of approximately 50 billion dollars per year.

Considering these facts, natural disaster risk management is crucial if the most appropriate mitigating actions are to be appropriately planned and taken. Solecki et al. (2011) point that climate change has a direct impact on such risks. Climate change such as temperature variations and oscillations in precipitation patterns can have a direct impact on the probability of extreme events taking place. Changes in the intensity and distribution of rainfall might well increase the occurrence of flooding or water rationing. High temperatures and melting glaciers may well lead to the sea level being raised, thus increasing the chance of severe flooding in coastal regions. Models in flooding context have also considered risk analysis (Hansson et al. 2013; Vari et al. 2003).

However, some observations need to be made about some obstacles that hinder the better management of risks from natural disasters and of assessing such risks. One of them concerns the availability of a reliable data base, since the dynamics of the social context (e.g. significant changes in the demographic occupation and use of land) associated with climate change often make data collected in previous periods of little or no current value. Keller and DeVecchio (2012) assert that, nowadays, there is a need for effective risk assessment under different scenarios for hazards that need to be associated with the analysis of natural disasters. Due to the occurrence of climate change past events often fail to provide adequate information on what may happen today or in future.

Moreover, population migration itself sees to it that different aspects, in distinct time intervals and in the same locality, should be recorded as well as different aspects related to vulnerability, an aspect that should be taken into account in the risk management process regarding natural disasters. Vulnerability, according to Pelling (2003), can be defined from the degree of exposure to natural hazards, and the ability of the area and community affected to prepare for and recover from given negative impacts.

Pine (2009) states that the changes in disasters frequency may be the result of natural climatic variations that occur over a time period or arise from changes of variables that impact the frequency or severity of environmental change. The intensification of human activity in hazardous areas such as the construction of residences without planning permission on hills subject to landslides or on land known to be occasionally subject to severe flooding are examples. Additionally, changes to the environment (such as those caused by buildings, technology and the infrastructure to support human habitation) that lead to the degradation of natural systems can also increase the severity of the hazard.

Thus, it is of fundamental importance to consider the dynamic nature of risk and vulnerability. Karimi and Hüllermeier (2007) reinforce this idea by stating that, due to there being all manner of uncertainty types, evaluating the risk of losses due to natural disasters is a complex activity, mainly for lack of sufficient physical knowledge and inadequate statistical data with respect to the origin, characteristics and consequences of each disaster that has actually taken place.

Bobrowsky (2013) states that risks related to natural hazards and climate change are not autonomous or externally generated. Therefore, society ought to be able to react, adapt or respond to them. These risks are the result of the interaction between society and the natural or built environment. Consequently, risk management requires a better understanding of this relationship and the factors influencing it.

Unlike the aspects considered in traditional risk management, in natural disaster environments, some additional concepts are important so that DMs can evaluate the situation more adequately. Hence there is a need to consider aspects such as vulnerability and resilience, besides the concept of risk. The concepts of risk, vulnerability and resilience are important in studies on natural disasters, and are used as an approach to understand the dynamics of natural disasters (Paul 2011).

According to Field et al. (2012), vulnerability is the result of different conditions and processes, which must include considering historical, social, political, cultural, institutional and environmental matters and natural resources. Resilience is defined in the context of natural disasters as a means of promoting sustainable livelihoods, which enables individuals or systems to be able to cope with an extreme event without using all the available resources (Paul 2011). Resilient systems tend to reduce physical damage, thereby providing time for the environment to recover after an extreme event has occurred. Therefore, it reflects the interest in improving the capacity of human and physical systems to respond to natural events that occur.

According to Field et al. (2012), the risk of a disaster can be understood as the possibility of adverse effects in the future arising from interaction between social and environmental processes, and a combination of physical hazards and vulnerabilities in the exposed elements. The simultaneous consideration of risk, vulnerabilities and dynamic changes in the different phases of crises and disasters produces a complex scenario out of which the degree of risk and vulnerability that this contains needs to be identified and assessed, as should the measures that need to be taken to mitigate risk and to adapt strategies. An understanding of extreme events and disasters is a prerequisite for drawing up adaptation strategies related to climate change and reducing risk in disaster risk management.

Natural disasters manifest themselves independent of the pre-existing states of economic, social and physical environmental. Therefore, infrastructure, services and organizations are prone to being affected by an event triggered by a natural phenomenon (such as an earthquake or a flood) or a technical event like an explosion or gas leakage (De-Leon 2006; Guikema 2009). Thus, it is observed that a disaster is preceded by at least two aspects: the possibility that an initiated event occurs, usually termed a danger from this potential state, and; a pre-existing vulnerability. In other words, there is a predisposition for people, processes, infrastructure, services, organizations or systems to be affected, damaged or destroyed when an event occurs.

Considering the coupled with vulnerability and danger as a prerequisite of there being a risk of a disaster, the exposure can be considered as another prerequisite. The exposure is understood as the number of people and/or other elements at risk that may be affected by a particular event (Thywissen 2006). Among other definitions, risk must be understood as a function of hazard, vulnerability, exposure and resilience.

Another issue discussed in studies related to natural disasters is consequence analysis. The occurrence of an event or combination of two or more events can cause different impacts in different dimensions.

The potential effects of climate change on natural hazards are an input to the formulation of strategies to adapt risk management practices using knowledge developed about the risks associated with people and with economic impacts (Zischg et al. 2013).

Extreme natural events can induce higher losses especially when they occur in vulnerable and/or areas that are densely populated (Huttenlau and Stotter 2011). Risk analysis in natural hazards is used for estimating the consequences in order to provide information to the public and DM. It is observed in this type of analysis that different perceptions of the concept of risk are seen to have different goals and to take different approaches, as given in Chap. 3, there are different approaches when considering individual or social risks. Therefore, depending on the approach given to the risk evaluation, different perspectives may be considered and depending on how the consequences are evaluated an MCDM/A approach is a more suitable manner to address the problem of evaluating multidimensional consequences. Considering the complexity of consequence evaluation activity, MCDM/A approaches allow to include different types of losses.

There are some aspects that may be considered for measuring the effects of natural disasters, amongst which are economic and social disruption and environmental impacts. Social disruption can, for example, include the number of people made homeless or the incidence of crime such as the number of homicides, arrests, the extent of civil disorder, including riots and street-fighting (Pine 2009).

Economic disruption may be associated with unemployment, lost work days, loss of production volume, decrease in sales and in tax collection. The environmental impacts can be evaluated for cost recovery, re-establishing water or sewer systems, the number of days of unhealthy air, or the number of warnings that involve not eating fish or restrict the use of water (Pine 2009).

These aspects can be adjusted depending on the type of situation analyzed, taking into account, for example, the kind of loss resulting from the natural disaster having occurred. Therefore, losses can be classified into: direct tangible losses and indirect losses. In the first type of loss, the losses considered are those that occur immediately after the event such as deaths, injuries and repair costs. Indirect losses involve loss of income due to unemployment, sales losses, productivity losses, disease and increase in the crime rate (Pine 2009).

Impacts can also be broadly classified by making distinctions between social and physical impacts, where physical impacts include property damage, deaths and injuries. Social impacts can be more difficult to measure once it develops over a long time run. A better understanding of social impacts is important to enable appropriate contingency plans to be drawn up to prevent and/or minimize adverse effects from extreme events. The social impacts of natural disasters are often broken down into demographic, economic, political, institutional, psychological and health impacts (Paul 2011).

A more critical situation is the possibility of a natural disaster occurring in industrial areas which can increase the chance of occurrence of events with extremely catastrophic consequences. According to Krausmann et al. (2011) the threat of natural disasters impacting on chemical industries, refineries, nuclear power plants and pipelines and the consequent leakage of hazardous substances have been recognized as an emerging risk in today's society. Industrial accidents from natural events such as earthquakes and floods are mentioned in many studies on *Natech* accidents. *Natech* accidents can generate leaks of hazardous substances leading to deaths, injury to persons, environmental pollution and economic losses. *Natech* risks differ from technological and natural risks by requiring a risk management approach that is integrated and more complex. One of the main problems with this type of scenario is the simultaneous occurrence of a natural disaster and a technological accident, both requiring simultaneous response efforts. In addition, the leakage of hazardous materials can be inducted by a single source or multiple sources simultaneously, from various hazardous installations in the area impacted by a natural disaster.

According to Krausmann and Cruz (2013), a practical example can be found in the earthquake and tsunami that hit Japan on March 11, 2011, damaging and destroying many industrial plants and killing more than 16,000 people, with more evidence due to the effects of the Fukushima's nuclear power plant facility. This event shows that even well-prepared countries are subject to the occurrence of *Natech* events. In the case of natural disasters that hit a wide impact area, multiple and simultaneous leaks of hazardous materials may occur, these being more severe in areas which are close to residential areas.

Girgin and Krausmann (2013) highlight that *Natech* risks is likely to be more frequent in the future due to industrial growth, changes in the patterns of occurrence (due to climate change) of natural disasters and the fact that society is becoming increasingly vulnerable, the more interconnected it becomes – something that is happening with each passing day.

In conclusion, organizations and countries should regard the search for more effective risk management as a fundamental goal. Adequate control of risk and monitoring should be done in the best way as possible, thereby minimizing the occurrence of catastrophic consequences.

To illustrate how an MCDM/A model may address such aspects, the following section structures an MCDM/A model for multidimensional risk evaluation considering flooding as an example of natural disaster.

12.4.1 An MCDM/A Model that Evaluates the Risk of Flooding

There are several natural hazards related to climate changes and global warming. In this section, an illustrative example is presented of using MCDM/A to evaluate risk considering multidimensional consequences specifically for one of the most frequent natural disasters, namely flooding.

Therefore, an MCDM/A model is described (Priori Jr. et al. 2015) focusing on specific aspects, including the occurrence of different events/scenarios, the choice of criteria, different methods and the distinct rationality required from the DM`s preference structure.

Some steps of the general procedure for building MCDM/A models proposed in Chap. 2 are mentioned throughout the presentation of the model.

This risk evaluation considers urban areas located in coastal regions at sea level, or even below sea level, for example, in the Netherlands.

In underdeveloped countries, there are poor communities that are more exposed than others to flooding due to the lack of infrastructure. This vulnerability puts people at risk from landslides as a consequence of rainfall even without flooding affecting the safety of such communities.

A probabilistic background is necessary for this type of evaluation. Thus, a risk hierarchy can be built, based on Utility Theory, for the most critical areas by assigning priorities to risks with a view to reducing or mitigating them in order to allocate the available resources better and to a level above the local safety standards applied. Therefore, when considering step 2 of the general procedure for building MCDM/A models, the overall objective is to assign priorities to risk so as to guide how resources will be allocated (Lins and de Almeida 2012). According to step 3, human, environmental, financial and infrastructure are considered consequence dimensions. The hierarchical structure of these dimensions and attributes is shown in Fig. 12.6.

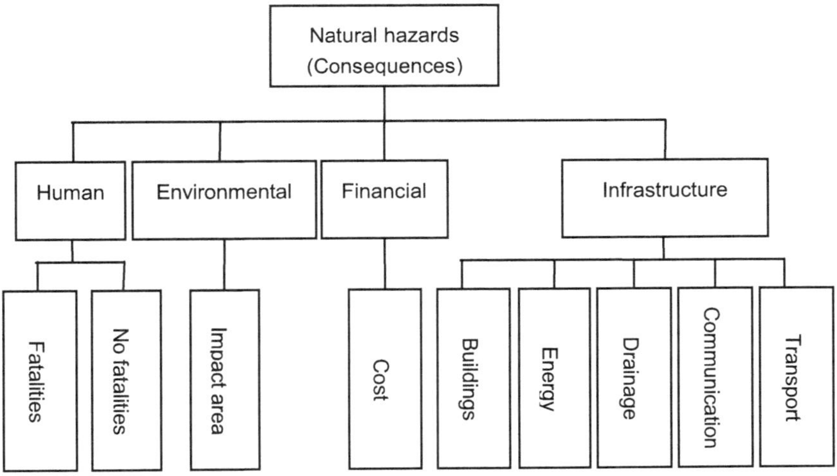

Fig. 12.6 Hierarchical structure of the consequence dimensions and attributes

Specifically, it is important to note that for the infrastructure dimension there are different attributes. The definition of these attributes is based on a World Bank report (Jha et al. 2013). Additional comments are presented for each consequence dimension and attributes considered from after the occurrence of the natural hazard:

- Human consequences (*h*): This dimension considers fatalities and injuries (no fatalities) as possible consequences;
- Financial consequences (*f*): This dimension considers the financial losses that arise from the occurrence of the event such as the production losses;
- Environmental consequences (*e*): This dimension considers the area impacted (including the area covered by vegetation area, fauna and flora);
- Infrastructure consequences (*s*): This dimension considers different attributes. In the building attribute, the current state of the edification is considered, taking into account the possibility of its structural collapse. In the Energy attribute, the physical structure of the power grid is analyzed. With regard to the drainage attribute, aspects of the drainage facility are taken into account. As to the communication attribute, the communication facility is studied. Finally, with respect to the transport attribute, the operation of existing transports systems is evaluated.

Based on step 4, there is a discrete set of elements $A = \{a_1, a_2, a_3, \ldots, a_n\}$, defined as limited urban areas, in which the extent of the urban area is established including some factors, such as infrastructure, number of inhabitants, topography and climate.

Step 5 deals with identifying the state of nature. The DM has to consider a set of controllable and uncontrollable inputs that impacts the problem analyzed. Additionally, the DM makes decisions that take account of states of nature and probabilities for several consequences. Therefore, utility functions are defined so as to represent the DM's preferences for different consequences (Cox 2009).

Elements of decision theory are used to evaluate the consequences. The consequences are represented by c and the set of alternatives by A. The state of nature θ, represents the uncertainty related to the problem, measured by the magnitude of the rainfall. The state of nature is represented by a continuous set denoted by real numbers regarding the rate of rainfall in a given region per hour (*mm/h*), in a determined rainfall event. Lognormal or Gamma probability density functions could be applied to represent the *mm/h* for a determined rainfall event in a specific location (Cho et al. 2004).

As to considering consequences of a rainfall, a probabilistic approach can be introduced to incorporate the associated uncertainties in A, considering a probability distribution over consequences given the state of nature. By eliciting utility functions for these consequences, DM's preferences are represented in the model. The prior probability $\pi(\theta)$ is introduced as the probability of each state of nature. Therefore, the expected utility $E[U(\theta,a_i)]$ is used to represent the risk associated with each given alternative (Berger 1985).

The utility is calculated by combining the probability of the consequences c in A, the consequence function $P(c|\theta,a_i)$. The effects of the tide on run-off rainwater can be included in this probabilistic mechanism. Therefore, the expected utility $E[U(\theta, a_i)]$ of these consequences is represented by (12.4).

$$E[U(\theta,a_i)] = \int_c P(c|\theta,a_i)U(c)dc \tag{12.4}$$

According to Berger (1985), the loss function is defined as the negative of the utility function $L(\theta, a_i) = - E[U(\theta, a_i)]$. Thus, the losses are computed for each criterion ($L_{(h)}$; $L_{(f)}$; $L_{(e)}$; $L_{(s)}$) considering the urban area analyzed and the state of nature. Moreover, the risk to an urban area is defined by (12.5).

$$r(a_i) = \int_\theta \pi_i(\theta)L(\theta,a_i)d\theta \tag{12.5}$$

Step 6 from Chap. 2 on preference modeling is required in order to evaluate the DM's preferences. This model considers that the DM's preferences satisfy MAUT axiomatic requirements for an additive utility function.

Thus, the intra-criterion and inter-criteria evaluations consider steps 7 and 8 respectively. The intra-criterion evaluation is based on the conditional utility function, defined for each dimension. The inter-criteria evaluation relies on the additive utility function, achieved by elicitation procedures through lotteries (Keeney and Raiffa 1976). Therefore, (12.6) represents the additive utility function.

$$U(a_i) = k_h U(h) + k_e U(e) + k_f U(f) + k_s U(s) \tag{12.6}$$

where k_h, k_e, k_f, k_s are scale constants for the human, environmental, financial and infrastructure dimensions respectively.

From the hierarchical structure of the attributes, the infrastructure dimension s and the human dimension h present specific attributes. These attributes are considered in the MAU function as given by (12.7).

$$\begin{aligned} U(a_i) = k_h\left[k_{h1} U(h_1) + k_{h2} U(h_2)\right] + k_e U(e) \\ + k_f U(f) + k_s\left[k_{s1} U(h_{s1}) + \ldots + k_{s5} U(h_{s5})\right] \end{aligned} \tag{12.7}$$

where:

k_h, k_e, k_f, k_s are a scale constants that represent the value of the tradeoff (dimensions);

k_{h1}, k_{h2}, k_{s1},…, k_{s5} are a scale constants that represent the value of the tradeoff (specific attributes).

Step 9 consists of applying an algorithm in the decision model in order to evaluate the set of alternatives. In this model, the interval scale of the utility function is applied to provide additional information based on comparing the utility values and ratios of the increments of utility between alternatives.

Finally, step 10 of the procedure for resolving problems and building MCDM/A models presented in Chap. 2 consolidates step 9 by conducting a sensitivity analysis to verify the robustness of the model, incorporating the data and parameters analyzed.

12.5 Operation Planning of a Power System Network

The demand for electric power has increased and all current forecasts indicate even higher growth due to improvements in the quality of life, besides population growth. As mentioned in Sect. 12.4, when discussing the effects and trends of changes in the climate, consequences in generating and consuming electric power can also be observed. The effects of climate changes can limit power generation capacity, when considering renewable sources, and in terms of environmental constraints for generation when considering other sources such as coal and oil. On the other hand, the demand for electric power rises due to severe weather such as colder winters or higher temperatures in summer. Therefore, the impact of power outages becomes more and more critical due to the great dependence of modern society on power always functioning.

Thus, since there are many social and economic issues that are rising in importance, there are many MCDM/A problems regarding decisions on the reliability of power systems.

From this perspective, due to the threat of irreversible damages to the environment, there is a need to consider various forms of energy production in order to satisfy the aspects related to demand, environment and cost (Jebaraj and Iniyan 2006).

Besides environmental and social changes, electric power systems have evolved and have been restructured according to the country's needs, and these depend on each country's specific energy policies. Therefore, there is a need to emerge from the unique view of provision of power at "minimum cost" to a broader perspective that allows consideration of multiple aspects, and this may consider the different interests of the actors involved in dealing with planning energy systems (Diakoulaki et al. 2005).

For power systems predominantly based on hydropower generation, such as Canada and Brazil, additional complexity is introduced when depending on this power source. Such complexity concerns planning the power system due to the dynamics of river flows and precipitation patterns in such energy producing systems.

As to step 5 of the general procedure for building MCDM/A models given in Chap. 2, the state of nature θ reflects the potential energy generation stored in water reservoirs over time. Depending on the context, different levels of uncertainty are found. For example, while in Canada river flows are more predictable due to the relation with the accumulated volume of snow layers; in Brazil the prediction of potential energy generation stored in water reservoirs is more difficult since there is no such relation (Albuquerque et al. 2009).

Considering the importance of potential energy generation forecasts, the operational planning of these power systems is much more complex than in power systems higher percentages of coal, oil and nuclear generation sources. The operational planning of generating power in such systems, which are not predominantly dependent on hydropower, is not highly associated with matching temporal to spatial data (Diniz and Maceira 2008).

Therefore, many MCDM/A problems arise in order to assure the supply reliability. Another aspect that must be considered is how the electric power system is designed, as this may result in different kinds of constraints and consequences over the system.

The objective of the system operator is to assure that the demand will be met with minimum cost and the maximum reliability of system supply. Thus the planning of such systems must be formulated with an MCDM/A paradigm that takes into account besides cost and the reliability of system supply, other aspects related both to the quality of service (such as voltage, power-frequency and harmonics) and to environmental impacts.

Thus, a reliability decision problem in power systems includes several objectives and a variety of constraints which reflect the physical system (Pinto et al. 2013). In the classical optimization approach, it is usual to remove important objectives and consider doing so as one of the problem constraints. The MCDM/A approach enriches the decision process by allowing a compromise solution, beyond constraint levels defined when modeling an objective as a constraint, to be evaluated. Therefore a DM can make tradeoffs and find the most suitable recommendation if he/she uses an appropriate methodology for modeling.

What is more and more observed is the need to consider other criteria regarding environmental issues due to the atmosphere being increasingly polluted which has already been experienced in China. As a result, the emission of greenhouse gases is another aspect that has been covered when considering MCDM/A approaches for power system planning (Diakoulaki et al. 2005; Batista et al. 2011).

Therefore, in order to meet the exhortations of environmental regulations, environmental aspects become a new criterion for planning and operating of power systems (Farag et al. 1995; Yokoyama et al., 1988; Wong et al., 1995).

12.6 Integrated Production and Maintenance Scheduling

Production scheduling models can support decision making on allocating jobs in manufacturing systems in order to optimize a given objective function. Generally, objective functions are related to the productivity of the system, such as: maximum tardiness, total tardiness, total weighted tardiness, total weighted completion time, maximum lateness, number of tardy jobs and makespan (Pinedo 2012). However, machine breakdowns can result in losses in terms of the productivity performance measured by these objective functions. In other words, the solution found for a specific problem, which assumes that failures are not possible, can be unrealistic when breakdowns occur.

In order to deal with breakdowns, a maintenance policy can recommend preventive actions with the objective of reducing the probability of machine failures. This means that performing preventive maintenance necessarily incurs a cost and time must be set aside for this. Besides having to take stochastic features of failures and repair of machines into account, maintenance performance can be in conflict with production performance. In this context, the main issue of this problem is how to balance maintenance and production objectives.

According to Aghezzaf and Najid (2008) most of the time, a contingency review of the production plan, due to a failure, is very expensive and also impacts the quality of products. Therefore, preventive maintenance has an essential role to play, not only to ensure the production plan is fulfilled by reducing the number of failures, but also to ensure quality and service within appropriate levels.

Independently of the kind of system, an appropriate production scheduling enables production systems to achieve strategic objectives, which might range from achieving minimum cost and tardiness. Therefore, it seems that, in most cases, dealing separately with the production schedule and maintenance plan does not work in practice. Indeed, it is not possible to ensure long lasting results for the production perspective, since production scheduling does not last for long due to failures. Nevertheless, a very common hypothesis considers that equipment is always available during the scheduling period, even though the probability of failure in intensively used production systems has a significant value (Allaoui et al. 2008).

Thus, it is interesting to include the activity of preventive maintenance in a production schedule in an integrated form (Ángel-Bello et al. 2011). According to Allaoui et al. (2008), in the literature, there are two particularly prominent approaches with regard to the problem of integrating production and preventive maintenance. For the first kind, the optimum maintenance schedule in the production system can be determined. The second approach comprises optimizing the scheduling of production by considering a preventive maintenance plan. By doing so, the maintenance schedule decision could be drawn up in advance of the production schedule. The problem with this approach is that the dynamic nature of the problem is overlooked.

Despite there being some interesting papers dealing simultaneously with a maintenance and a production schedule, most of them consider only one decision criterion (Alardhi et al. 2007; Benmansour et al. 2011; Ji et al. 2007; Sortrakul and Cassady 2007; Su and Tsai 2010). In fact, since these integrated models derive from the original problem of the production schedule, some of these papers still consider only the original objectives such as total weighted expected tardiness. Therefore, maintenance features that influence joint scheduling are dealt with as secondary aspects, mostly as elements of constraint.

It is worth stating that maintenance aspects are completely different from the common criteria used to define the production schedule. Instead of simply having some rules based on the strategy of client satisfaction, such as, expected tardiness and makespan; maintenance aspects are related to the performance of the equipment, such as availability, probability of finishing by the end of the schedule, the total cost, considering preventive maintenance and interruption; and so on. Thus, it is not difficult to realize that the operational and maintenance aspects are complementary.

An integrated decision model that takes into account two objectives to be optimized simultaneously: minimizing total weighted expected tardiness and minimizing expected maintenance costs is developed. The conflicts of the maintenance function and production are dealt with under the approach of MCDM/A. Some results give evidence that on applying NSGA-II (Deb et al. 2002), satisfactory solutions can be found for the integrated scheduling problem.

It is assumed that there are a number of jobs to be scheduled in a single machine in a production system. Each job has a fixed processing time, due date and importance weight. In addition to production scheduling, it is assumed that this machine may be unavailable due to preventive maintenance or repairs that are needed due to failures. These features imply a conflict between the production and maintenance objectives. Whereas the production objective may be related to minimizing tardiness in finishing jobs, the maintenance objective may be related to minimizing time losses incurred by unnecessary maintenance actions and is mainly characterized by the expected cost of maintenance. To estimate the latter, it is assumed that the time to failure of this machine is governed by a Weibull probability distribution (Sortrakul and Cassady 2007). Replacements should be recommended when the expected cost of replacement is lower than the cost of preventive maintenance and additional costs including production losses.

It is assumed that jobs cannot be preempted by preventive maintenance activity, and only one failure can occur during the processing of a job. The basic decision variables to be determined are: what the sequence of the jobs and when preventive maintenance actions should be performed with the objective of minimizing the total weighted expected tardiness and expected maintenance cost (Cassady and Kutanoglu 2003; Sortrakul and Cassady 2007).

The mathematical model is defined by (12.8), in order to minimize two objective functions. Let F_1 be the total weighted expected tardiness and F_2, the total expected cost of maintenance.

$$\begin{aligned} &\text{minimize } F_1(x_{ij}, y_i) = \sum_{i=1}^{n} w_{[i]} \left(\sum_{k=0}^{i} \theta_{[i,k]} \pi_{[i,k]} \right) \\ &\text{minimize } F_2(x_{ij}, y_i) = \sum_{i=1}^{n} \sum_{k=0}^{i} cm_{[i,k]} \pi_{[i,k]} \end{aligned} \tag{12.8}$$

The maintenance cost is given by (12.9), job completion time is given by (12.10) and tardiness is given by (12.11).

$$cm_{[i,k]} = c_b \sum_{l=1}^{i} y_{[l]} + c_a \cdot k \quad k = 0,1,\ldots,i;\ i = 1,\ldots,n \tag{12.9}$$

$$c_{[i,k]} = t_p \sum_{l=1}^{i} y_{[l]} + \sum_{l=1}^{i} p_{[l]} + kt_r \quad k = 0,1,\ldots,i;\ i = 1,\ldots,n \tag{12.10}$$

$$\theta_{[i,k]} = \max(0, c_{[i,k]} - d_{[i]})\ k = 0,1,\ldots,i;\ \ i = 1,\ldots,n \tag{12.11}$$

where:
n - Total number of jobs to be scheduled;
$p_{[i]}$ - Processing time for the *i-th* job performed;
$d_{[i]}$ - Due date for the *i-th* job performed;
$w_{[i]}$ - Weight for the *i-th* job performed;

$c_{[i]}$ - Completion time for the *i-th* job performed;
$\theta_{[i]}$ - Tardiness for the *i-th* job performed;
β - Shape parameter of the Weibull distribution;
η - Scale parameter of the Weibull distribution;
$a_{[i]}$ - The age of machine immediately after finishing the *i-th* job;
$\bar{a}_{[i-1]}$ - The age of machine immediately before processing the *i-th* job;
$y_{[i]}$ - Binary variable decision when preventive maintenance is performed prior to the *i-th* job;
$\pi_{[i,k]}$ - probability mass function of *k* failures during the *i-th* job;
t_p - Duration of preventive maintenance action;
t_r - Duration of corrective repairs to a machine;
c_b - Cost of preventive maintenance action;
c_a - Cost of corrective repairs to a machine.

The result of a simulation of NSGA-II is presented in Fig. 12.7. As can be seen, nine solutions obtained with the algorithm were found in PF_{true}.

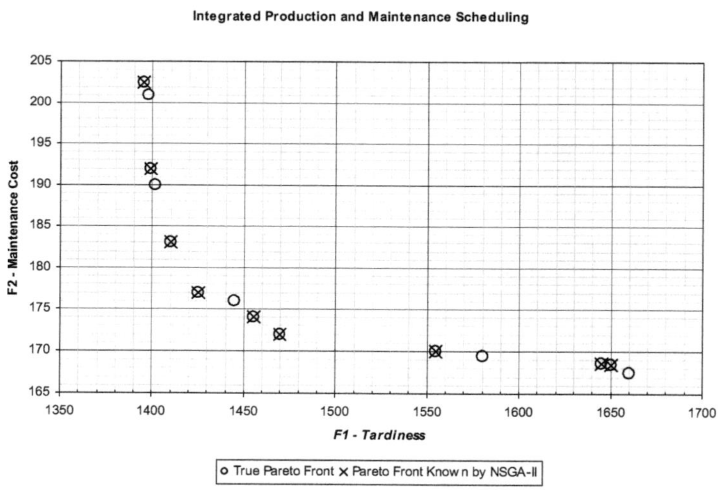

Fig. 12.7 Result of NSGA-II in the integrated production and maintenance scheduling

The integrated model proposed may well be of interest to industry in order to tackle production and maintenance needs which take into account two conflicting objectives related to tardiness and maintenance cost with regard to scheduling production jobs.

12.7 Maintenance Team Sizing

Maintenance team sizing is a topic that involves various types of methodologies. Simulation and queuing theory are examples of approaches that can be used to determine the best number of maintenance personnel. Such approaches aim to minimize the waiting time and service costs due to losses associated with an inappropriate team size, high investments and strategic reasons. What should be taken into account are both: the cost of hiring personnel and the estimated cost of the consequences of the unavailability of the system, which can be represented by the cost of production losses. Hillier (1963) proposes economic models that minimize the total cost, which comprise of the expected costs and service costs.

Queuing theory allows the DM to analyze the problem using a structure that can incorporate the probabilistic mechanism present in the reliability and maintainability of systems. A maintenance system can consist of several queues. Customers are devices that need repairs and the servers are personnel who perform repair services, which may eventually form a virtual queue waiting for service.

Some queuing system indicators show stochastic features that can support decisions about maintenance team sizing. Examples are the utilization factor of the personnel; the probability of finding n customers in the system, the probability that all the servers are busy, the average number of items in the queue and the average time spent in the queue waiting for the equipment and system.

Two maintenance models are investigated in a flexible manufacturing system using queuing theory to study features of the system utilization. The failure of a machine requires the activation of a stand-by, while the failed unit goes to repair. A stand-by is required to perform a certain level of service. For these types of systems to minimize the cost of loss of production, which includes the cost of customer dissatisfaction, these costs can be minimized by maximizing system availability (Lin et al. 1994). A bi-objective formulation to solve a maintenance workforce sizing problem is found by using a branch and bound algorithm (Ighravwe and Oke 2014).

The sizing of the maintenance team can be defined for corrective or preventive maintenance. In some situations, estimating failure and repair rates from historical data may be difficult and an expert's knowledge could be useful. The use of prior knowledge deals with the uncertainty in a more appropriate way. Therefore, a decision model for maintenance team sizing with use of prior knowledge is developed.

The model considers a system with p maintenance teams, denoted by MT_i, where $i = 1, \ldots, p$. Each maintenance team MT_i is responsible for the repair of q_i different items of the equipment j. Each piece of equipment, denoted by Eq_{ij}, has

n_{ij} items. Each MT_i has team size s_i. Each maintenance team has similar characteristics with respect to repair time. A generic representation of the maintenance system is shown in Fig. 12.8.

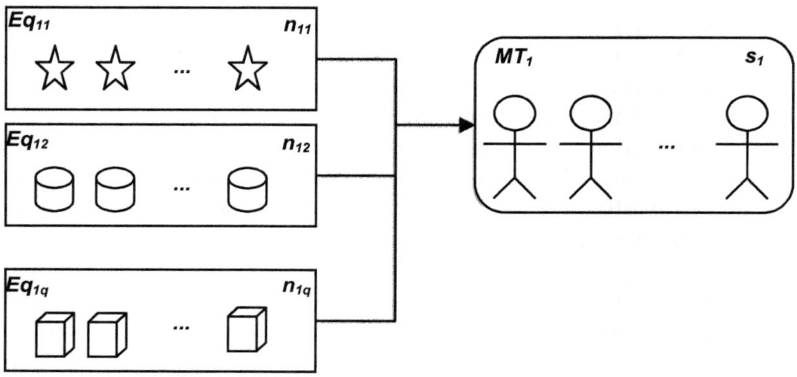

Fig. 12.8 Example of a maintenance system

In general, a combination of reliability, maintainability and cost directly influences the system performance measures. In some situations, it may be interesting to acquire more reliable items than to increase the size of maintenance teams. Reliability and maintainability features of the system are represented by the failure and repair rate, respectively.

Assuming an exponential probability distribution for the equipment reliability function, λ represents the equipment failure rate, with different values for each type of equipment and this corresponds to the arrivals of the customers in queuing system of the maintenance team.

The repair times are modeled by an exponential distribution, where the constant μ is the repair rate, which may have different values for each maintenance team.

The objective of this problem is to determine the required number of maintenance personnel to achieve satisfactory levels in the performance indicators and associated costs.

The system can be represented by p models of one-queue, s servers and an infinite population. Based on the Kendall notation, the decision model developed for the maintenance team sizing problem is represented by type *M*/*M*/*s*. A structure of the decision model proposed is shown in Fig. 12.9.

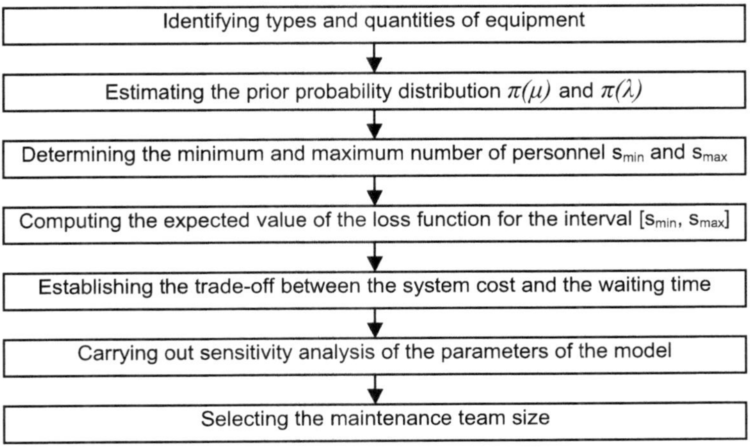

Fig. 12.9 Structure of decision model with use of prior knowledge, $\theta = [\lambda, \mu]$

The objective of the decision model is to define the team size while dealing with the tradeoff between service level and cost. Thus, the problem is solved by maximizing the MAU function given by (12.12).

$$U(s_i) = k_1 U_1\left(\int_{-\infty}^{\infty} \int_{-\infty}^{\infty} L(\lambda, \mu, s) \cdot \pi(\lambda) d\lambda \cdot \pi(\mu) d\mu \right) + k_2 U_2 \left(c_p s \right) \tag{12.12}$$

The expected queue length, $L(\lambda, \mu, s)$, is given by (12.13) and the maximum number of servers, s_{max}, is given by (12.14).

$$L(\lambda, \mu, s) = \frac{\left(\frac{\lambda}{\mu}\right)^s \cdot (\lambda \cdot \mu)}{(s-1)! \cdot (s \cdot \mu - \lambda)^2} \cdot \frac{1}{\sum_{n=0}^{s-1} \frac{1}{n!} \cdot \left(\frac{\lambda}{\mu}\right)^n + \frac{1}{s!} \cdot \left(\frac{\lambda}{\mu}\right)^s \cdot \left(\frac{s \cdot \mu}{s \cdot \mu - \lambda}\right)} + \left(\frac{\lambda}{\mu}\right) \tag{12.13}$$

$$s_{\max} = \frac{1}{2} + \frac{1}{2} \cdot \sqrt{1 + \frac{4 \cdot \lambda}{\mu \cdot \Delta\rho_{critical}}} \tag{12.14}$$

where:
λ - failure rate;
μ - repair rate;
s - number of servers;
c_p - cost of personnel
$\pi(\lambda)$ - prior probability on failure rate;
$\pi(\mu)$ - prior probability on repair rate;
$\Delta\rho_{critical}$ - critical value of difference in terms of factor utilization.

Considering an illustrative example with $\pi(\lambda)$ and $\pi(\mu)$ given by Weibull probability distributions with β=11.3 and η=0.05154 for $\pi(\lambda)$; and β=10.7 and η=0.01153 for $\pi(\mu)$. In addition, an additive function with scale constants $k_1 = 0.10$ and $k_2 = 0.90$ for the MAU function given by (12.12).

The minimum number of servers s_{min} to satisfy the stationary condition of 99% is found by considering the λ_{max} and μ_{min} from the inverse Weibull function. Therefore, $s_{min} = 8$.

Based on (12.14), the maximum number of servers is defined as $s_{max} = 23$. Thus, for this scenario, when maximizing the MAU function, the best maintenance team size would be $s^*=10$. This result is illustrated in Fig. 12.10.

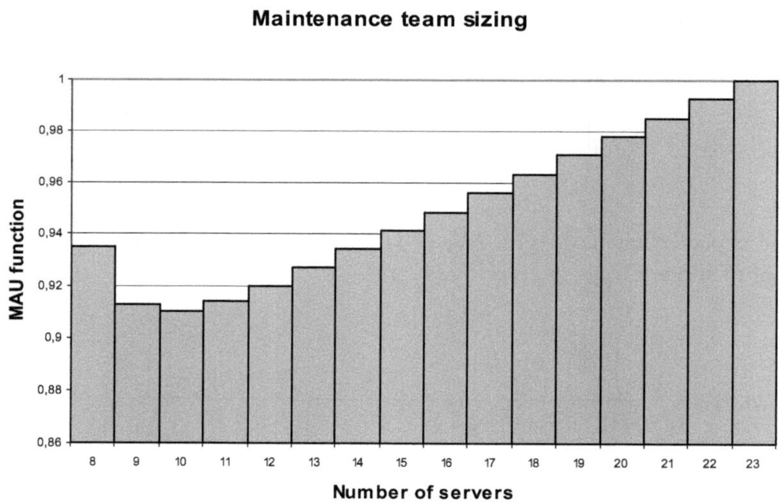

Fig. 12.10 $U(s_i)$ for s_i as $[s_{min}, s_{max}]$

To evaluate the robustness of the solution recommended, a sensitivity analysis is presented in Table 12.1. From the achieved results, it is possible to realize that the scale parameters of prior distributions are the most sensitive parameters in the model.

Table 12.1 Sensitivity analysis of parameters of the model

β of π(λ)	η of π(λ)	β of π(μ)	η of π(μ)	k_1	k_2
+20%	+5%	+20%	+20%	+20%	+20%
-20%	-20%	-20%	-5%	-20%	-20%

The result observed from the sensitivity analysis indicates that the analyst should give more attention to the elicitation process that obtained the scale parameters from the experts. As these are the most sensitive parameters, this elicitation process has to be more accurate than other parameters not so sensitive.

12.8 Bayesian Reliability Acceptance Test Based on MCDM/A

The operations and maintenance planning of systems requires the use of information on the reliability of equipment, which is usually provided by the manufacturers. A concern about the state of nature (θ_λ) that reflects the real reliability of this equipment has triggered the need to ensure, by contract, that reliability acceptance testing takes place (de Almeida and Souza 2001; de Almeida and Souza 1986).

The number of equipment failures during the phase of operation trials that can occur in the set of equipment ordered is limited, so that the actual failure rate θ_λ is in accordance with the specified λ_0.

The problem considered in this section regards the decision about the acceptance of θ_λ for given equipment during the phase of operation trials, in order to decide whether or not to return the equipment. If it is, this may delay the start of the industrial plant project. If the decision is to return the equipment to the manufacturer, this implies delaying the completion time of the project. Therefore, the DM has to consider the tradeoff between delivering the industrial plant project on time which may adversely affect the reliability of the project by accepting equipment not in accordance with the specified λ_0; and delaying the project conclusion so as to assure the reliability requirements of the project. Therefore, this is an MCDM/A problem with two clear objectives (de Almeida and Souza 2001; de Almeida and Souza 1986).

Thus, the decision is more than just testing the hypothesis on θ_λ with regard to λ_0 , thus requiring the DM's preferences to be evaluated for each specific situation which may lead to different decisions depending on the specific priorities or aspects involved. Thus a DM may decide to delay the conclusion of a project if safety requirements are compromised due to the θ_λ , for example, or else he/she may decide to accept equipment with lower reliability in order to conclude the project on time.

As an illustrative example, consider a study consisting of certain sampling restrictions considering N=36 new items and an observation time, Δt, of 3 months in the phase of operation trials. Thus, for this observation time, x failures are observed in the 36 items, representing θ_λ. This considers that the unit failure rate λ_u specified by the equipment's manufacturer as $\lambda_u = 5.88x10^{-6}$ per hour; $\lambda_0 = \lambda_u N \Delta t = 0.457$ for the set of 36 items, in a time interval of 3 months.

The function $P(x|\ \theta_\lambda)$ corresponds to the probability of x failures occurring, given a true failure rate θ_λ. Thus, the number of failures x is explained by a Poisson process.

Thus, the DM seeks to find out the number of failures that can occur in the population of 36 items during the phase of operation trials, so that the actual failure rate θ_λ is compatible with the specified λ_0.

There are three approaches for addressing this problem: Hypothesis testing under the Neymann-Pearson approach; a decision problem formulation with a Bayesian criterion; and a definition of minimax estimators with a Bayesian estimator for the failure rate θ_λ (de Almeida and Souza 2001; de Almeida and Souza 1986).

A model using the specified MTBF considers that the x failures observed are random, and not caused by improper operation or by external effects or by faults in the manufacturing process. Furthermore, it is considered that in the beginning of the phase of operation trials, premature failure have been removed after burn-in testing and debugging. Thus, it is possible to conclude that the failure rate is in the operational phase of the bathtub curve that is constant over time.

Considering a hypothesis test, the state of nature becomes the equipment's failure rate, which can be represented by a discrete set such as $\theta = \{\theta_0 \text{ and } \theta_1\}$, in which θ_0 means that $\theta_\lambda \leq \lambda_0$ and θ_1 means that $\theta_\lambda > \lambda_0$. Moreover, it is assumed that there is no initial knowledge about the state of nature θ_λ, and the Neyman-Pearson approach can be applied.

By using this Neyman-Pearson approach, the problem is reduced to the choice of the best decision rule that minimizes the risk R_b for a given θ, subject to the constraint that the R_b risk to the other θ is less than or equal to a predetermined level α as given by (12.15).

$$\begin{aligned} &\min\ R_b(\theta_1) \\ &s.t.\ \ R_b(\theta_0) \leq \alpha \end{aligned} \tag{12.15}$$

This formulation corresponds to testing the null hypothesis H_0: that the equipment has a lower failure rate than or equal to a specified against the alternative hypothesis H_1. This means that H_0: $\theta_\lambda \leq \lambda_0$ and H_1: $\theta_\lambda > \lambda_0$.

There are two errors involved in the hypothesis test that should be minimized. The probability α of rejecting the null hypothesis when it is true, $R_b(\theta_0)$, and the probability β of incorrectly accepting the null hypothesis, $R_b\ (\theta_1)$. These errors are known as error Type I and error Type II, respectively.

While the DM prefers to increase the probability α to reduce the β probability, the equipment manufacturer seeks to reduce probability α in order to increase the β probability.

Usually, when considering statistical hypothesis tests a value of 0.05 is adopted by convention. However, for this specific case, if a DM chooses an α level by convention, that DM's preferences have not been considered and therefore, neither has the context of the problem.

Using the Bayesian approach the DM considers an α level that meets his/her expectations. To solve this problem a decision rule has to be defined (de Almeida and Souza 2001; de Almeida and Souza 1986) in order to minimize risk r_d as given by (12.16).

$$\min_{i}\left\{\sum_{x=0}^{i}\frac{1}{x!}\int_{a}^{b}\pi(\theta)[L(\theta,a_0)-L(\theta,a_1)]\cdot e^{-\theta}\cdot e^{x}d\theta\right\} \tag{12.16}$$

where:
θ - state of nature;
$\pi(\theta)$ - prior probability distribution;
$L(\theta,a_i)$ - Loss function.

The interval *[a, b]* corresponds to the given range of θ_λ in a prior distribution. The solution that minimizes (12.16) represents the maximum number of failures that there are for not rejecting the null hypothesis. If during the phase of operation trials, there occurs a greater number of failures than the solution of (12.16), the null hypothesis must be rejected.

This procedure using a Bayesian approach can support a DM to find the maximum number of failures that would be acceptable, considering the objectives and knowledge available.

Thus, when considering reliability acceptance for equipment there are conflicting objectives as discussed previously regarding the system's reliability and the project being delivered on time.

Therefore, in some situations a DM can be concerned with evaluating of the tradeoff between the error Type I dimension and the delay in delivering the project due to rejecting the equipment.

For example, if a DM is interested in the evaluating whether the reliability defined in the contract is consistent with the actual reliability of the purchased items, he/she may consider a tolerance level. He/she does so in order to include an upper limit greater than the equipment's nominal reliability value, which is usually the one set out on the contract, for the null hypothesis. Thus, this tolerance level represents how much the DM is willing to tradeoff in terms of reliability in order to succeed in delivering the project on time. Therefore, this means that the DM may decide to accept the null hypothesis since the reliability is lower than the one defined by contract, but respects the tolerance level accepted by the DM in order to deliver the project on time.

This decision may involve a substantial delay in project execution time from several months to years due to the specificities of the items that may be rejected. For these kinds of items, the lead time involved with the purchasing process and delivery is sufficiently long to compromise delivering the project on time. This is because the manufacturer usually adopts a make-to-order strategy for this kind of equipment which means that in most cases, it is not available in the short term i.e. immediately after ordering the equipment.

Thus, for dealing with a decision that has such relevant and strategic objectives, the MCDM/A approach provides techniques and methods for modeling the DM's preferences in order to give a recommendation on the reliability acceptance test considering the broader aspects involved in the problem. Therefore, the general procedure defined for building MCDM/A models in Chap. 2 gives the directions to build a suitable decision model for a problem with these features.

12.9 Some Multiobjective Optimization Models on Reliability and Maintenance

This section is divided into two topics based on the generation of MOEAs (Multiobjective Evolutionary Algorithms).

12.9.1 Approaches in the 1980s and 1990s

The first papers using multiobjective formulation to find Pareto solutions were published in the 1980s. Dewispelare (1984) formulated a non-linear multiple objective problem with regard to a pre-production decision on an airborne tactical missile where the reliability, survivability, combat effectiveness, cost and flight area were considered as objective functions. Feasible space was explored for all non-dominated solutions obtained by a constrained optimization technique. Although non-dominated solutions should be found, a scalar scoring function is recommended when the DM is not able to make a choice due to the incomplete ordering of the Pareto solutions set.

Soltani and Corotis (1988) constructed a trade-off curve of a design for structural systems as a result of using multiobjective linear programming obtained and a constrained optimization technique to formulate objective functions of cost of failure versus initial cost.

Fu and Frangopol (1990) found Pareto optimal solutions in a multiobjective formulation of structural systems considering three objectives: weight, system reliability and redundancy. They used the ε-constraint method to find Pareto solutions.

Misra and Sharma (1991), Dhingra (1992) and Rao and Dhingra (1992) used MOEAs for redundancy allocation, as discussed in Chap. 9.

12.9.2 Approaches in the 2000s and 2010s

With the development of the Second Generation of MOEAs close to 2000, several studies have developed with a view to evaluating the effectiveness of these techniques and more often from this point on, in the field of maintenance and reliability, such as: NSGA-II, SPEA2 and other approaches.

The use of the second generation of MOEAs in Reliability and Maintenance problems has become one of the most common approaches across the Pareto-front approaches. Some cases from the literature are highlighted in Table 12.2.

Table 12.2 Some Pareto-front approaches used in Reliability and Maintenance problems

Optimization Method	References
Constrained optimization technique; ε-constraint method; goal programming; goal-attainment	Dewispelare (1984), Soltani and Corotis (1988), Fu and Frangopol (1990), Dhingra (1992), Rao and Dhingra (1992), Barakat et al. (2004), Azaron et al. (2009), Moghaddam (2013)
Min-max concept; Exact algorithm; PSO; GPSIA	Misra and Sharma (1991), Certa et al. (2011), Chou and Le (2011)
MOEA; MOGA; NSGA-II	Ramirez-Rosado and Bernal-Agustin (2001), Marseguerra et al. (2002), Marseguerra et al. (2004), Kumar et al. (2006), Kumar et al. (2008), Cadini et al. (2010), Moradi et al. (2011), Wang and Hoang (2011), Chiang (2012), Torres-Echeverria et al. (2012), Zio et al. (2012), Gjorgiev et al. (2013), Jin et al. (2013), Li et al. (2013), Lins et al. (2013), Rathod et al. (2013), Trivedi et al. (2013), Zidan et al. (2013)

These approaches have been applied to several reliability and maintenance problems, such as:

- Design selection (Ramirez-Rosado and Bernal-Agustin 2001; Marseguerra et al. 2004; Barakat et al. 2004; Azaron et al. 2009; Chiang 2012; Torres-Echeverria et al. 2012; Rathod et al. 2013);
- Maintenance strategy selection (Marseguerra et al. 2002);
- Service restoration (Kumar et al. 2006; Kumar et al. 2008);
- Power system planning (Cadini et al. 2010; Zio et al. 2012; Gjorgiev et al. 2013; Jin et al. 2013; Li et al. 2013; Trivedi et al. 2013; Zidan et al. 2013);
- Preventive maintenance (Certa et al. 2011; Chou and Le 2011; Moradi et al. 2011; Wang and Hoang 2011; Moghaddam 2013).

Ramirez-Rosado and Bernal-Agustin (2001) applied a multiobjective evolutionary algorithm to determine the set of non-dominated solutions in the design of distribution systems for two objective functions: economic costs and reliability.

Marseguerra et al. (2002) considered a continuously monitored multi-component system and used a genetic algorithm and Monte Carlo simulation to determine the optimal degradation level beyond which preventive maintenance has to be performed in order to optimize two objective functions: profit and availability. A multiobjective genetic algorithm approach was also applied in nuclear safety system by Marseguerra et al. (2004). They considered two objectives: unavailability and the variance of its estimate.

Barakat et al. (2004) proposed the use of an ε-constraint method when designing pre-stressed concrete beams, and set minimizing the overall cost and maximizing the reliability of the system and of its flexural strength as objectives. The ε-constraint method decomposes the multiobjective optimization into a series of single objective optimizations. The procedure involves minimizing a primary objective, and expressing the other objectives in the form of inequality constraints. Consequently, the entire Pareto set can be obtained by varying the ε value.

Kumar et al. (2006) introduced an NSGA-II model for service restoration in a distribution system using three objectives: out-of-service area, number of switch operations and losses. Kumar et al. (2008) used an NSGA-II model for service restoration considering various practical operational issues in a distribution system, such as priority customers, presence of remotely controlled, as well as manually controlled switches. The same objective functions as those defined in Kumar et al. (2006) were used.

Azaron et al. (2009) found Pareto solutions in a cold-standby redundancy scheme using genetic algorithms and the goal attainment method in order to minimize the initial purchase cost of the system, to maximize its MTTF (mean time to failure), to minimize its VTTF (variance of time to failure) and also to maximize its reliability during the mission time.

Cadini et al. (2010) studied the optimal expansion of an existing electrical power transmission network using multiobjective genetic algorithms with two objectives: maximizing reliability and minimizing cost.

Certa et al. (2011) evaluated when maintenance actions should be undertaken in order to assure the required reliability level until the next fixed stop for maintenance, thereby minimizing the global maintenance cost and the total maintenance time. They proposed an exact algorithm that is able to find the whole optimal Pareto frontier.

Chou and Le (2011) used a multiobjective particle swarm optimization (MOPSO) technique in order to optimize the reliability and cost of roadway pavement maintenance.

Moradi et al. (2011) investigated an integrated flexible job shop problem with preventive maintenance activities, thereby optimizing two objectives: minimizing makespan and system unavailability. Four evolutionary algorithms are compared,

NSGA-II, NRGA, CDRNSGA-II and CDRNRGA. A composite dispatching rule (CDR) was included in the last two.

Wang and Hoang (2011) used an NSGA-II approach in order to optimize availability and the cost of an imperfect preventive maintenance policy for dependent competing risk systems with hidden failure.

Chiang (2012) discussed a multiobjective genetic algorithm integrated with a DEA approach to create an optimal design chain partner combination with total expected cost, total expected time for product development and product reliability as objective functions.

Konak et al. (2012) dealt with a multi-state multiple sliding window system problem and used NSGA-II where each failure type constitutes a minimization objective.

Torres-Echeverria et al. (2012) used a multiobjective genetic algorithm approach to design and test safety instrumented systems using NSGA-II and set three objectives: those of calculating the average probability on demand of dangerous failure, the spurious trip rate and the lifecycle cost.

Zio et al. (2012) analyzed the vulnerability of the Italian high-voltage electrical transmission network in which the most critical groups of links were identified. A multiobjective genetic algorithm approach was carried out. Two objective functions are considered: the betweenness centrality of a group of edges and the cardinality of the group of edges.

Gjorgiev et al. (2013) recommended a multiobjective genetic algorithm for scheduling the optimal generation from a power system for which they set three objectives: those of minimizing cost, emissions and unavailability.

Jin et al. (2013) proposed a multicriteria model based on genetic algorithms to design and operate a wind-based distributed generation with two objective functions: cost and reliability.

Li et al. (2013) formulated a multiobjective optimization model for protecting against cascading failures in complex networks based on the principles of NSGA-II with three objective functions: those of minimizing global connectivity loss, local connectivity loss, number of lines switched-off.

Lins et al. (2013) evaluated a multiobjective genetic algorithm to select the design for a security system which had two objectives: those of calculating the probability of a successful defense and of minimizing the acquisition and operational costs.

Moghaddam (2013) used a goal programming technique integrated with a Monte Carlo simulation to determine Pareto-optimal preventive maintenance and replacement schedules for a repairable multi-workstation manufacturing system which had been experiencing an increasing rate of the occurrence of failures. Three objective functions were evaluated: costs, reliability and availability.

Rathod et al. (2013) proposed a multiobjective genetic algorithm for a reliability-based robust design optimization problem where seven specific objective functions were defined using the first version of the NSGA.

Trivedi et al. (2013) addressed day-ahead thermal generation based on genetic algorithms using three objective functions: scheduling operation cost, emission cost and reliability. The population is ranked using the constrained-domination principle of the constrained NSGA-II.

Zidan et al. (2013) modeled how to plan a distribution network using NSGA II with two objective functions: an economic function involving costs of line upgrades, energy losses, switching operations required for network reconfiguration, and distributed generation capital, operation and maintenance costs, and an environmental function involving emissions from both grid and distributed generation units. Decision variables are defined such as switch status, line to be upgraded, distributed generation size, location and type, and year in which each decision is to be implemented.

References

Aghezzaf E-H, Najid NM (2008) Integrated production planning and preventive maintenance in deteriorating production systems. Inf Sci (Ny) 178:3382–3392

Alardhi M, Hannam RG, Labib AW (2007) Preventive maintenance scheduling for multi - cogeneration plants with production constraints. J Qual Maint Eng 13:276–292

Albuquerque LL de, Almeida AT de, Cavalcante CAV (2009) Aplicabilidade da programação matemática multiobjetivo no planejamento da expansão de longo prazo da geração no Brasil (Multiobjective mathematical programming applicability in long-term expansion planning of generation in Brazil). Pesqui Operacional 29 :153–177

Allaoui H, Lamouri S, Artiba A, Aghezzaf E (2008) Simultaneously scheduling n jobs and the preventive maintenance on the two-machine flow shop to minimize the makespan. Int J Prod Econ 112:161–167

Almeida-Filho AT de, Ferreira RP, de Almeida AT (2013) A DSS Based on Multiple Criteria Decision Making for Maintenance Planning in an Electrical Power Distributor. In: Purshouse R, Fleming P, Fonseca C, et al. (eds) Evol. Multi-Criterion Optim. SE - 58. Springer Berlin Heidelberg, pp 787–795

ANEEL (2012) Agência Nacional de Energia Elétrica, Brazil. Qualidade do serviço (Quality of service). Available at http://www.aneel.gov.br/area.cfm?idArea=79&idPerfil=2. Accessed 05 Mai 2012

Ángel-Bello F, Álvarez A, Pacheco J, Martínez I (2011) A heuristic approach for a scheduling problem with periodic maintenance and sequence-dependent setup times. Comput Math with Appl 61:797–808

Azaron A, Perkgoz C, Katagiri H, et al. (2009) Multi-objective reliability optimization for dissimilar-unit cold-standby systems using a genetic algorithm. Comput Oper Res 36:1562–1571

Barakat S, Bani-Hani K, Taha MQ (2004) Multi-objective reliability-based optimization of pre-stressed concrete beams. Struct Saf 26:311–342

Batista FRS, Geber de Melo AC, Teixeira JP, Baidya TKN (2011) The Carbon Market Incremental Payoff in Renewable Electricity Generation Projects in Brazil: A Real Options Approach. Power Syst IEEE Trans 26:1241–1251

Benmansour R, Allaoui H, Artiba A, et al. (2011) Simulation - based approach to joint production and preventive maintenance scheduling on a failure - prone machine. J Qual Maint Eng 17:254–267

Berger J (1985) Statistical Decision Theory and Bayesian Analysis (Springer Series in Statistics). Springer, New York

Bobrowsky PT (ed) (2013) Encyclopedia of Natural Hazards. Springer, Dordrecht

Bornstein CT, Maculan N, Pascoal M, Pinto LL (2012) Multiobjective combinatorial optimization problems with a cost and several bottleneck objective functions: An algorithm with reoptimization. Computers and Operations Research, 39 (9): 1969-1976

Brandeau ML, Chiu SS (1989) An Overview of Representative Problems in Location Research. Manage Sci 35:645–674

Brans JP, Mareschal B (1984) PROMETHEE: a new family of outranking methods in multicriteria analysis. Operational Research 84. Brans JP (eds). Amsterdam: North-Holland, pp. 408–421

Brans JP, Mareschal B (2002) Prométhée-Gaia: une méthodologie d'aide à la décision en présence de critères multiples. Éditions de l'Université de Bruxelles

Cadini F, Zio E, Petrescu CA (2010) Optimal expansion of an existing electrical power transmission network by multi-objective genetic algorithms. Reliab Eng Syst Saf 95:173–181

Cassady CR, Kutanoglu E (2003) Minimizing Job Tardiness Using Integrated Preventive Maintenance Planning and Production Scheduling. IIE Trans 35:503–513

Certa A, Galante G, Lupo T, Passannanti G (2011) Determination of Pareto frontier in multi-objective maintenance optimization. Reliab Eng Syst Saf 96:861–867

Chiang T-A (2012) Multi-objective decision-making methodology to create an optimal design chain partner combination. Comput Ind Eng 63:875–889

Cho H-K, Bowman KP, North GR (2004) A Comparison of Gamma and Lognormal Distributions for Characterizing Satellite Rain Rates from the Tropical Rainfall Measuring Mission. J Appl Meteorol 43:1586–1597

Chou J-S, Le T-S (2011) Reliability-based performance simulation for optimized pavement maintenance. Reliab Eng Syst Saf 96:1402–1410

Cox LA Jr (2009) Risk analysis of complex and uncertain systems. Springer Science & Business Media

de Almeida AT, Cavalcante CAV, Ferreira RJP, et al. (2006) Location of Back-up Transformers. Eng. Manag. Conf. 2006 IEEE Int. IEEE, Salvador, Bahia, pp 300–302

de Almeida AT, Ferreira RJP, Cavalcante CAV (2015) A review of multicriteria and multi-objective models in maintenance and reliability problems. IMA Journal of Management Mathematics 26(3):249–271

de Almeida AT, Souza FMC (1986) Bayes-Like Decisions in Reliability Engineering. In: Proceedings of International Conference on Information Processing and Management of Uncertainty in Knowledge-Based Systems, Paris, 30 June-4 July, 87-90

de Almeida AT, Souza FMC (2001) Gestão da Manutenção: na Direção da Competitividade (Maintenance Management: Toward Competitiveness) Editora Universitária da UFPE. Recife

De León JCV (2006) Vulnerability: A conceptual and methodological review. 'Studies of the University: Research, Counsel, Education. No. 4/2006, Institute for Environment and Human Security (UNU-EHS), Bonn, Available at https://www.ehs.unu.edu/file/get/8337.pdf. Accessed 16 Jan 2014

Deb K, Pratap A, Agarwal S, Meyarivan T (2002) A fast and elitist multiobjective genetic algorithm: NSGA-II. IEEE Trans Evol Comput 6:182–197

DeWispelare AR (1984) A computer based application of non-linear multiple objective optimization. Comput Ind Eng 8:143–152

Dhingra AK (1992) Optimal apportionment of reliability and redundancy in series systems under multiple objectives. IEEE Trans Reliab 41:576–582

Diakoulaki D, Antunes CH, Gomes Martins A (2005) MCDA and Energy Planning. Mult. Criteria Decis. Anal. State Art Surv. SE - 21. Springer New York, pp 859–890

Diniz AL, Maceira MEP (2008) A Four-Dimensional Model of Hydro Generation for the Short-Term Hydrothermal Dispatch Problem Considering Head and Spillage Effects. Power Syst IEEE Trans 23:1298–1308

Drezner Z, Hamacher HW (2004) Facility location: applications and theory. Springer Science & Business Media, New York

Farag A, Al-Baiyat S, Cheng TC (1995) Economic load dispatch multiobjective optimization procedures using linear programming techniques. Power Syst IEEE Trans 10:731–738

Ferreira RJP, de Almeida AT, Ferreira HL (2010) Multi-attribute p-median model for location of back-up transformers. Brazilian J Oper Prod Manag 7(2):09–28

Ferreira RJP, Ferreira HL (2012) Decision support system for location of back-up transformers based on a multi-attribute p-median model. Syst. Man, Cybern. (SMC), 2012 IEEE Int. Conf. IEEE, Seoul, pp 629–631

Field CB, Barros V et al. (eds) (2012) IPCC – Managing the risks of extreme events and disasters to advance climate change adaptation. A Special Report of Working Groups I and II of the Intergovernmental Panel on Climate Change. Cambridge University Press, New York

Fu G, Frangopol DM (1990) Balancing weight, system reliability and redundancy in a multi-objective optimization framework. Struct Saf 7:165–175

Girgin S, Krausmann E (2013) RAPID-N: Rapid natech risk assessment and mapping framework. J Loss Prev Process Ind 26(6):949–960

Gjorgiev B, Kančev D, Čepin M (2013) A new model for optimal generation scheduling of power system considering generation units availability. Int J Electr Power Energy Syst 47:129–139

Guikema SD (2009) Natural disaster risk analysis for critical infrastructure systems: An approach based on statistical learning theory. Reliab Eng Syst Saf 94:855–860

Hansson K, Danielson M, Ekenberg L, Buurman J (2013) Multiple Criteria Decision Making for Flood Risk Management. In: Amendola A, Ermolieva T, Linnerooth-Bayer J, Mechler R (eds) Integr. Catastr. Risk Model. SE - 4. Springer Netherlands, pp 53–72

Hillier FS (1963) Economic Models for Industrial Waiting Line Problems. Manage Sci 10:119–130

Huttenlau M, Stötter J (2011) The structural vulnerability in the framework of natural hazard risk analyses and the exemplary application for storm loss modelling in Tyrol (Austria). Nat Hazards 58:705–729

IEEE (2012) Std 1366-2012: IEEE Guide for Electric Power Distribution Reliability Indices. IEEE, New York

Ighravwe DE, Oke SA (2014) A non-zero integer non-linear programming model for maintenance workforce sizing. Int J Prod Econ 150:204–214

Jebaraj S, Iniyan S (2006) A review of energy models. Renew Sustain Energy Rev 10:281–311

Jha AK, Miner TW, Stanton-Geddes Z (2013) Building Urban Resilience: Principles, Tools, and Practice. 1–180. © World Bank. http://elibrary.worldbank.org/doi/book/10.1596/978-0-8213-8865-5

Ji M, He Y, Cheng TCE (2007) Single-machine scheduling with periodic maintenance to minimize makespan. Comput Oper Res 34:1764–1770

Jiang R, Ji P (2002) Age replacement policy: a multi-attribute value model. Reliab Eng Syst Saf 76(3):311–318

Jin T, Tian Y, Zhang CW, Coit DW (2013) Multicriteria Planning for Distributed Wind Generation Under Strategic Maintenance. Power Deliv IEEE Trans 28:357–367

Karimi I, Hüllermeier E (2007) Risk assessment system of natural hazards: A new approach based on fuzzy probability. Fuzzy Sets Syst 158:987–999

Keeney RL, Raiffa H (1976) Decisions with multiple objectives: Preferences and Value Trade-Offs. Wiley Series in Probability and Mathematical Statistics. Wiley and Sons, New York

Keller EA, DeVecchio DE (2012) Earth's Processes as Hazards, Disasters, and Catastrophes. Natural Hazards. Pearson Prentice Hall, New Jersey

Konak A, Kulturel-Konak S, Levitin G (2012) Multi-objective optimization of linear multi-state multiple sliding window system. Reliab Eng Syst Saf 98(1):24–34

Krausmann E, Cozzani V, Salzano E, Renni E (2011) Industrial accidents triggered by natural hazards: an emerging risk issue. Nat Hazards Earth Syst Sci 11:921–929

Krausmann E, Cruz A (2013) Impact of the 11 March 2011, Great East Japan earthquake and tsunami on the chemical industry. Nat Hazards 67:811–828

Kumar Y, Das B, Sharma J (2006) Service restoration in distribution system using non-dominated sorting genetic algorithm. Electr Power Syst Res 76:768–777

Kumar Y, Das B, Sharma J (2008) Multiobjective, multiconstraint service restoration of electric power distribution system with priority customers. Power Deliv IEEE Trans 23:261–270

Li F, van Gelder PHAJM, Ranasinghe R, et al. (2014) Probabilistic modelling of extreme storms along the Dutch coast. Coast Eng 86:1–13

Li YF, Sansavini G, Zio E (2013) Non-dominated sorting binary differential evolution for the multi-objective optimization of cascading failures protection in complex networks. Reliab Eng Syst Saf 111:195–205

Lin C, Madu CN, Chien TW, Kuei C-H (1994) Queueing Models for Optimizing System Availability of a Flexible Manufacturing System. J Oper Res Soc 45(10):1141–1155

Linnekamp F, Koedam A, Baud ISA (2011) Household vulnerability to climate change: Examining perceptions of households of flood risks in Georgetown and Paramaribo. Habitat Int 35:447–456

Lins ID, Rêgo LC, Moura M das C, Droguett EL (2013) Selection of security system design via games of imperfect information and multi-objective genetic algorithm. Reliab Eng Syst Saf 112:59–66

Lins PHC, de Almeida AT (2012) Multidimensional risk analysis of hydrogen pipelines. Int J Hydrogen Energy 37:13545–13554

Mareschal B, De Smet Y, Nemery P (2008) Rank reversal in the PROMETHEE II method: Some new results. Ind Eng Eng Manag 2008 IEEM 2008 IEEE Int Conf 959–963

Marseguerra M, Zio E, Podofillini L (2002) Condition-based maintenance optimization by means of genetic algorithms and Monte Carlo simulation. Reliab Eng Syst Saf 77:151–165

Marseguerra M, Zio E, Podofillini L (2004) A multiobjective genetic algorithm approach to the optimization of the technical specifications of a nuclear safety system. Reliab Eng Syst Saf 84:87–99

Misra KB, Sharma U (1991) An efficient approach for multiple criteria redundancy optimization problems. Microelectron Reliab 31:303–321

Moghaddam KS (2013) Multi-objective preventive maintenance and replacement scheduling in a manufacturing system using goal programming. Int J Prod Econ 146:704–716

Moradi E, Fatemi Ghomi SMT, Zandieh M (2011) Bi-objective optimization research on integrated fixed time interval preventive maintenance and production for scheduling flexible job-shop problem. Expert Syst Appl 38:7169–7178

Moubray J (1997) Reliability-centered maintenance. Industrial Press Inc., New York

Paul, BK (2011) Environmental hazards and disasters: contexts, perspectives and management. Wiley-Blackwell, Chichester

Pelling M (2003) Vulnerability of cities. Earthscan Publications Ltd., London

Pine JC (2009) Natural hazards analysis: reducing the impact of disasters. CRC Press, Taylor & Francis Group, Florida

Pinedo ML (2012) Scheduling: theory, algorithms, and systems. Fourth Edition. Springer, New York.

Pinto RJ, Borges CLT, Maceira MEP (2013) An Efficient Parallel Algorithm for Large Scale Hydrothermal System Operation Planning. Power Syst IEEE Trans 28:4888–4896

Priori Jr L, Alencar MH, de Almeida AT (2015) Adaptations to possible climate change impacts: problem structuring based on VFT methodology. CDSID Working report

Ramirez-Rosado IJ, Bernal-Agustin JL (2001) Reliability and costs optimization for distribution networks expansion using an evolutionary algorithm. Power Syst IEEE Trans 16:111–118

Rao SS, Dhingra AK (1992) Reliability and redundancy apportionment using crisp and fuzzy multiobjective optimization approaches. Reliab Eng Syst Saf 37:253–261

Rathod V, Yadav OP, Rathore A, Jain R (2013) Optimizing reliability-based robust design model using multi-objective genetic algorithm. Comput Ind Eng 66:301–310

Shafiee M, Finkelstein M (2015) An optimal age-based group maintenance policy for multi-unit degrading systems. Reliab Eng Syst Saf 134:230–238

Solecki W, Leichenko R, O'Brien K (2011) Climate change adaptation strategies and disaster risk reduction in cities: connections, contentions, and synergies. Curr Opin Environ Sustain 3:135–141

Soltani M, Corotis RB (1988) Failure cost design of structural systems. Struct Saf 5:239–252

Sortrakul N, Cassady CR (2007) Genetic algorithms for total weighted expected tardiness integrated preventive maintenance planning and production scheduling for a single machine. J Qual Maint Eng 13:49–61

Su L, Tsai H (2010) Flexible preventive maintenance planning for two parallel machines problem to minimize makespan. J Qual Maint Eng 16:288–302

Thywissen K (2006) Components of Risk: A Comparative Glossary. Institute for Environment and Human Security (UNU-EHS), Bonn

Torres-Echeverría AC, Martorell S, Thompson HA (2012) Multi-objective optimization of design and testing of safety instrumented systems with MooN voting architectures using a genetic algorithm. Reliab Eng Syst Saf 106:45–60

Trivedi A, Srinivasan D, Sharma D, Singh C (2013) Evolutionary Multi-Objective Day-Ahead Thermal Generation Scheduling in Uncertain Environment. Power Syst IEEE Trans 28:1345–1354

Vari A, Linnerooth-Bayer J, Ferencz Z (2003) Stakeholder Views on Flood Risk Management in Hungary's Upper Tisza Basin. Risk Anal 23:585–600

Wang Y, Pham H (2011) A Multi-Objective Optimization of Imperfect Preventive Maintenance Policy for Dependent Competing Risk Systems with Hidden Failure. Reliab IEEE Trans 60:770–781

Wong KP, Fan B, Chang CS, Liew AC (1995) Multi-objective generation dispatch using bi-criterion global optimisation. Power Syst IEEE Trans 10:1813–1819

Yokoyama R, Bae SH, Morita T, Sasaki H (1988) Multiobjective optimal generation dispatch based on probability security criteria. Power Syst IEEE Trans 3:317–324

Zidan A, Shaaban MF, El-Saadany EF (2013) Long-term multi-objective distribution network planning by DG allocation and feeders' reconfiguration. Electr Power Syst Res 105:95–104

Zio E, Golea LR, Rocco S. CM (2012) Identifying groups of critical edges in a realistic electrical network by multi-objective genetic algorithms. Reliab Eng Syst Saf 99:172–177

Zischg A, Schober S, Sereinig N, et al. (2013) Monitoring the temporal development of natural hazard risks as a basis indicator for climate change adaptation. Nat Hazards 67:1045–1058

Index

A.T. de Almeida et al., *Multicriteria and Multiobjective Models for Risk, Reliability and Maintenance Decision Analysis*, International Series in Operations Research & Management Science 231, DOI 10.1007/978-3-319-17969-8

MIX
Papier aus verantwortungsvollen Quellen
Paper from responsible sources
FSC® C105338

If you have any concerns about our products,
you can contact us on
ProductSafety@springernature.com

In case Publisher is established outside the EU,
the EU authorized representative is:
Springer Nature Customer Service Center GmbH
Europaplatz 3, 69115 Heidelberg, Germany

Printed by Libri Plureos GmbH
in Hamburg, Germany